55837

541.345

02

STUDIES IN INTERFACE SCIENCE

Ultrasound for Characterizing Colloids
Particle Sizing, Zeta Potential, Rheology

STUDIES IN INTERFACE SCIENCE

SERIES EDITORS
D. Möbius and R. Miller

Vol. 1 Dynamics of Adsorption at Liquid Interfaces. *Theory, Experiment, Application.* By S.S. Dukhin, G. Kretzschmar and R. Miller

Vol. 2 An Introduction to Dynamics of Colloids. By J.K.G. Dhont

Vol. 3 Interfacial Tensiometry. By A.I. Rusanov and v.A. Prokhorov

Vol. 4 New Developments in Construction and Functions of Organic Thin Films. Edited by T. Kajiyama and M. Aizawa

Vol. 5 Foam and Foam Films. By D. Exerowa and P.M. Kruglyakov

Vol. 6 Drops and Bubbles in Interfacial Research. Edited by D. Möbius and R. Miller

Vol. 7 Proteins at Liquid Interfaces. Edited by D. Möbius and R. Miller

Vol. 8 Dynamic Surface Tensiometry in Medicine. By V.M. Kazakov, O.V. Sinyachenko, V.B. Fainerman, U. Pison and R. Miller

Vol. 9 Hydrophile-Lipophile Balance of Surfactants and Solid Particles. *Physicochemical Aspects and Applications.* By P.M. Kruglyakov

Vol. 10 Particles at Fluid Interfaces and Membranes. *Attachment of Colloid Particles and Proteins to Interfaces and Formation of Two-Dimensional Arrays.* By P.A. Kralchevsky and K. Nagayama

Vol. 11 Novel Methods to Study Interfacial Layers. By D. Möbius and R. Miller

Vol. 12 Colloid and Surface Chemistry. By E.D. Shchukin, A.V. Pertsov, E.A. Amelina and A.S. Zelenev

Vol. 13 Surfactants: Chemistry, Interfacial Properties, Applications. Edited by V.B. Fainerman, D. Möbius and R. Miller

Vol. 14 Complex Wave Dynamics on Thin Films. By H.-C. Chang and E.A. Demekhin

Vol. 15 Ultrasound for Characterizing Colloids. *Particle Sizing, Zeta Potential, Rheology* By A.S. Dukhin and P.J. Goetz

Ultrasound for Characterizing Colloids

Particle Sizing, Zeta Potential, Rheology

Andrei S. Dukhin

and

Philip J. Goetz

Dispersion Technology Inc., NY, USA

2002
ELSEVIER
Amsterdam – Boston – London – New York – Oxford – Paris – San Diego
San Francisco – Singapore – Sydney – Tokyo

ELSEVIER SCIENCE B.V.
Sara Burgerhartstraat 25
P.O. Box 211, 1000 AE Amsterdam, The Netherlands

© 2002 Elsevier Science B.V. All rights reserved.

This work is protected under copyright by Elsevier Science, and the following terms and conditions apply to its use:

Photocopying
Single photocopies of single chapters may be made for personal use as allowed by national copyright laws. Permission of the Publisher and payment of a fee is required for all other photocopying, including multiple or systematic copying, copying for advertising or promotional purposes, resale, and all forms of document delivery. Special rates are available for educational institutions that wish to make photocopies for non-profit educational classroom use.

Permissions may be sought directly from Elsevier Science Global Rights Department, PO Box 800, Oxford OX5 1DX, UK; phone: (+44) 1865 843830, fax: (+44) 1865 853333, e-mail: permissions@elsevier.co.uk. You may also contact Global Rights directly through Elsevier's home page (http://www.elsevier.nl), by selecting 'Obtaining Permissions'.

In the USA, users may clear permissions and make payments through the Copyright Clearance Center, Inc., 222 Rosewood Drive, Danvers, MA 01923, USA; phone: (+1) (978) 7508400, fax: (+1) (978) 7504744, and in the UK through the Copyright Licensing Agency Rapid Clearance Service (CLARCS), 90 Tottenham Court Road, London W1P 0LP, UK; phone: (+44) 207 631 5555; fax: (+44) 207 631 5500. Other countries may have a local reprographic rights agency for payments.

Derivative Works
Tables of contents may be reproduced for internal circulation, but permission of Elsevier Science is required for external resale or distribution of such material.
Permission of the Publisher is required for all other derivative works, including compilations and translations.

Electronic Storage or Usage
Permission of the Publisher is required to store or use electronically any material contained in this work, including any chapter or part of a chapter.

Except as outlined above, no part of this work may be reproduced, stored in a retrieval system or transmitted in any form or by any means, electronic, mechanical, photocopying, recording or otherwise, without prior written permission of the Publisher.
Address permissions requests to: Elsevier Science Global Rights Department, at the mail, fax and e-mail addresses noted above.

Notice
No responsibility is assumed by the Publisher for any injury and/or damage to persons or property as a matter of products liability, negligence or otherwise, or from any use or operation of any methods, products, instructions or ideas contained in the material herein. Because of rapid advances in the medical sciences, in particular, independent verification of diagnoses and drug dosages should be made.

First edition 2002

Library of Congress Cataloging in Publication Data
A catalog record from the Library of Congress has been applied for.

ISBN: 0 444 51164 4
ISSN: 1383 7303

⊖ The paper used in this publication meets the requirements of ANSI/NISO Z39.48-1992 (Permanence of Paper).
Printed in The Netherlands.

PREFACE

The roots of this book go back twenty years. In the early 1980's Philip Goetz, then President of Pen Kem, Inc., was looking for new ways to characterize ζ-potential, especially in concentrated suspensions. His scientific consultants, Bruce Marlow, Hemant Pendse and David Fairhurst, pointed towards utilizing electroacoustic phenomena. Several years of effort resulted in the first commercial electroacoustic instrument for colloids, the PenKem-7000. In the course of their work they learned much about the potential value of ultrasound and collected a large number of papers published over the last two centuries on the use of ultrasound as a technique to characterize a diverse variety of colloids. From this it became clear that electroacoustics was only a small part of an enormous new field. Unfortunately, by that time Pen Kem's scientific group broke apart; Bruce Marlow died, David Fairhurst turned back to design his own electrophoretic instruments, and Hemant Pendse concentrated his efforts on the industrial on-line application of ultrasound. In order to fill the vacuum of scientific support, Philip Goetz invited me to the US motivated by my experience and links to the well established group of my father, Prof. Stanislav Dukhin, then the Head of the Theoretical Department in the Institute of Colloid Science of the Ukrainian Academy of Sciences.

Whatever the reasons, I was fortunate to become involved in this exciting project of developing new techniques based on ultrasound. I was even more lucky to inherit the vast experience and scientific base created by my predecessors at Pen Kem. Their hidden contribution to this book is very substantial, and I want to express here my gratitude to them for their pioneering efforts.

It took ten years for Phil and I to reach the point when, overwhelmed with the enormous volume of collected results, we concluded the best way to summarize them for potential users was in the form of a book. There were many people who helped us during these years of development and later in writing this book. I would like to mention here several of them personally.

First of all, Vladimir Shilov helped us tremendously with the electroacoustic theory. I believe that he is the strongest theoretician alive in the field of electrokinetics and related colloidal phenomena. Although he is very well known in Europe, he is less recognized in the United States.

More recently, David Fairhurst joined us again. He brought his extensive expertise in the field of particulates, various colloids related applications and techniques; his comments were very valuable.

In our company, Dispersion Technology Inc., Ross Parrish, Manufacturing and Service Manager, and Betty Rausa, Office Manager, extended their responsibilities last year thus allowing Phil and I time to write this book. We are very grateful for their patience, reliability and understanding.

We trust that publishing this book will truly mark the end of ultrasound's "childhood" in the field of colloids; it is no longer a "new" technique for colloids. Currently, there are more than 100 groups in the US, Germany, Japan, Taiwan, China, UK, Belgium, Finland, Singapore, South Korea, Mexico, and Canada using Dispersion Technology ultrasound based instruments. We do not know the total numbers of competitive instruments from Malvern, Matec, Colloidal Dynamics and Sympatec that are also being used in the field, but we estimate that there are over 200 groups world-wide. We strongly believe that this is only the beginning. Ultrasonics has "come of age" and we hope that you will also share this view after reading this book and learning of the many advantages ultrasound can bring to the characterization of colloid systems.

Andrei Dukhin
President of Dispersion Technology Inc.

Table of Contents

Preface v

Chapter 1. INTRODUCTION 1
1.1 Historical overview. 5
1.2 Advantages of ultrasound over traditional characterization techniques. 9
References 13

Chapter 2. FUNDAMENTALS OF INTERFACE AND COLLOID SCIENCE 17
2.1 Real and model dispersions. 18
2.2 Parameters of the model dispersion medium. 20
    2.2.1 Gravimetric parameters. 21
    2.2.2 Rheological parameters. 21
    2.2.3 Acoustic parameters. 22
    2.2.4 Thermodynamic parameters. 22
    2.2.5 Electrodynamic parameters. 24
    2.2.6 Electroacoustic parameters. 24
    2.2.7 Chemical composition. 25
2.3 Parameters of the model dispersed phase. 26
    2.3.1 Rigid vs. soft particles. 27
    2.3.2 Particle size distribution. 28
2.4. Parameters of the model interfacial layer. 33
    2.4.1. Flat surfaces. 35
    2.4.2 Spherical DL, isolated and overlapped. 36
    2.4.3 Electric Double Layer at high ionic strength. 38
    2.4.4 Polarized state of the Electric Double Layer 39
2.5. Interactions in Colloid and Interface science 42
    2.5.1. Interactions of colloid particles in equilibrium. Colloid stability 43
    2.5.2 Interaction in a hydrodynamic field. Cell and core-shell models. Rheology. 46
    2.5.3 Linear interaction in an electric field. Electrokinetics and dielectric spectroscopy. 52
    2.5.4 Non-linear interaction in the electric field. Electrocoagulation and electro-rheology. 58
2.6. Traditional particle sizing. 62
    2.6.1 Light Scattering. Extinction=scattering+absorption. 63
References 68

Chapter 3. FUNDAMENTALS OF ACOUSTICS IN LIQUIDS 75
3.1. Longitudinal waves and the wave equation. 75
3.2. Acoustics and its relation to Rheology. 77
3.3. Acoustic Impedance. 82

3.4 Propagation through phase boundaries - Reflection.	84
3.5 Propagation in porous media.	86
3.6 Chemical composition influence.	89
References	94

Chapter 4. ACOUSTIC THEORY FOR PARTICULATES — 101

4.1 Extinction=absorption + scattering. Superposition approach.	104
4.2 Acoustic theory for a dilute system.	114
4.3 Ultrasound absorption in concentrates.	117
4.3.1 Coupled phase model.	118
4.3.2 Viscous loss theory.	122
4.3.3 Thermal loss theory.	126
4.3.4 Structural loss theory.	130
4.3.5 Intrinsic loss theory.	133
4.4. Ultrasound scattering	135
4.4.1 Rigid sphere.	140
4.4.2 Rigid Cylinder.	141
4.4.3 Non-rigid sphere.	141
4.4.4 Porous sphere.	142
4.4.5 Scattering by a group of particles.	143
4.4.6 Ultrasound resonance by bubbles.	144
4.5 Input parameters.	145
References	149

Chapter 5. ELECTROACOUSTIC THEORY — 153

5.1 The Theory of Ion Vibration Potential (IVP).	157
5.2 The Low frequency electroacoustic limit - Smoluchowski limit, (SDEL).	158
5.3 The O'Brien theory.	160
5.4 The Colloid Vibration Current in concentrated systems.	163
5.4.1 CVI and Sedimentation Current.	164
5.4.2 CVI for polydisperse systems.	169
5.4.3 Surface conductivity.	171
5.4.4 Maxwell-Wagner relaxation. Extended frequency range.	172
5.5 Qualitative analysis.	173
References	176

Chapter 6. EXPERIMENTAL VERIFICATION OF THE ACOUSTIC AND ELECTROACOUSTIC THEORIES 181

6.1 Viscous losses. 181
6.2 Thermal losses. 187
6.3 Structural losses. 189
6.4 Scattering losses. 193
6.5 Electroacoustic phenomena. 196
References 201

Chapter 7. ACOUSTIC AND ELECTROACOUSTIC MEASUREMENT TECHNIQUES 205

7.1 Historical Perspective. 205
7.2 Difference between measurement and analysis. 206
7.3 Measurement of attenuation and sound speed using Interferometry. 207
7.4 Measurement of attenuation and sound speed using the transmission technique. 208
    7.4.1 Historical development of the transmission technique. 208
    7.4.2 Detailed Description of the Dispersion Technology DT-100 Acoustic Spectrometer. 210
7.5 Precision, accuracy, and dynamic range for transmission measurements. 221
7.6 Analysis of Attenuation and Sound Speed to yield desired outputs. 224
    7.6.1 The ill-defined problem. 224
    7.6.2 Precision, accuracy, and resolution of the analysis. 230
7.7 Measurement of Electroacoustic properties. 234
    7.7.1 Electroacoustic measurement of CVI. 234
    7.7.2 CVI measurement using energy loss approach. 237
7.8 Zeta potential calculation from the analysis of CVI. 239
7.9 Measurement of acoustic Impedance. 240
References 243

Chapter 8. APPLICATIONS OF ACOUSTICS FOR CHARACTERIZING PARTICULATE SYSTEMS 247

8.1 Characterization of aggregation and flocculation. 247
8.2 Stability of emulsions and microemulsions. 256
8.3 Particle sizing in mixed colloids with several dispersed phases. 264
    8.3.1 High density contrast - Ceramics, oxides, minerals, pigments. 267
    8.3.2 Cosmetics - Sunscreen. 278
    8.3.3 Composition of mixtures. 282
8.4. Chemical-mechanical polishing. Large particle resolution. 287

8.5. Titration using Electroacoustics.	296
8.5.1 pH titration.	296
8.5.2 Time titration, kinetic of the surface-bulk equilibration.	298
8.5.3 Surfactant titration.	300
8.6. Colloids with high ionic strength - Electroacoustic background.	305
8.7 Effect of air bubbles.	311
8.8 Table of Applications.	312
References	321

List of symbols   331

Bibliography alphabetical   339

Index   367

Chapter 1. INTRODUCTION

Two key words define the scope of this book: "ultrasound" and "colloids". In turn, each word is a key to a major scientific discipline, Acoustics on one hand and Colloid Science on the other. It is a rather curious situation that, historically, there has been little real communication between disciples of these two fields. Although there is a large body of literature devoted to ultrasound phenomena in colloids, mostly from the perspective of scientists from the field of Acoustics, there is little recognition that such phenomena may be of real importance for both the development, and application, of Colloid Science. From the other side, colloid scientists have not embraced acoustics as an important tool for characterizing colloids. The lack of any serious dialog between these scientific fields is perhaps best illustrated by the fact that there are no references to ultrasound or Acoustics in the major handbooks on Colloid and Interface Science [1,2], nor any reference to colloids in handbooks on acoustics [3,4,5].

One might ask "Perhaps this link does not exist because it is not important to either discipline?" In order to answer this question, let us consider the potential place of Acoustics within an overall framework of Colloid Science. For this purpose, it is helpful to classify non-equilibrium colloidal phenomena in two dimensions; the first determined by whether the relevant disturbances are electrical, mechanical, or electro- mechanical in nature, and the second based on whether the time domain of that disturbance can be described as stationary, low frequency, or high frequency. Table 1.1 illustrates this classification of major colloidal phenomena. The low and high frequency ranges are separated based on the relationship between either the electric or mechanical wavelength λ, and some system dimension L.

Clearly, light scattering represents electrical phenomena in colloids at high frequency (the wavelength of light is certainly smaller than the system dimension). There was, however, no mention in colloid textbooks, until very recently, of any mechanical or electro-mechanical phenomena in the region where the mechanical or electrical wavelength is shorter than the system dimension. This would appear to leave two empty spaces in Table 1.1. Such mechanical wavelengths are produced by "Sound" or, when the frequency exceeds our hearing limit of 20 KHz, by "Ultrasound". For reference, ultrasound wavelengths lie in the range from 10 microns to 1 mm, whereas the system dimension is usually in the range of centimeters. For this reason, we consider ultrasound related effects to lie within the high frequency range. One of the empty spaces can be filled by acoustic measurements at ultrasound frequencies, which characterize colloidal phenomena of a mechanical nature at high frequency. The second empty space can be filled by electroacoustic measurements, which allow us to characterize electro-mechanical phenomena at high frequency. This book will help fill these gaps and demonstrate that acoustics (and electroacoustics) and can bring much useful knowledge to Colloid Science. As an aside, we do not consider here the use of high power ultrasound for modifying colloidal systems, just the use of low power sound as a non-invasive investigation tool that has very unique capabilities.

Table 1.1 Colloidal phenomena

	Electrical nature	Electro-mechanical	Mechanical nature
Stationary	Conductivity, Surface conductivity.	Electrophoresis, Electroosmosis, Sedimentation potential, Streaming current/potential, Electro-viscosity	Viscosity, Stationary colloidal hydrodynamics, Osmosis, Capillary flow.
Low frequency ($\lambda > L$)	Dielectric spectroscopy.	Electro-rotation, Dielectrophoresis.	Oscillatory rheology.
High frequency ($\lambda < L$)	Optical scattering, X-ray spectroscopy.	Empty? Electroacoustics!	Empty? Acoustics!

There are several questions that one might ask when starting to read this book. We think it is important to deal with these questions right away, at least giving some preliminary answers, which will then be clarified and expanded later in the main text. Here are these questions and the short answers.

Why should one care about Acoustics if generations of colloid scientists worked successfully without it?

While it may be true at present that the usefulness of Acoustics is not widely understood, it seems that earlier generations had a somewhat better appreciation. Many well-known scientists applied Acoustics to colloidal systems, as will be described in a detailed historical overview in the next section. Briefly, we can mention the names of Stokes, Rayleigh, Maxwell, Henry, Tyndall, Reynolds, and Debye, all of whom considered acoustic phenomena in colloids as deserving of their attention. The first colloid-related acoustic effect to be studied was the propagation of sound through fog; contributions by Henry, Tyndall and Reynolds made more than century ago between 1870-80. Another interesting, but not so well known fact, is that Lord Rayleigh, the first author of a scattering theory, entitled his major books "Theory of Sound". He developed the mathematics of scattering theory mostly for sound, not for light as is often assumed by those not so familiar with the history of Colloid Science. In fact, the main reference to light in his work was a paragraph or two on "why the sky is blue".

If Acoustics is so important, why has it remained almost unknown in Colloid Science for such a long time?

We think that the failure to exploit acoustic methods might be explained by a combination of factors: the advent of the laser as a convenient source of monochromatic light, technical problems with generating monochromatic sound beams within a wide frequency range, the mathematical complexity of the theory, and complex statistical analysis of the raw data. In addition, acoustics is more dependent on mathematical calculations than other traditional instrumental techniques. Many of these problems have now been solved mostly due to the advent of fast computers and the development of new theoretical approaches. As a result there are a number commercially available instruments utilizing ultrasound for characterizing colloids, produced by Matec, Malvern, Sympatec, Colloidal Dynamics, and Dispersion Technology.

What information does ultrasound based instruments yield?

For colloidal systems, ultrasound provides information on three important areas of particle characterization: Particle sizing, Rheology, and Electrokinetics.

In addition, ultrasound can be used as a tool for characterizing properties of pure liquids and dissolved species like ions or molecules, but we will cover this aspect only briefly.

An Acoustic spectrometer may measure the attenuation of ultrasound, the propagation velocity of this sound, and/or the acoustic impedance, in any combination depending on the instrument design. The measured acoustic properties contain information about the particle size distribution, and volume fraction, as well as structural and thermodynamic properties of the colloid. One can extract this information by applying the appropriate theory in combination with a certain set of *a'priori* known parameters. Hence, an Acoustic spectrometer is not simply a particle-sizing instrument. By applying sound we apply stress to the colloid and consequently the response can be interpreted in rheological terms, as will be shown below.

In addition to acoustics there is one more ultrasound-based technique, which is called Electroacoustics. The Electroacoustic phenomenon, first predicted by Debye in 1933, results from coupling between acoustic and electric fields. There are two ways to produce such an Electroacoustic phenomenon depending on which field is the driving force. When the driving force is the electric field and we observe an acoustic response we speak of ElectroSonic Amplitude (ESA). Alternatively, when the driving force is the acoustic wave we speak, instead, of the Colloid Vibration Potential (CVP) if we observe an open circuit potential, or a Colloid Vibration Current (CVI) if we observe a short circuit current. Such electroacoustic techniques yield information about the electrical properties of colloids. In principle, it can also be used for particle sizing.

Where can one apply ultrasound?

The following list gives some idea of the existing applications for which the ultrasound based characterization technique is appropriate:

Aggregative stability, Cement slurries, Ceramics, Chemical-Mechanical Polishing, Coal slurries, Coatings, Cosmetic emulsions, Environmental protection, Flotation, Ore enrichment, Food products, Latex, Emulsions and micro emulsions, Mixed dispersions, Nanosized dispersions, Non-aqueous dispersions, Paints, Photo Materials.

This list is not complete. A table in Chapter 8 summarizes all experimental works currently known to us.

What are the advantages of ultrasound over traditional characterization techniques?

There are so many advantages of ultrasound. The last section of this chapter is devoted to describing the relationship between ultrasound based and traditional colloidal characterization techniques.

Finally, we would like to stress that this book targets primarily scientists who consider colloids as their major object of interest. As such we emphasize those aspects of acoustics that are important for colloids, and thereby neglect many others.

On the other hand, scientist working with ultrasound will already be familiar with many of the theoretical and experimental developments presented in this book. At the same time they will find several important new developments. In particular we would like to mention:

- a general approach to acoustics in colloids by combination of ultrasound absorption and ultrasound scattering;
- a general solution for eliminating multiple scattering;
- a theory of ultrasound absorption in concentrated systems;
- an electroacoustic theory for concentrates;
- experimental verification of theses theories for concentrated systems;
- multiple existing applications.

1.1 Historical overview

The roots of our current understanding of sound go back more than 300 years to the first theory for calculating sound speed suggested by Newton [6]. Newton's work is still interesting for us today because it illustrates the importance of thermodynamic considerations in trying to adequately describe ultrasound phenomena. Newton assumed that sound propagates while maintaining a constant temperature, i.e. an isothermal case. Laplace later corrected this misunderstanding by showing that it was actually adiabatic in nature [6].

This thermodynamic aspect of sound provides a good example of the importance of keeping a historical perspective. At least twice during the past 200 years the thermodynamic contribution to various sound-related phenomena was initially neglected, and only later found to be quite important. This thermodynamic neglect happened first in the 19th century, when Stokes's purely hydrodynamic theory for sound attenuation [9, 10] was later corrected by Kirchhoff [7, 8]. Then again, in the 20th century, Sewell's hydrodynamic theory for sound absorption in heterogeneous media [12] was later extended by Isakovich [11] by the introduction of a mechanism for thermal losses.

We have now a very similar situation concerning electroacoustics. Until quite recently all such theories neglected any thermodynamic contribution [13, 14, 15, 16, and 17]. Based on historical perspective, we might reasonably inquire about the potential importance of thermodynamic considerations for electroacoustics. Shilov and others [18] have addressed this query and revealed a new interesting feature of the electroacoustic effect.

Table 1.2 lists important steps in the development of our understanding of sound. From the very beginning sound was considered as a rather simple example that allowed development of a general theory of "wave" phenomena. Then, later, the new understanding achieved for sound was extended to other wave phenomena, such as light. Tyndall, for example, used reference to sound to explain the wave nature of the light [19, 20]. Newton's Corpuscular Theory of Light was first opposed both by the celebrated astronomer Huygens and the, no less celebrated, mathematician Euler. They each held that light, like *sound*, was a product of wave-motion. In the case of *sound*, the velocity depends upon the relation of elasticity to density in the body that transmits the sound. The greater the elasticity the greater is the velocity, and the less the density the greater is the velocity. To account for the enormous velocity of propagation in the case of light, the substance that transmits it is assumed to have both extreme elasticity and extreme density.

This dominance of sound over light as examples of the wave phenomena continued even with Lord Rayleigh, who developed his theory of scattering mostly for sound and paid much less attention to light [6,21-24]. At the end of the 19th century sound and light parted because further investigation was directed more on the physical roots of each phenomenon instead of on their common wave nature.

The history of light and sound in Colloid Science is very different. Light has been an important tool since the first microscopic observations of Brownian motion and the first electrophoretic measurements. It became even more important in middle of the 20th century through the use of light scattering for the determination of particle size.

In contrast, sound remained unknown in Colloid Science, despite a considerable amount of work in the field of Acoustics using fluids that were essentially of a colloidal nature. The goal of these studies was to learn more about Acoustics, but not about colloids. This is the spirit in which the ECAH theory (Epstein-Carhart-Allegra-Hawley [25, 26]) for ultrasound propagation through dilute colloids was developed.

Although Acoustics was not used specifically for colloids, it was a powerful tool for other purposes. For instance, it was used to learn more about the structure of pure liquids and the nature of chemical reactions in liquids. These studies, associated with the name of Prof. Eigen, who received a Nobel Prize in 1968 [27-29], are described in more detail in the chapter "Fundamentals of Acoustics".

It is curious that the penetration of ultrasound into Colloid Science began with electroacoustics, which is more complex than traditional acoustics. An Electroacoustic effect was predicted for ions by Debye in 1933 [30], and later extended to colloids by Hermans and, independently, Rutgers in 1938 [31]. The early experimental electroacoustic work is associated with Yeager and Zana, who conducted many experiments in the 1950's and 60's with various co-authors [32-35]. Later this work was continued by Marlow, O'Brien, Ohshima, Shilov, and the authors of this book [13, 15, 16, 17, 18, 36, and 37]. As a result, there are now several commercially available electroacoustic instruments for characterizing ζ-potential.

Acoustics only attained some recognition in the field of colloid science very recently. It was first suggested as a particle sizing tool by Cushman and others [77] in 1973, and later refined by Uusitalo and others [77], and for large particles by Riebel [38]. Development as a commercial instrument having the capability to measure a wide particle size range, was begun by Goetz, A.Dukhin, and Pendse in the 90's [39, 40, 41,42]. At the same time a group of British scientists, including McClements, Povey, and others [43, 44, 45, 46, 47, and 48], actively promoted it, especially for emulsions. There are now four commercially available acoustic spectrometers, manufactured by Malvern, Sympatec, Matec, and Dispersion Technology.

To conclude this short historical review we would like to mention a development that we consider of great importance for the future, namely the combination of both acoustic and electroacoustic spectroscopies. The synergism of this combination is described in papers and patents by A.Dukhin and P.Goetz [17, 37, 40, 41, 42, 49, 50, 72, and 73].

Table 1.2 Key Steps in understanding Sound related to Colloids.

Year	Author	Topic
1687	Newton [6]	Sound speed in fluid, theory, erroneous isothermal assumption
Early 1800's	Laplace [6]	Sound speed in fluid, theory, adiabatic assumption
1820	Poisson [51,52]	Scattering by atmosphere arbitrary disturbance, first successful theory
1808	Poisson [51,52]	Reflection from rigid plane, general problem
1845-1851	Stokes [9,10]	Sound attenuation in fluid, theory, viscous losses
1842	Doppler [53]	Alternation of pitch by relative motion
1868	Kirchhoff [7,8]	Sound attenuation in fluid, theory, thermal losses

1866	Maxwell [54]	Kinetic theory of viscosity
1870-80	Henry, Tyndall, Reynolds [19,20,55,56]	First application to colloids - sound propagation in fog
1871	Rayleigh [21,22]	Light scattering theory
1875-80	Rayleigh [6,23,24]	Diffraction and scattering of sound, Fresnel zones in Acoustics
1878	Rayleigh [6]	Theory of Sound, Vol. II
1910	Sewell [12]	Viscous attenuation in colloids, theory
1933	Debye [30]	Electroacoustic effect, introduction for ions
1936	Morse [3]	Scattering theory for arbitrary wavelength-size ratio
1938	Hermans [31]	Electroacoustic effect, introduced for colloids
1944	Foldy [57,58]	Acoustic theory for bubbles
1948	Isakovich [11]	Thermal attenuation in colloids, theory
1946	Pellam, Galt [59]	Pulse technique
1947	Bugosh, Yaeger [32]	Electroacoustic theory for electrolytes
1951-3	Yeager, Hovorka, Derouet, Denizot [33-35,60]	First electroacoustic measurements
1951-2	Enderby, Booth [61-62]	First electroacoustic theory for colloids
1953	Epstein and Carhart [25]	General theory of sound attenuation in dilute colloids
1958-9	Happel, Kuwabara [63-65]	Hydrodynamic cell models
1962	Andreae et al [66,67]	Multiple frequencies attenuation measurement
1967	Eigen et al [27-28]	Nobel price, acoustics for chemical reactions in liquids
1972	Allegra, Hawley [26]	ECAH theory for dilute colloids
1973	Cushman [77]	First patent for acoustic particle sizing
1974	Levine, Neale [68]	Electrokinetic cell model
1978	Beck [36]	Measurement of ζ-potential by ultrasonic waves

1981	Shilov, Zharkikh [69]	Corrected electrokinetic cell model
1983	Marlow, Fairhurst, Pendse [36]	First electroacoustic theory for concentrates
1983	Uusitalo [76]	Mean particle size from acoustics, patent
1983	Oja, Peterson, Cannon [70]	ESA electroacoustic effect
1988	Harker, Temple [71]	Coupled phase model for acoustics of concentrates
1987	Riebel [38]	Particle size distribution, patent for the large particles
1988-9	O'Brien [13,15]	Electroacoustic theory, particle size and ζ-potential from electroacoustics
1990	Anson, Chivers [47]	Materials database
1999	Shilov and others [16,17]	Electroacoustic theory for CVI in concentrates
1990 to present	McClements, Povey [43,44,45]	Acoustics for emulsions
1996 to present	A.Dukhin, P.Goetz [39,72,73]	Combining together acoustics and electroacoustics for Particle sizing, Rheology and Electrokinetics

1.2 Advantages of ultrasound over traditional characterization techniques

There is one major advantage of ultrasound-based techniques compared to traditional characterization methods. Ultrasound can propagate through concentrated suspensions and consequently allows one to characterize concentrated dispersions as is, without any dilution. This feature of ultrasound is applicable to both particle size and ζ-potential measurement. Dilution required by traditional techniques can destroy aggregates or flocs and the corresponding measured particle size distribution for that dilute system would not be correct for the original concentrated sample.

Elimination of dilution is especially critical for ζ-potential characterization, because this parameter is a property of both the particle and the surrounding liquid; dilution changes the suspension medium and, as a result, ζ-potential.

The many advantages of ultrasound for characterizing particle size are summarized in Table 1.3. Detailed analysis of ultrasound-based techniques is given later in this book. The following is a short summary.

Table 1.3 Features and benefits of acoustics over traditional particle sizing techniques.

Feature	Benefit
No dilution required.	Less sensitive to contamination
No calibration with the known particle size	More accurate
Particle size range from 5 nm to 1000 microns with the same sensor.	Simpler hardware, more cost effective
Simple decoupling of sound adsorption and sound scattering	Simplifies theory
Possible to eliminate multiple scattering even at high volume fractions up to 50%vol	Simplifies theory for large particle size
Existing theory for ultrasound absorption in concentrates with particles interaction	Possible to treat small particles in concentrates
Data available over wide range of wavelength	Allows use of simplified theory and reduces particle shape effects
Innate weight basis, lower power of the particle size dependence	Better for polydisperse systems
Particle sizing in dispersions with several dispersed phases (mixed dispersions)	Real world, practical systems
Particle sizing in structured dispersions.	

Acoustic methods are very robust and precise [72,73]. They are much less sensitive to contamination compared to traditional light-based techniques, because the high concentration of particles in a fresh sample dominates any small residue from the previous sample. It is a relatively fast technique. Normally a single particle size measurement can be completed in a few minutes. This feature, together with the ability to measure flowing systems, makes acoustic attenuation very attractive for monitoring particle size on-line.

There are several advantages of ultrasound over light based instruments because of the longer wavelength used. The wavelength of ultrasound in water, at the highest frequency typically used (100 MHz), is about 15 microns, and it increases even further to 1.5 millimeters at the lowest frequency (1 MHz). In contrast, light based instruments typically use wavelengths on the order of 0.5 microns. If the particles are small compared to the wavelength we say that this satisfies the Rayleigh long wavelength requirement. It is known that particle sizing in this long wavelength range is more desirable than in the intermediate or short wavelength range, because of lower sensitivity to shape factors and also a simpler theoretical interpretation. As a result, using the longer wavelengths available through acoustics allows us to characterize a much wider range of particle size, while still meeting this long wavelength requirement.

Nature provided one more significant advantage of ultrasound over light, and that is related to the wavelength dependence. As the wave travels through the colloid, it is known that the extinction of both ultrasound and light occurs due to the combined effects of both scattering and absorption [3, 74]. Since most light scattering experiments are performed at a single wavelength it is not possible to experimentally separate these two contributions to the total extinction. In fact, more often than not, the absorption of light is simply neglected in most light scattering experiments, and this can lead to errors.

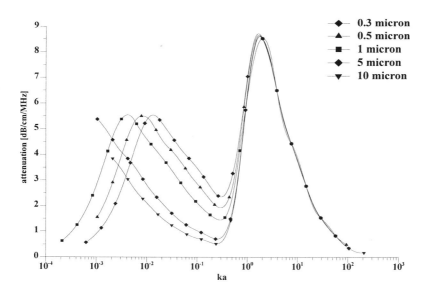

Figure 1.1 Scattering attenuation and viscous absorption of ultrasound.

In the case of ultrasound, the absorption and scattering are distinctively separated on the wavelength scale. Figure 1.1 illustrates the dependence of ultrasound attenuation as a function of relative wavelength ka defined by:

$$ka = \frac{2\pi a}{\lambda} \tag{1.1}$$

where a is the particle radius and λ is the wavelength of ultrasound.

It is seen that attenuation curve has two prominent ranges. The low frequency region corresponds to absorption; the higher frequency region corresponds to scattering. It is obvious from inspection of Figure 1.1 that it is a simple matter to separate both contributions because there is very little, indeed almost negligible, overlap.

This peculiar aspect of ultrasound frequency dependence allows one to simplify the theory tremendously. Indeed, in the wide majority of cases absorption and scattering can be considered separately. This simplification is valid except for very high volume fractions and for some special non-aqueous systems with soft particles [75].

Electroacoustics is a relatively new technique compared to acoustics. In principle it can provide information for both particle sizing and ζ-potential characterization. However, we believe that acoustics is much more suited to particle sizing than electroacoustics. For this reason, justified later in Chapters 4, 5, and 7, we consider electroacoustics as primarily a technique for characterizing only the electric surface properties like ζ-potential. In this sense electroacoustics competes with microelectrophoresis and other traditional electrokinetic methods. However, electroacoustics has many advantages over traditional electrokinetic methods that can be summarized as:

- no dilution required, volume fraction up to 50%vl;
- less sensitive to contamination, easier to clean;
- higher precision (± 0.1 mV);
- low surface charges (down to 0.1 mV);
- electrosmotic flow is not important;
- convection is not important;
- faster.

In addition, electroacoustic probes can be used for various titration experiments, as it will be shown later in Chapter 8.

The third field where ultrasound competes with traditional colloid characterization is the field of rheology. This is relatively new area of ultrasound application. We can count two obvious advantages of ultrasound over traditional rheometers. First, ultrasound measurements are non-destructive and allow us to obtain information about the high frequency rheological properties while keeping the sample intact. The second advantage is related to the ability to characterize volume viscosity in addition to shear viscosity. This was already known to Stokes 150 years ago [9]. Volume viscosity is a more sensitive probe of any structural features in a system but it is impossible to measure using shear-based techniques. Ultrasound attenuation is the only known technique able to characterize this important rheological parameter. This aspect will be dealt with in detail later in Chapter 3.

In conclusion, we think that the combination of acoustics and electroacoustics enhances each of them [39]. In addition, there is certain overlap in their nature, that offers a way to create various consistency tests to verify the reliability of the data.

REFERENCES

1. Lyklema, J. "Fundamentals of Interface and Colloid Science", Volumes 1, Academic Press, (1993)

2. Hunter, R.J. "Foundations of Colloid Science", Oxford University Press, Oxford, (1989)

3. Morse, P. and Ingard, U. "Theoretical Acoustics", McGraw-Hill, NY, (1968)

4. Kinsler, L., Frey, A., Coppens, A. and Sanders, J. "Fundamentals of Acoustics", J. Wiley & Sons, NY, (2000)

5. Blackstock, D. "Fundamentals of Physical Acoustics", J. Wiley & Sons, NY, (2000)

6. Rayleigh, L. "The Theory of Sound", Vol.2, Macmillan and Co., NY, second edition 1896, first edition (1878).

7. Kirchhoff, Pogg. Ann., vol. CXXXIV, p.177, (1868)

8. Kirchhoff, "Vorlesungen uber Mathematische Physik", (1876)

9. Stokes, "On a difficulty in the Theory of Sound", Phil. Mag., Nov. (1848)

10. Stokes, "Dynamic Theory of Diffraction", Camb. Phil. Trans., IX, (1849)

11. Isakovich, M.A. Zh. Experimental and Theoretical Physics, 18, 907 (1948)

12. Sewell, C.T.J., "The extinction of sound in a viscous atmosphere by small obstacles of cylindrical and spherical form", PhilTrans. Roy. Soc., London, 210, 239-270 (1910)

13. O'Brien, R.W. "Electro-acoustic Effects in a dilute Suspension of Spherical Particles", J. Fluid Mech., 190, 71-86 (1988)

14. Hunter, R.J. "Review. Recent developments in the electroacoustic characterization of colloidal suspensions and emulsions", Colloids and Surfaces, 141, 37-65 (1998)

15. O'Brien, R.W. "Determination of Particle Size and Electric Charge", US Patent 5,059,909, Oct.22, (1991)

16. Dukhin, A.S., Shilov, V.N., Ohshima, H., Goetz, P.J "Electroacoustics Phenomena in Concentrated Dispersions. New Theory and CVI Experiment", Langmuir, 15, 20, 6692-6706, (1999)

17. Dukhin, A.S., Shilov, V.N, Ohshima, H., Goetz, P.J "Electroacoustics Phenomena in Concentrated Dispersions. Effect of the Surface Conductivity", Langmuir, 16, 2615-2620 (2000)

18. Shilov, V.N. and Dukhin A.S. "Sound-induced thermophoresis and thermodiffusion in electric double layer of disperse particles and electroacoustics of concentrated colloids." Langmuir, submitted.

19. Tyndall, J. "Light and Electricity", D. Appleton and Com., NY (1873).

20. Tyndall, J. "Sound", Phil. Trans., 3rd addition, (1874)

21. Rayleigh, J.W. "On the Light from the Sky", Phil. Mag., (1871)

22. Rayleigh, J.W. "On the scattering of Light by small particles", Phil. Mag., (1871)

23. Rayleigh, J.W. "Acoustical Observations", Phil. Mag., vol IX, p. 281, (1880)

24. Rayleigh, J.W. "On the Application of the Principle of Reciprocity to Acoustics", Royal Society Proceedings, vol XXV, p. 118, (1876)

25. Epstein, P.S. and Carhart R.R., "The Absorption of Sound in Suspensions and Emulsions", J. of Acoust. Soc. Amer., 25, 3, 553-565 (1953)

26. Allegra, J.R. and Hawley, S.A. "Attenuation of Sound in Suspensions and Emulsions: Theory and Experiments", J. Acoust. Soc. Amer., 51, 1545-1564 (1972)

27. Eigen., "Determination of general and specific ionic interactions in solution", Faraday Soc. Discussions, , No.24, p.25 (1957)

28. Eigen, M. and deMaeyer, L. in "Techniques of Organic Chemistry", (ed. Weissberger) Vol. VIII Part 2, Wiley (1963)

29. DeMaeyer, L., Eigen, M., and Suarez, J. "Dielectric Dispersion and Chemical Relaxation", J. Of the Amer. Chem. Soc., 90, 12, 3157-3161(1968)

30. Debye, P. J. Chem. Phys., 1, 13 (1933)

31. Hermans, J. Philos. Mag., 25, 426 (1938)

32. Bugosh, J., Yeager, E. and Hovorka, F. J. Chem. Phys. 15, 592 (1947)

33. Yeager, E. and Hovorka, F. J. Acoust. Soc. Amer., 25, 443 (1953)

34. Yeager, E., Dietrick, H. and Hovorka, F. J. Acoust. Soc. Amer., 25, 456 (1953)

35. Zana, R. and Yeager, E. J. Phys. Chem, 71, 4241 (1967)

36. Beck et al., "Measuring Zeta Potential by Ultrasonic Waves", Tappi, vol.61, 63-65, (1978)

37. Dukhin, A.S., Shilov, V.N. and Borkovskaya. Yu. "Dynamic Electrophoretic Mobility in Concentrated Dispersed Systems. Cell Model.", Langmuir, 15, 10, 3452-3457 (1999)

38. Riebel, U. et al. "The Fundamentals of Particle Size Analysis by Means of Ultrasonic Spectrometry" Part. Part. Syst. Charact., vol.6, pp.135-143, (1989)

39. Dukhin, A.S. and Goetz, P.J. "Acoustic and Electroacoustic Spectroscopy", Langmuir, 12, 19, 4336-4344 (1996)

40. Dukhin, A.S. and Goetz, P.J. "Characterization of aggregation phenomena by means of acoustic and electroacoustic spectroscopy", Colloids and Surfaces, 144, 49-58 (1998)

41. Dukhin, A.S. and Goetz, P.J. "Method and device for characterizing particle size distribution and zeta potential in concentrated system by means of Acoustic and Electroacoustic Spectroscopy", patent USA, 09/108,072, (2000)

42. Dukhin, A.S. and Goetz. P.J. "Method and device for Determining Particle Size Distribution and Zeta Potential in Concentrated Dispersions", patent USA, pending

43. Povey, M. "The Application of Acoustics to the Characterization of Particulate Suspensions", in Ultrasonic and Dielectric Characterization Techniques for Suspended Particulates, ed. V. Hackley and J. Texter, Am. Ceramic Soc., Ohio, (1998)

44. McClements, J.D. "Ultrasonic Determination of Depletion Flocculation in Oil-in-Water Emulsions Containing a Non-Ionic Surfactant", Colloids and Surfaces, 90, 25-35 (1994)

45. McClements, D.J. "Comparison of Multiple Scattering Theories with Experimental Measurements in Emulsions" The Journal of the Acoustical Society of America, vol.91, 2, pp. 849-854, February (1992)

46. Holmes, A.K., Challis, R.E. and Wedlock, D.J. "A Wide-Bandwidth Study of Ultrasound Velocity and Attenuation in Suspensions: Comparison of Theory with Experimental Measurements", J. Colloid and Interface Sci., 156, 261-269 (1993)

47. Anson, L.W. and Chivers, R.C. "Thermal effects in the attenuation of ultrasound in dilute suspensions for low values of acoustic radius", Ultrasonic, 28, 16-25 (1990)

48. Holmes, A.K., Challis, R.E. and Wedlock, D.J. "A Wide-Bandwidth Ultrasonic Study of Suspensions: The Variation of Velocity and Attenuation with Particle Size", J. Colloid and Interface Sci., 168, 339-348 (1994)

49. Dukhin, A.S. and Goetz, P.J. "Acoustic Spectroscopy for Concentrated Polydisperse Colloids with High Density Contrast", Langmuir, 12, [21] 4987-4997 (1996)

50. Wines, T.H., Dukhin A.S. and Somasundaran, P. "Acoustic spectroscopy for characterizing heptane/water/AOT reverse microemulsion", JCIS, 216, 303-308 (1999)

51. Poisson, "Sur l'integration de quelques equations lineaires aux differnces prtielles, et particulierement de l'equation generalie du mouvement des fluides elastiques", Mem., de l'Institut, t.III, p.121, (1820)

52. Poisson, Journal de l'ecole polytechnique, t.VII, (1808)

53. Doppler, "Theorie des farbigen Lichtes der Doppelsterne", Prag, (1842)

54. Maxwell, "On the Viscosity or Internal Friction of Air and other Gases", Phil. Trans. vol 156, p.249, (1866)

55. Henry, Report of the Lighthouse Board of the United States for the year 1874.

56. Reynolds, O. Proceedings of the Royal Society, vol. XXII, p.531, (1874)

57. Foldy, L.L "Propagation of sound through a liquid containing bubbles", OSRD Report No.6.1-sr1130-1378, (1944)

58. Carnstein, E.L. and Foldy, L.L "Propagation of sound through a liquid containing bubbles", J. of Acoustic Society of America, 19, 3, 481- 499 (1947)

59. Pellam, J.R. and Galt, J.K. "Ultrasonic propagation in liquids: Application of pulse technique to velocity and absorption measurement at 15 Megacycles", J. of Chemical Physics, 14, 10 , 608-613 (1946)

60. Derouet, B. and Denizot, F. C. R. Acad. Sci., Paris, 233,368 (1951)

61. Booth, F. and Enderby, J. "On Electrical Effects due to Sound Waves in Colloidal Suspensions", Proc. of Amer. Phys. Soc., 208A, 32 (1952)

62. Enderby, J.A. "On Electrical Effects Due to Sound Waves in Colloidal Suspensions", Proc. Roy. Soc., London, A207, 329-342 (1951)

63. Happel J. and Brenner, H, "Low Reynolds Number Hydrodynamics", Martinus Nijhoff Publishers, Dordrecht, The Netherlands, (1973)

64. Happel J., "Viscous flow in multiparticle systems: Slow motion of fluids relative to beds of spherical particles", AICHE J., 4, 197-201 (1958)

65. Kuwabara, S. "The forces experienced by randomly distributed parallel circular cylinders or spheres in a viscous flow at small Reynolds numbers", J. Phys. Soc. Japan, 14, 527-532 (1959)

66. Andreae, J. and Joyce, P. "30 to 230 Megacycle Pulse Technique for Ultrasonic Absorption Measurements in Liquids", Brit. J. Appl. Phys., v.13, p.462-467 (1962)

67. Andreae, J., Bass, R., Heasell, E., and Lamb, J. "Pulse Technique for Measuring Ultrasonic Absorption in Liquids", Acustica, v.8, p.131-142 (1958)

68. Levine, S. and Neale, G.H. "The Prediction of Electrokinetic Phenomena within Multiparticle Systems.1.Electrophoresis and Electroosmosis.", J. of Colloid and Interface Sci., 47, 520-532 (1974)

69. Shilov, V.N., Zharkih, N.I. and Borkovskaya, Yu.B. "Theory of Nonequilibrium Electrosurface Phenomena in Concentrated Disperse System.1.Application of Nonequilibrium Thermodynamics to Cell Model.", Colloid J., 43,3, 434-438 (1981)

70. Oja, T., Petersen, G. and Cannon, D. "Measurement of Electric-Kinetic Properties of a Solution", US Patent 4,497,208, (1985)

71. Harker, A.H. and Temple, J.A.G., "Velocity and Attenuation of Ultrasound in Suspensions of Particles in Fluids", J.Phys.D.:Appl.Phys., 21, 1576-1588 (1988)

72. Dukhin, A.S. and Goetz, P.J. "Acoustic and Electroacoustic Spectroscopy for Characterizing Concentrated Dispersions and Emulsions", Adv. In Colloid and Interface Sci., 92, 73-132 (2001)

73. Dukhin, A.S. and Goetz. P.J. "New Developments in Acoustic and Electroacoustic Spectroscopy for Characterizing Concentrated Dispersions", Colloids and Surfaces, 192, 267-306 (2001)

74. Bohren, C. and Huffman, D. "Absorption and Scattering of Light by Small Particles", J. Wiley & Sons, (1983)

75. Babick, F., Hinze, F. and Ripperger, S. "Dependence of Ultrasonic Attenuation on the Material Properties", Colloids and Surfaces, 172, 33-46 (2000)

76. Uusitalo, S.J., von Alfthan, G.C., Andersson, T.S., Paukku, V.A., Kahara, L.S. and Kiuru, E.S. "Method and apparatus for determination of the average particle size in slurry", US Patent 4,412,451 (1983)

77. Cushman and oth. US Patent 3,779,070 (1973)

Chapter 2. FUNDAMENTALS OF INTERFACE AND COLLOID SCIENCE

The goal of this section is to provide a general description of the objects that go to make up the term "colloids" referred to in the title of this book, and related phenomena. We will introduce here some terms that will be used to characterize these objects, together with a brief overview of corresponding theories and traditional measuring techniques.

In writing this section, we follow J. Lyklema's definition of a colloid as "an entity, having at least in one direction a dimension between 1 nm and 1 µm, i.e. between 10^{-9} and 10^{-6} m. The entities may be solid or liquid or, in some cases, even gaseous. They are dispersed in the medium" [1]. This medium may be also solid or liquid or gaseous. We have found it more useful in this work to refer to these "entities", instead, as the "dispersed phase" and the "medium" as the "dispersion medium".

There is an enormous variety of systems satisfying this wide definition of a colloid. Given three states for the dispersed phase and three states for the dispersion medium allows, altogether, for nine possible varieties of colloids. For the purpose of this book, we restrict our concern to just three of these possibilities; namely, a collection of solid, liquid, or gas particles dispersed in a single liquid media. All of our colloids can, therefore, be defined as "a collection of particles immersed in a liquid"; the particles can be solid (a suspension or dispersion), liquid (an emulsion), or gas (a foam). These three types of dispersed systems play an important role in all kinds of applications: paints, latices, food products, cements, minerals, ceramics, blood, and many others.

All these disperse systems have one common feature; because of their small size, they all have a high surface area relative to their volume. It is surface related phenomena, therefore, that primarily determine their behavior in various processes and justify consideration of colloids as effectively a different state of matter.

What follows are the combined terms and methods that have been created over the course of two centuries to characterize these very special systems. In addition to Lyklema's book, we also incorporate terminology from "Nomenclature in Dispersion Science and Technology" published by the US National Institute of Standards and Technology [93].

2.1 Real and model dispersions

Real colloidal systems, by their very nature, are quite complex. Two of these complexities are the variation of particle size and shape within a particular dispersion. Furthermore, the particles might not even be separate, but linked together to form some network, or the liquid might contain complex additives. This inherent complexity makes it difficult, perhaps impossible, to fully characterize a real colloidal system except in the most simple cases. However, we do not always need a complete detailed characterization; a somewhat simplified picture can often be sufficient for a given particular purpose. There are two approaches to obtaining a simplified picture of any real system under investigation.

The first approach is "phenomenological" [1], in that it relates only experimentally measured variables, and usually is based on simple thermodynamic principles. This can be a very powerful approach, but, typically, it does not yield very much insight concerning any underlying structure within the system.

The second approach is "modeling", in which we replace the practical system with some approximation based on an imaginary "model colloid". This model colloid is an attempt to describe a real colloid in terms of a set of simplified model parameters including, of course, the characteristics which we hope to determine. The model, in effect, makes a set of assumptions about the real world in order to reduce the complexity of the colloid, and thereby also simplifies the task of developing a suitable prediction theory.

It is important to recognize that in substituting a model colloid for the real colloid we introduce a certain "modeling error" into the characterization process. Sometimes this error may be so large as to make this substitution inadequate or, perhaps, even worthless. For example, most particle size measuring instruments substitute a model that assumes that the particles are spherical. In this case, a complete geometrical description of a single particle is specified using just one parameter, its diameter d. Obviously, such a model would not adequately describe a dispersion of long thin carpet fibers that have a high aspect ratio, and any theory based on this over-simplified model might well give very misleading results. This is an example of an extreme case where the modeling error makes the model system completely inappropriate.

It is important to acknowledge that some "modeling error" almost always exists. We can reduce it by adopting a more complicated model in response to the complexities of the real-world colloid, but we can seldom make it vanish. In turn, the selected model then limits, or increases, the complexity of the theory that must be brought into play to adequately predict any acoustic properties based on the chosen model. Having then chosen a model for our colloid, one can then select and develop a more appropriate theory. In the course of such a development, certain theoretical assumptions, or approximations, may be made which introduce additional "theoretical errors". These theoretical errors reflect our inability to completely predict even the behavior of our already simplified model system. However, it would be pointless to use some overly elaborate and complex theory in an effort to reduce these theoretical errors to values much less than the already unavoidable modeling errors.

In doing experimental tests to confirm various theories, it is of course necessary to minimize the difference between the real colloid used for the tests and the model colloid on which the proposed theory is based. For example, the model colloid mentioned by Lyklema [1], (a collection of spherical particles with a certain size in the liquid), describes quite precisely a real colloid consisting of a suspension of monodisperse latex particles.

The design of any model colloid is also strongly related to the notion of a "dimensional hierarchy", which we introduce here to generalize the rather widely accepted idea of macroscopic and microscopic levels of characterization. For instance, Lyklema described macroscopic laws as dealing on a large scale with "gigantic numbers of molecules". In contrast, the microscopic level seeks an *ab initio* interpretation based on a very small scale or a molecular picture.

In dealing with colloids we further need to expand this simple picture of a merely two-dimensional hierarchy. Certainly, we can speak of a molecular, microscopic level, and the colloid as a whole can be viewed on a macroscopic level. But there are intermediate levels of importance as well. For instance, the size of a typical colloid particle, which lies in between the molecular and macroscopic dimension ranges, is also an important dimension in its own right, and we refer to this as the "single particle level". For many colloids, other key dimensions might also be important. For example, in a polymer system we may wish to speak of a dimension level related to the length of the polymer chains. In a flocculated system of particles, a dimension related to the floc size may be useful. For a given colloid, we refer to the complete set of all important dimensions as its "dimensional hierarchy".

The design of a suitable "model colloid", and the related theory, must somehow reflect the "dimensional hierarchy" of the real system. The model might put more emphasis on one dimension than on another; it depends on the final goal of the characterization.

As an example, consider a dispersion of biological cells in saline. We might create an elaborate model at the "single particle level" by modeling the cells as multilayer spheres with built-in ionic pumps, etc., while very simply at the "microscopic level" modeling the saline dispersion medium as a continuum with a certain viscosity and density.

However, there is one important dimension common to all colloids, namely the thickness of the interfacial layer. The behavior of this boundary layer plays a key role in many colloid-related phenomena and must also be adequately reflected in the design of the "model colloid".

The complexity of real colloids, and the corresponding diversity in the dimensional hierarchy necessary to depict these colloids, makes it practically impossible to provide a description of all the colloid models that have been proposed over the years. Historically, the more common models include the microscopic level to describe the dispersion medium, the single particle level to characterize the disperse phase, and the interfacial level to describe the interface boundary layers. We present below some general models that have been developed during last two centuries for each of these three levels.

2.2 Parameters of the model dispersion medium

Although we have restricted our attention to colloids having only liquid media, we need not limit ourselves to pure liquids. The dispersion media might be a mixture of different liquids, a liquid with many dissolved ion species, or some dissolved polymers. We can even include in our media definition the possibility that the media itself might be a colloid. For example, consider the case of kaolin clay dispersed in water and stabilized by the addition of latex particles. In reality this suspension contains two types of dispersed particles: kaolin and latex. However, an alternative view would be to consider the kaolin particles dispersed in a new latex/water "effective dispersion medium". This reduces a complex three phase colloid to a much simpler two phase colloid, but the price that we must pay is a more complex model for this "effective dispersion medium".

Basically, we are playing here with two concepts: a "homogeneous media" and a "heterogeneous media". Depending on the dimensional scale of the effect of interest, we can consider everything with lesser dimensions as the "homogeneous media". At the same time the colloid system is described as a "heterogeneous media".

Let us assume first that the dispersion medium is a simple pure liquid. For clarity, we will organize the parameters that characterize this pure liquid into six classes: gravimetric, rheological, acoustic, thermodynamic, electrodynamic and electroacoustic. To some extent all of these parameters are temperature dependent; this dependence is strong for some parameters and negligible for others. Some of these parameters also depend on the frequency of the ultrasound, which we will discuss in the following text.

The effect of the liquid chemical composition on these properties is considered in the last section of this chapter.

2.2.1 Gravimetric parameters

The density of the dispersing media, ρ_m, is an intensive (i.e. independent on the sample volume), parameter that characterizes the liquid mass per unit volume and is expressed in units of kg/m^3. Since the volume of the media changes with temperature, the density is also temperature dependent. There are some empirical expressions for this temperature dependency [3]. The density of the media is assumed to be independent of the ultrasound frequency.

2.2.2 Rheological parameters

Viscosity and elasticity are rheological parameters of general interest for any material. Most pure liquids are primarily viscous, having little elasticity. Many traditional rheological instruments can not measure the elastic component, or if possible, can measure it only at low frequency. Elasticity yields information about structure. A small elasticity implies a weak structure. Acoustics provide a means to measure the elastic properties of liquids and to characterize their structure at much higher frequencies than those available with more traditional methods [4, 5, 6].

The elastic properties of the liquid are defined by a bulk storage modulus, M_m, that is the reciprocal of the liquid compressibility, β^P_m. This compressibility is the change in the liquid volume due to a unit variation of pressure. It is usually assumed that the liquid exhibits no yield stress. The SI units of the bulk modulus are [Pascal] (Pa), which is equal to 0.1 [dyne cm^2] in CGS units.

The viscous properties of the liquid are characterized by two dynamic viscosity coefficients: a shear dynamic viscosity, η_m, and a volume dynamic viscosity, η_m^v [7]. The volume viscosity only appears in phenomena related to the compressibility of the liquid. All effects that can be described assuming an incompressible liquid are independent of the volume viscosity. This is why the volume viscosity is hard to measure and, therefore, is usually assumed to be zero.

Ultrasonic attenuation is the only known way to measure volume viscosity [6]. It turns out that all liquids exhibit a volume viscosity. For example, the volume viscosity of water is almost three times larger than the shear viscosity. The volume viscosity depends on the structure of the liquid. Hence acoustic attenuation measurements potentially provide important information about liquid structure.

The SI units for these dynamic viscosities are [Pa s] which is equal to 1000 [centipoise] (cP) in CGS units.

Rheological parameters usually depend on temperature; some empirical expressions for this dependency can be found in the Handbook of Chemistry and Physics [3]. These parameters are independent of the frequency of the ultrasound for simple Newtonian liquids.

2.2.3 Acoustic parameters

Rheological and acoustic properties are closely related; each provides a different perspective of the viscoelelastic nature of the fluid. There are two key acoustic parameters: the attenuation coefficient, α_m, and the sound speed, c_m. We have already described two key rheological parameters: the bulk storage modulus, M_m, and the dynamic viscosity, η_m. The bulk modulus and sound speed both characterize the elastic nature of the liquid; the dynamic viscosity and attenuation both characterize its viscous, or dissipative, nature.

Elastic and dissipative characteristics can be combined into one complex parameter. In acoustics this complex parameter is referred to as a "compression complex wave number", $k_{c,m}$. The attenuation and sound speed are related to the imaginary and real parts of this complex wave number by the expressions:

$$\alpha_m = -\operatorname{Im} k_{c,m} \tag{2.1}$$

$$c_m = \frac{\omega}{\operatorname{Re} k_{c,m}} \tag{2.2}$$

The units of sound speed are [m s^{-1}]. The attenuation can be expressed in several different ways as discussed in Chapter 3. In all of the following text, we define attenuation in units of dB cm^{-1} MHz^{-1}.

Another complex parameter also widely used in acoustics is the acoustic impedance, Z_m. It is defined by:

$$Z_m = c_m \rho_m (1 - j\frac{c_m \alpha_m}{\omega}) = c_m \rho_m (1 - j\frac{\alpha_m \lambda}{2\pi}) \tag{2.3}$$

where λ is a wavelength of the ultrasound in the liquid.

The attenuation is almost independent of variations in temperature, whereas the sound speed is very sensitive to temperature. For example, the sound speed of water changes 2.4 m/sec per degree Celsius at room temperature. This is an important distinction between these two parameters.

The frequency dependence of these two parameters is also quite different. For simple Newtonian liquids, the sound speed is practically independent of frequency, whereas the attenuation is roughly proportional to frequency, at least over the frequency range of 1-100 MHz (see Chapter 3 for details).

2.2.4 Thermodynamic parameters

There are three thermodynamic parameters that are important in characterizing the thermal effects in a colloid: thermal conductivity, τ_m [mW/cm C^0], heat capacity Cp^m [J/g C^0], and thermal expansion β_m [10^{-4} 1/C^0]. Fortunately, it turns out that for almost all liquids, except water, that τ_m is virtually the same, and that Cp^m can be approximated by a simple function of the material density. Figures 2.1 and 2.2 illustrate the variation of both properties for more than 100 liquids; the data is taken from the paper of Anson and Chivers [94] and our own database. This reduces the number of required thermodynamic properties to just one, namely the thermal expansion coefficient.

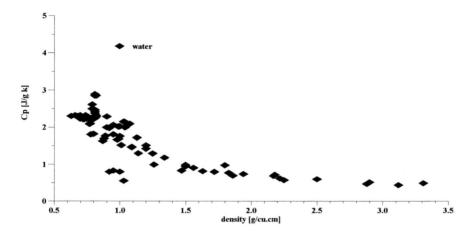

Figure 2.1 Heat capacity at constant pressure for various liquids, Anson and Chivers [94].

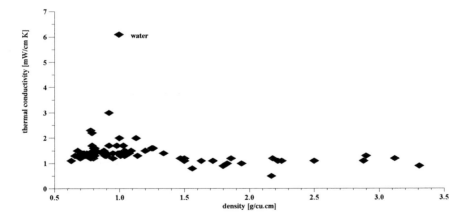

Figure 2.2 Thermal conductance for various liquids, Anson and Chivers [94].

In principle, all thermodynamic parameters are temperature dependent, but independent of the ultrasound frequency.

2.2.5 Electrodynamic parameters

The two essential electrodynamic parameters which characterize the response of the liquid to an applied electric field, are conductivity, K_m, and dielectric permittivity, $\varepsilon_o \varepsilon_m$.

The SI units of conductivity are S m^{-1} [Cs^{-1}V^{-1} m^{-1}], and of dielectric permittivity are [C^2 N^{-1} m^{-2}].

The relative dielectric permittivity, ε_m, (dielectric constant) of the liquid is the dielectric permittivity of the media divided by that of a vacuum, ε_o, and is dimensionless. The dielectric permittivity of a vacuum, ε_o, is 8.8542×10^{12} C^2 N^{-1} m^{-2}.

In principle, both the conductivity and dielectric permittivity are frequency dependent, even for a pure liquid. However, for the typical frequency range of acoustic measurements, (1 to 100 MHz), both parameters are usually assumed to be independent of frequency.

The temperature dependence of the dielectric permittivity for some liquids is given in The Handbook of Chemistry and Physics [3].

These two parameters can be combined into one complex number (either a complex dielectric permittivity, ε^*_m, or a complex conductivity, K^*_m). This simplifies the calculations with respect to an applied alternating electric field:

$$K_m^* = K_m - j\omega\varepsilon_o\varepsilon_m \qquad (2.4)$$

$$\varepsilon_m^* = \varepsilon_m(1+j\frac{K_m}{\omega\varepsilon_m\varepsilon_0}) \qquad (2.5)$$

2.2.6 Electroacoustic parameters

There are two reciprocal Electroacoustic effects that can be differentiated, depending on the nature of the driving force. For ultrasound as the driving force we speak of the Ion Vibration Current as representative of the electroacoustic effect. The Ion Vibration Current is an alternating current generated by the motion of ions in the sound field and is characterized by a magnitude, IVI, and phase, Φ_{ivi}. The IVI and phase, when normalized by the magnitude and phase of the acoustic pressure, can be considered to be the electroacoustic properties of the dispersion medium.

In the reverse case, when an electric field is the driving force, the resultant ultrasound signal, called Electrosonic Amplitude [8, 9], represents the electroacoustic effect. It is also characterized by a magnitude, ESA, and phase, Φ_{esa}. These also can be used as electroacoustic properties of the dispersion medium.

The magnitude of both electroacoustic effects is much less sensitive to temperature than to phase. Both IVI and ESA can be considered to be frequency independent within the megahertz frequency range. Both electroacoustic effects are strongly dependent on the chemical composition of the liquid.

2.2.7 Chemical composition

The dispersion media may contain many different chemical species. The collection of specified amounts of these species is called its chemical composition. Each specie is characterized with an electrochemical potential, μ_i, given by:

$$\mu_i = \mu_i^0 + 2RT \ln y_i c_i + z_i F\phi \qquad (2.6)$$

where μ_i^0 is a standard chemical potential that describes the interaction between specie, i, and the liquid, c_i is the concentration of that specie, y_i is an activity coefficient which depends on the interaction between species, R is the Faraday constant, T is the absolute temperature, z_i is the valence of that specie if electrically charged, and ϕ is an electric potential.

Finally, there are three additional parameters that characterize the dynamic response of the solution. The best known is a diffusion coefficient, D_i. In addition we will need the mass, m_i, and volume, V_i, of any solvated particles, especially for electrically charged species. These last two parameters play an important role in the electroacoustics of solvated ions.

Although all of the dispersion medium properties depend on this chemical composition, the degree of this dependence is not the same for each property. In many cases we can neglect the chemical composition.

For instance, the viscosity of a liquid depends on the concentration of ions. The continuous adjustment of the hydration layer upon ion displacement leads to friction and hence to energy dissipation. This process is known as dielectric drag. For very small ions this friction may well be stronger than for larger ions. Consequently, the viscosity, η_s, of an aqueous electrolyte solution increases with ion concentration, c, and generally follows the Jones-Dole law:

$$\frac{\eta_s}{\eta_w} = 1 + Ac^{0.5} + Bc + \ldots \qquad (2.7)$$

where η_w is the viscosity of water.

However, the parameters A and B are rather small and we can usually neglect this effect when the concentration of ions is low. For example, the viscosity of 1M KCl (aq) decreases only 1% compared to that of pure water [3].

This does not mean that we can always neglect the influence of the chemical composition on the viscosity of the medium. Many chemicals strongly affect the viscosity of the media, and the dimension of these viscosity modifying species is very important. In calculating particle size from acoustic data, the relevant viscosity to be used is the one experienced by the particle as it moves in response to the sound wave. In the case of gels, or other structured systems, this "microviscosity" may be significantly less than the "macroviscosity" measured with conventional rheometers.

This example of viscosity thus highlights the complexity regarding the influence of chemical composition on a particular property of the dispersion medium. In general, there is no theoretically justified answer that will be valid in all cases. However, in many situations empirical experience does provide a guide. From our experience, the chemical composition is always important when considering: sound speed, conductivity, and the electroacoustic signal. As a result, these are good candidates for studying the properties of ions, molecules, and other simple species in liquids. Conversely, attenuation is the least sensitive to chemical composition. This makes the attenuation measurement a very good candidate for a robust particle sizing technique.

2.3 Parameters of the model dispersed phase

First, any model for the dispersed phase particles must describe the properties of a single particle: its shape, physical and chemical properties, etc. The model must also reflect any possible polydispersity, i.e. the variation in properties from one particle to another. This is a rather difficult task. Fortunately, at least with particle shape, there is one factor that helps - Brownian motion.

Colloidal particles are generally irregular in shape as illustrated in Figure 2.3. There are of course a small number of exceptions such as emulsions or latices. All colloidal particles experience Brownian motion that constantly changes their position. A natural averaging occurs over time; the resulting time averaged particle looks like a sphere with a certain "equivalent diameter", d. It is for this reason that a sphere is the most widely and successfully used model for particle shape.

Obviously a spherical model does not work well for very asymmetric particles. Here an ellipsoid model can yield more useful results [10-12]. This adds another geometrical property: an aspect ratio, the ratio of the longest dimension to the shortest dimension. For such asymmetric particles the orientation also becomes important, bringing yet further complexity to the problem.

Here we will simply assume that the particles can be adequately represented by spheres. This assumption affects ultrasound absorption much more than ultrasound scattering, as is the case with light scattering. According to both our experience and preliminary calculations for ultrasound absorption [12] this spherical assumption is valid for aspect ratios less than 5:1.

The material of the dispersed phase particles is characterized with the same set of gravimetrical, rheological, thermodynamic, electrodynamic, acoustic and electroacoustic parameters as the material of the dispersion medium described in the previous section. We will use the same symbols for these parameters, changing only the subscript index to p.

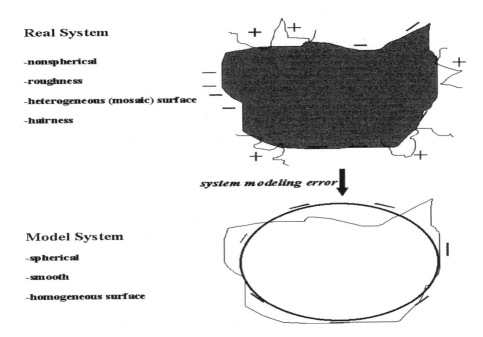

Figure 2.3 Real and model spherical particle.

The fact that we define so many properties for both the media and the particles does not mean that, in order to derive useful information from the measured acoustic or electroacoustic measurement, we need to know this complete set for a particular dispersion. Later, in Chapter 4, we will outline the set of input properties that are necessary to constructively use ultrasound measuring techniques. The exact composition of this required set depends on the nature of the particles, particularly on their rheology. In this regard, it turns out that it is very useful to divide all particles into just two classes, namely rigid and soft.

2.3.1 Rigid vs. soft particles

Historically, Morse and Ingard [5] introduced these two classes of rigid and non-rigid (soft) particles to the acoustic theory for describing ultrasound scattering. In the case of scattering, the notion of a rigid particle serves as an extreme but not real case. It turns out that this classification is even more important in the case of ultrasound absorption.

In the case of ultrasound absorption, the notion of rigid particles corresponds to real colloids, such as all kinds of minerals, oxides, inorganic and organic pigments; indeed, virtually everything except for emulsions, latices and biological material. Rigid particles dissipate (absorb) ultrasound without changing their shape in either an applied sound field or an electric field. For interpreting ultrasound absorption caused by rigid particles, we need to know, primarily, the density of the material. There is no need to know any other gravimetric or acoustic properties or, in fact, any rheological, thermodynamic, electrodynamic or electroacoustic properties at all. Furthermore, the sound speed of these rigid particles is, usually, close to 6000 m/sec. Thus, in most cases, we can simply assume that this same value is appropriate. The sound attenuation inside the body of these particles is usually negligible. We do not need to know the electrodynamic properties, except in rare cases of conducting and semi-conducting particles.

It is important to mention here that sub-micron metal particles are not truly conducting. There is not enough difference in electric potential between particle poles to conduct an electrochemical reaction across the small distance between these poles. Such sub-micron particles represent a special case of so-called "ideally-polarized particles" [33].

These empirical observations reduce the number of parameters, required to characterize rigid particles, to just one, namely the density. It is also important to mention that most often the density of these particles significantly exceeds that of the liquid media.

The situation with soft particles is very different. First, the density of these softer particles quite often closely matches the density of the dispersion media. As a result, all effects which depend on the density contrast are minimized. Instead, thermodynamic effects are much more important and the thermal expansion coefficient of the particle and suspending media replace the density contrast as a critical parameter for ultrasound absorption.

So far we have talked about particles that are internally homogeneous, constructed from the same material. However, there are some particles that are heterogeneous inside, consisting of different materials at different points within their bodies. Usually a "shell model" is used to describe the properties of these complex particles [13, 16]. Occasionally a simple averaging of properties over the particle volume suffices, as for instance, using an average density.

2.3.2 Particle size distribution

Colloidal systems are generally of a polydisperse nature - i.e. the particles in a particular sample vary in size. Aggregation, for instance, usually leads to the formation of polydisperse sols, mainly because the formation of new nuclei and the growth of established nuclei occur simultaneously, and so the particles finally formed are grown from nuclei formed at different times.

Table 2.1. Representation of PSD Data

Measure of quantity Type of quantity	Cumulative Q_r	Density $q_r(X) = \dfrac{dQ_r(X)}{dX}$
r=0 : number	Cumulative PSD on number basis	Density PSD on number basis
r=1 : length	Figure 2.5.	Figure 2.4
r=2 : area	Cumulative PSD on area basis	Density PSD on area basis
r=3 : volume, weight	Cumulative PSD on weight basis	Density PSD on weight basis

In order to characterize this polydispersity we use the well-known conception of the "particle size distribution" (PSD). We define this here following Leschonski [14] and Irani [15].

There are multiple ways to represent the PSD depending on the physical principles and properties used to determine the particle size. A convenient general scheme created by Leschonski is reproduced in Table 2.1.

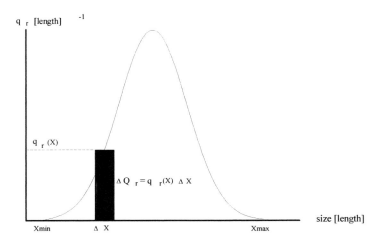

Figure 2.4. Density particle size distribution.

The independent variable (abscissa) describes the physical property chosen to characterize the size, whereas the dependent variable (ordinate) characterizes the type and measure of the quantity. The different relative amount of particles, measured in certain size intervals, form a so-called density distribution, $q_r(X)$ that represents the first derivative of the cumulative distribution, $Q_r(X)$. The subscript, r, indicates the type of the quantity chosen.

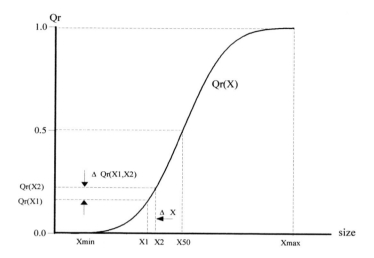

Figure 2.5. Cumulative particle size distribution.

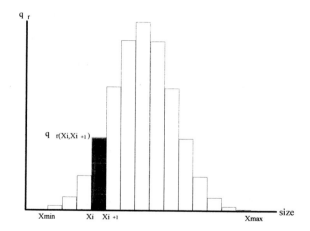

Figure 2.6. Discrete density particle size distribution

Figure 2.4 and 2.5 illustrate the density and cumulative particle size distributions. The parameter, X, characterizes a physical property uniquely related to the particle size. It is assumed that a unique relationship exists between the physical property and a one-dimensional property unequivocally defining "size".

This can be achieved only approximately. For irregular particles, the concept of an "equivalent diameter" allows one to characterize irregular particles. This is the diameter of a sphere that yields the same value of any given physical property when analyzed under the same conditions as the irregularly shaped particle.

The cumulative distribution, $Q_r(X)$ on Figure 2.5 is normalized. This distribution determines the amount of particles (in number, length, area, weight or volume depending on r) that are smaller than the equivalent diameter X.

Figure 2.6 shows the normalized discrete density distribution or histogram $q_r(X_i, X_{i+1})$. This distribution specifies the amount of particles having diameters larger than x_i and smaller than x_{i+1} and is given by:

$$q_r(X_i, X_{i+1}) = \frac{Q_r(X_{i+1}) - Q_r(X_i)}{X_{i+1} - X_i} \tag{2.7}$$

The shaded area represents the relative amount of the particles.

The histogram transforms to a continuous density distribution when the thickness of the histogram column limits to zero. It is presented in Figure 2.7 and can be expressed as the first derivative from the cumulative distribution:

$$q_r(X) = \frac{dQ_r(X)}{dX} \tag{2.8}$$

A histogram is suitable for presentation of the PSD when the value of each particle size fraction is known. It is the so-called "full particle size distribution". It can be measured using either counting or fractionation techniques. However, in many cases, full information about the PSD is either not available or not even required. For instance, Figure 2.7 represents histograms with a non-monotonic and non-smooth variation of the column size. This might be related to fluctuations and have nothing to do with the statistically representative PSD for this sample. That is why in many cases histograms are replaced with various analytical particle size distributions.

There are several analytical particle size distributions that approximately describe empirically determined particle size distributions. One of the most useful, and widely used distributions, is the log-normal. In general, it can be assumed that the size, X, of a particle grows, or diminishes, according to the relation[15]:

$$\frac{dX}{dt} = const * \frac{(X - X_{min})(X_{max} - X)}{(X_{max} - X_{min})} \tag{2.9}$$

This equation reflects the obvious fact that when d approaches either X_{max}, or X_{min}, it becomes independent of time. Using assumptions of the normal distribution of growth and destruction times, combined with Equation 2.9, leads to:

$$q_r(X) = \frac{1}{\sqrt{2\pi}\ln\sigma_l} EXP(-[\frac{\ln\frac{(X-X_{min})(X_{max}-X_{min})}{(X_{max}-X)\bar{X}}}{\sqrt{2}\ln\sigma_l}]^2) \qquad (2.10)$$

where $\bar{X}$ and σ_l are two unknown constants that can have some certain physical meaning, especially for the standard log-normal PSD (Equation 2.11). The distribution given with Equation 2.10 is called the "modified log-normal". It is not symmetrical on a logarithmic scale of particle sizes. One example of a modified log-normal PSD is shown in Figure 2.7.

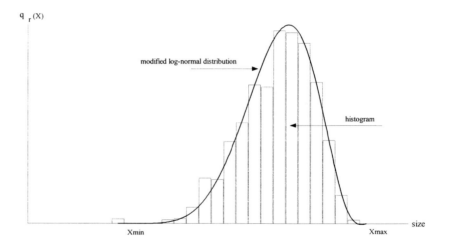

Figure 2.7. Transition from the discrete to the contineous distribution.

For the case when $X_{min} = 0$ and $X_{max} = $ infinity, the modified log-normal PSD reduces to the standard log-normal PSD:

$$q_r(X) = \frac{1}{\sqrt{2\pi}\ln\sigma_l} EXP\{-[\frac{\ln\frac{X}{\bar{X}}}{\sqrt{2}\ln\sigma_l}]^2\} \qquad (2.11)$$

Parameters $\bar{X}$ and σ_l are then the geometrical median size and geometrical standard deviation. The median size corresponds to the 50% point on the cumulative curve. The standard deviation characterizes the width of the distribution. It is smallest ($\sigma_l = 1$) for a monodisperse PSD. It can be defined as the ratio of size at 15.87% cumulative probability to that at 50%, or the ratio at 50% probability to that at 84.13%. These points, at 16% and 84% roughly, are usually reported together with the median size at 50%.

Log-normal and modified log-normal distributions require that we extract only a few parameters (2 and 4 respectively) from the experimental data. This is a big advantage in many cases when amount of the experimental data is restricted. These two distributions plus a bimodal PSD as the combination of two lognormals are usually sufficient for characterizing the essential features of the vast majority of practical dispersions.

2.4. Parameters of the model interfacial layer

In the previous two sections we have introduced a set of properties to characterize both the dispersion medium and the dispersed particles. It is clear that the same properties might have quite different values in the particle as compared to the medium. For example, in an oil-in-water emulsion, the viscosity and density inside the oil droplet would be quite different than the values in the bulk of the water medium. However, this does not mean that these parameters change stepwise at the water-oil interface. There is a certain transition, or interfacial layer, where the properties vary smoothly from one phase to the other. Unfortunately, there is no thermodynamic means to decide where one phase or the other begins. A convention suggested by Gibbs resolves this issue [1], but the details of this rather complicated thermodynamic problem are beyond the scope of this book.

We wish to discuss here only the electrochemical aspects of this interfacial layer and these are generally combined as the concept of the "electric double layer" (DL). Lyklema [1] made a most comprehensive review of this concept, which plays such an important role in practically all aspects of Colloid Science. We present here a short overview of the most essential features of the DL, with particular emphasis on those that can be characterized using ultrasound based technology.

We will consider the DL in two states: equilibrium and polarized. We use term "equilibrium" as a substitute for the term "relaxed" used by Lyklema [1] (The term relaxed is somewhat more general, and might include situations when the DL is created with non-equilibrium factors. For instance, the DL of living biological cells might have a component that is related to the non-equilibrium transmembrane potential [16, 17].)

For the first equilibrium state, the DL exists in an undisturbed dispersion characterized by a minimum value of the free energy. The second "polarized" state reflects a deviation from this equilibrium state due to some external disturbance such as an electric field.

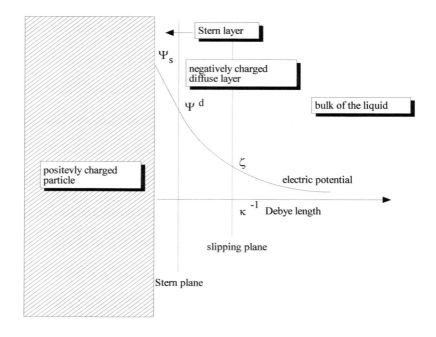

Figure 2.8. Illustration of the structure of the electric double layer.

According to Lyklema: "...the reason for the formation of a "relaxed" ("equilibrium") double layer is the non-electric affinity of charge-determining ions for a surface...". This process leads to the build up of an "electric surface charge", σ, expressed usually in $\mu C\ cm^{-2}$. This surface charge creates an electrostatic field that then affects the ions in the bulk of the liquid (Figure 2.8). This electrostatic field, in combination with the thermal motion of the ions, creates a countercharge, and thus screens the electric surface charge. The net electric charge in this screening diffuse layer is equal in magnitude to the net surface charge, but has the opposite polarity. As a result the complete structure is electrically neutral. Some of the counter-ions might specifically adsorb near the surface and build an inner sub-layer, or so-called Stern layer. The outer part of the screening layer is usually called the "diffuse layer".

What about the difference between the surface charge ions and the ions adsorbed in the Stern layer? Why should we distinguish them? There is a thermodynamic justification [1] but we think a more comprehensive reason is kinetic (ability to move). The surface charge ions are assumed to be fixed to the surface (immobile); they cannot move in response to any external disturbance. In contrast, the Stern ions, in principle, retain some degree of freedom, almost as high as ions of the diffuse layer [22, 24, 25, 26, 27].

We give below some useful relationships of the DL theory from Lyklema's review [1].

2.4.1. Flat surfaces

The DL thickness is characterized by the so-called Debye length κ^{-1} defined by:

$$\kappa^2 = F^2 \sum_i \frac{c_i z_i}{\varepsilon_0 \varepsilon_m RT} \qquad (2.12)$$

where the valences have sign included, for a symmetrical electrolyte $z_+ = -z_- = z$

For a flat surface and a symmetrical electrolyte, of concentration C_s, there is a straightforward relationship between the electric charge in the diffuse layer, σ^d, and the Stern potential, ψ^d, namely:

$$\sigma^d = -\sqrt{8\varepsilon_0 \varepsilon_m C_s RT} \sinh \frac{F\psi^d}{2RT} \qquad (2.13)$$

If the diffuse layer extends right to the surface, Equation 2.13 can then be used to relate the surface charge to the surface potential.

Sometimes it is helpful to use the concept of a differential DL capacitance. For a flat surface and a symmetrical electrolyte this capacitance is given by:

$$C_{dl} = \frac{d\sigma}{d\psi}\bigg|_{\psi=\psi_s} = \varepsilon_0 \varepsilon_m \kappa \cosh \frac{F\psi^d}{2RT} \qquad (2.14)$$

For a symmetrical electrolyte, the electric potential, ψ, at the distance, x, from the flat surface into the DL, is given by:

$$EXP(-\kappa x) = \frac{\tanh \frac{zF\psi(x)}{4RT}}{\tanh \frac{zF\psi^d}{4RT}} \qquad (2.15)$$

The relationship between the electric charge and the potential over the diffuse layer is given by:

$$\sigma^d = -(sign\psi^d)\sqrt{2\varepsilon_0 \varepsilon_m C_s RT}(v_+ \ell^{-z_+ \bar{\psi}_d} + v_- \ell^{-z_- \bar{\psi}_d} - v_+ - v_-)^{1/2} \qquad (2.16)$$

where $v_\pm$ is the number of cations and anions produced by dissociation of a single electrolyte molecule, and $\bar{\psi}^d$ is a dimensionless potential given by:

$$\bar{\psi}^d = \frac{F\psi^d}{RT} \qquad (2.17)$$

In the general case of an electrolyte mixture there is no analytical solution. However, some convenient approximations have been suggested [1, 22, 50, 68].

2.4.2. Spherical DL, isolated and overlapped

There is only one geometric parameter in the case of a flat DL, namely the Debye length $1/\kappa$. In the case of a spherical DL, there is an additional geometric parameter, namely the radius of the particle, a. The ratio of these two parameters (κa) is a dimensionless parameter that plays an important role in Colloid Science. Depending on the value of κa, two asymptotic models of the DL exist.

A "thin DL" model corresponds to colloids in which the DL is much thinner than particle radius, or simply:

$$\kappa a >> 1 \qquad (2.18)$$

The vast majority of aqueous dispersions satisfy this condition, except for very small particles in low ionic strength media. If we assume an ionic strength greater than 10^{-3} mol/l, corresponding to the majority of most natural aqueous systems, the condition $\kappa a >> 1$ is satisfied for virtually all particles having a size above 100 nm.

The opposite case of a "thick DL" corresponds to systems where the DL is much larger than the particle radius, or simply:

$$\kappa a << 1 \qquad (2.19)$$

The vast majority of dispersions in hydrocarbon media, having inherently very low ionic strength, satisfy this condition.

These two asymptotic cases allow one to picture, at least approximately, the DL structure around spherical particles. A general analytical solution exists only for low potential:

$$\psi^d << \frac{RT}{F} \approx 25.8 mV \qquad (2.20)$$

This so-called Debye-Hückel approximation yields the following expression for the electric potential in the spherical DL, $\psi(r)$, at a distance, r, from the particle center:

$$\psi(r) = \psi^d \frac{a}{r} EXP(-\kappa(r-a)) \qquad (2.21)$$

The relationship between diffuse charge and the Stern potential is then:

$$\sigma^d = -\varepsilon_o \varepsilon_m \kappa \psi^d (1 + \frac{1}{\kappa a}) \qquad (2.22)$$

This Debye-Hückel approximation is valid for any value of κa, but this is somewhat misleading, since it covers only isolated double layers. The approximation does not take into account the obvious probability of an overlap of double layers as in a concentrated suspension, i.e. high volume fraction. A simple estimate of this critical volume fraction, φ_{over}, is that volume fraction for which the Debye length is equal to the shortest distance between the particles. Thus::

$$\varphi_{over} \approx \frac{0.52}{(1+\frac{1}{\kappa a})^3} \qquad (2.23)$$

This dependence is illustrated in Figure 2.9.

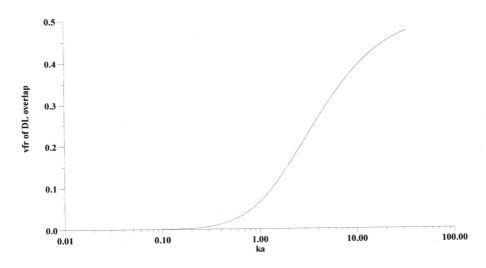

Figure 2.9. Estimate of the volume fraction of the overlap of the electric double layer

It is clear that for $\kappa a \gg 10$ (thin DL) we can consider the DL's as isolated entities even up to volume fractions of 0.4. However, the model for an isolated DL becomes somewhat meaningless for small κa (thick DL) because DL overlap then occurs even in very dilute suspensions. Theoretical treatments have been suggested for these two extreme cases of DL thickness. Briefly they are:

Thin DL ($\kappa a > 10$) Loeb-Dukhin-Overbeek [18, 19, 20] proposed a theory to describe the DL structure for a symmetrical electrolyte, applying a series expansion in terms of powers of $(\kappa a)^{-1}$. The final result is:

$$\sigma^d = -\frac{2FC_s z}{\kappa}(2\sinh\frac{z\overline{\psi}^d}{2} + \frac{4\tanh\frac{z\overline{\psi}^d}{4}}{\kappa a}) \qquad (2.24)$$

Thick DL ($\kappa a < 1$). No theory for an isolated DL is necessary, since DL overlap must be considered even in the dilute case. A theory that does include DL overlap has been proposed only very recently [21]:

$$\sigma^d = -\frac{2FC_s z}{\kappa}(2\sinh\frac{z\overline{\psi}^d}{2} + \frac{4\tanh\frac{z\overline{\psi}^d}{4}}{\kappa a} + \frac{3B}{\kappa a}EXP(-\frac{z\overline{\psi}^d}{2})\sinh(\frac{z\overline{\psi}^d}{2})) \qquad (2.25)$$

where
$$B = \frac{A\ EXP(\kappa(b-a))}{2\kappa a}(\kappa b - 1)$$

The constant, A, is obtained by matching the asymptotic expansions of the long distance potential distribution with the short distance distribution; b is the radius of the cell.

This new theory suggests, at last, a way to develop a general approach to the various colloidal-chemical effects in non- aqueous dispersions.

2.4.3. Electric Double Layer at high ionic strength

According to classical theory, the DL should essentially collapse and cease to exist when the ionic strength approaches 1 M. At this high electrolyte concentration, the Debye length becomes comparable with the size of the ions, implying that the counter ions should collapse onto the particle surface. However, there are indications [95-102] that, at least in the case of hydrophilic surfaces, the DL still exists even at ionic strengths exceeding 1 M.

The DL, at high ionic strength, is strongly controlled by the hydrophilic properties of the solid surfaces. The affinity of the particle surface for water creates a structured surface water layer, and this structure changes the properties of the water considerably from that in the bulk.

Figure 2.10 illustrates the DL structure near a hydrophilic surface. Note the difference between the much lower dielectric permittivity within the structured surface layer, ε_{sur}, compared to that of the bulk liquid, ε_m. There are two effects caused by this variation of the dielectric permittivity [100]

- The structured surface layer becomes insolvent, at least partially, to ions.
- The structured surface layer may gain electric charge due to the difference in the standard chemical potentials of the ions.

Both effects influence the electrokinetic behavior of the colloids at high ionic strength. The first effect is the more important because it leads to a separation of electric charges in the vicinity of the particle. The insolvent structured layer repels the screening electric charge from the surface to the bulk, and makes Electrokinetics possible, at high ionic strength, if the structured layer retains a certain hydrodynamic mobility in the lateral direction.

A theory of Electroosmosis for the insolvent and hydrodynamically mobile DL at high ionic strength has been proposed by Dukhin and Shilov in a paper which is not available in English. There is a description and reference to this theory in an experimental paper that is available in English [101]. We discuss these effects in more detail in Chapter 5.

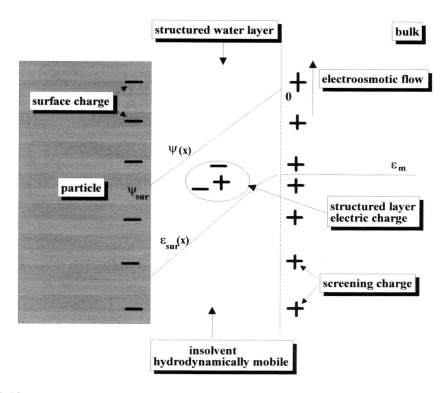

Figure 2.10. Electric Double Layer near hydrophilic surface with structured water layer.

2.4.4. Polarized state of the Electric Double Layer

External fields may affect the DL structure. These fields might be of various origins: electric, hydrodynamic, gravity, concentration gradient, acoustic, etc. Any of these fields might disturb the DL from its equilibrium state to some "polarized" condition. [1, 22].

Such external fields affect the DL by moving the excess ions inside it. This ion motion can be considered as an additional electric surface current, I_s (Figure 2.11). This current is proportional to the surface area of the particle, and to a parameter called the surface conductivity K^σ [1]. This parameter reflects the excess conductivity of the DL due to excess ions attracted there by the surface charge.

The surface current adds to the total current. But there is another component that depends on the nature of the particle, which has the opposite effect. For example, a non-conducting particle reduces the total current going through the conducting medium, simply because the current can not pass through the non-conducting volume of the particle. In the important case of charged non-conducting particles these two opposing effects influence the total current value. The surface current increases the total current, whereas the non-conducting character reduces it. The amplitude of the surface current is proportional to the surface conductivity K^σ. The amplitude current arising from the non-conductive character depends on the conductivity of the liquid, K_m; the higher the value the more current is lost, because the conducting liquid is replaced with non-conducting particles.

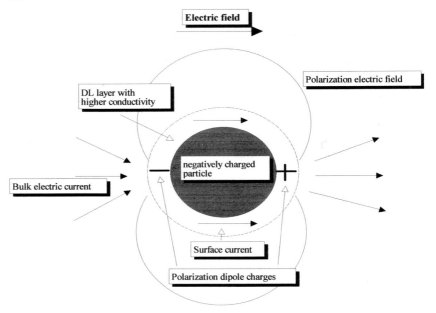

Figure 2.11. Mechanism of polarization of the electric double layer

The balance of these two effects depends on the dimensionless number, Du given by:

$$Du = \frac{K^\sigma}{aK_m} \quad (2.26)$$

The particle size, a, is in the denominator of Equation 2.26 because the surface current is proportional to the surface of the particle (a^2), whereas the reduction of the total current due to the non-conducting particles is proportional to the volume of the particles (a^3).

The abbreviation, *Du*, for this dimensionless parameter was introduced by J. Lyklema. He called it the "Dukhin number" [1] after S. S. Dukhin, who explicitly used this number in his analysis of electrokinetic phenomena [22,33].

The surface electric current redistributes the electric charges in the DL. One example is shown in Figure 2.11. It is seen that this current produces an excess of positive charge at the right pole of the particle and a deficit at the left pole. Altogether, this means that the particle gains an "induced dipole moment" (IDM) [22, 33] due to this polarization in the external field. A calculation of the IDM value for the general case is a rather complex problem. It can be substantially simplified in the case of the thin DL ($\kappa a > 10$).

There is a possibility to split the total electric field around the particle into two components: a "near field" component and a "far field" component. The "near field" is located inside of the DL. It maintains each normal section of the DL in a local equilibrium. This field is shown as a small arrow inside of the DL in Figure 2.11.

The "far field" is located outside of the DL, in the electro-neutral area. It is simply the field generated by the IDM.

Let us assume now that the driving external field is an electric field. The polarization of the DL affects all of the other fields around the particles: hydrodynamic, electrolyte concentration, etc. However, all other fields are secondary in relationship to the driving electric field applied to the dispersion. The IDM value, p_{idm}, is proportional to the value of the driving force, the electric field strength, E, in this case. In a static electric field these two parameters are related with the following equation:

$$p_{ind} = \gamma_{ind} E = -2\pi\varepsilon_0\varepsilon_m a^3 \frac{1-Du}{1+2Du} E \qquad (2.27)$$

where γ_{ind} is the polarizability of the particle.

The IDM value depends not only on the electric field distribution, but also on the distribution of the other fields generated near the particle. For example, the concentration of electrolyte shifts from its equilibrium value in the bulk of the solution, C_s. This effect, called concentration polarization, complicates the theory of electrophoresis, and is the main reason for the gigantic values found for dielectric dispersion in a low frequency electric field.

Fortunately, this concentration polarization effect is not important for ultrasound, since it appears only in the low frequency range where the frequency, ω, is smaller than the characteristic concentration polarization frequency, ω_{cp}:

$$\omega << \omega_{cp} = \frac{D_{eff}}{d^2} \qquad (2.28)$$

where D_{eff} is the effective diffusion coefficient of electrolyte.

Typically, this concentration polarization frequency is below 1 MHz, even for particles as small as ten nanometers.

The IDM value becomes a complex number in an alternating electric field, as well as polarizability. According to the Maxwell-Wagner-O'Konski theory [28-30], the IDM value for a spherical non-conducting particle is:

$$p^*_{idm} = \gamma^*_{idm} E = \frac{\varepsilon^*_p - \varepsilon^*_m}{2\varepsilon^*_m + \varepsilon^*_p} E \qquad (2.29)$$

where the complex conductivity of the media is defined by Equation 2.4.

The complex conductivity of the charged non-conducting particle is given by:

$$\varepsilon^*_p = \varepsilon_p + j\frac{2K_m Du}{\omega\varepsilon_0} \qquad (2.30)$$

The influence of the DL polarization on the IDM decreases as a function of the frequency (following from Equation 2.29). There is a certain frequency above which the DL influence becomes negligible because the field changes faster than the DL can react. This is the so-called Maxwell-Wagner frequency, ω_{MW}, which is defined as:

$$\omega_{MW} = \frac{K_m}{\varepsilon_0 \varepsilon_m} \qquad (2.31)$$

So far we have described the polarization of the DL mostly as it relates to an external electric field. Indeed, most theoretical efforts over last 150 years have been expended in this area. However, we have already mentioned that other fields can also polarize the DL. The major interest of this book is to describe in some detail the polarization caused, not by this electric field, but a hydrodynamic field, since this gives rise to the electroacoustic phenomena in which we are interested.

2.5. Interactions in Colloid and Interface science

The characterization of interactions is strongly connected to the idea of a "dimension hierarchy". We can speak of several specific dimensional levels in the colloid and each level has its own specific interaction features.

Interactions that originate at the microscopic level, where we deal with molecules and ions, are so-called "chemical interactions". They are usually related to the exchange of electron density between various microscopic bodies. There are many indications that ultrasound techniques are well suited for characterizing such chemical interactions. Although not the main subject of this book, this aspect will be covered briefly in Chapter 3.

We are most interested in effects that occur at the "single particle level" or, in other words, effects with a characteristic spatial parameter on the order of the size of a typical particle. This means that many thousands of molecules and ions are involved in this interaction. We observe the result of some averaging of individual chemical interactions. This averaging masks, and practically eliminates, any recognizable features of the original chemical interactions and microscopic objects. It justifies the introduction of a completely new set of concepts to describe effects at this higher structural level. We call all interactions that happen on this level "colloidal-chemical interactions", and can further subdivide these reactions into two categories: equilibrium and non-equilibrium.

Equilibrium interactions, as the name suggests, corresponds to interactions where the particles are in equilibrium with the dispersion medium. These interactions, subject to the classical DLVO theory, include:
1. Dispersion Van der Waals forces.
2. Electrostatic forces of the overlapped double layers.
3. Various steric interactions.

Non-equilibrium interactions correspond to cases where the equilibrium between the particle and the dispersed phase is disturbed. For example, any external field, whether it is electric, hydrodynamic, gravitational or magnetic, might create such a disturbance. Living biological cells present a special case in which the equilibrium is disturbed by an electrochemical exchange between the cell and the dispersion medium.

We briefly consider here both equilibrium and non-equilibrium colloidal-chemical interactions. The purpose of this overview is merely to determine what features, of these colloidal-chemical interactions, may affect ultrasound in the dispersions and, consequently, to learn which of these features might possibly be characterized with ultrasound-based techniques.

2.5.1. Interactions of colloid particles in equilibrium. Colloid stability.

The affinity between the particles and the dispersion medium allows us to separate colloids into two broad classes. "Lyophobic" colloids represent systems with very low affinity between particle and the liquid, whereas for "lyophilic" colloids this affinity is high. From this definition, it follows that particles of a lyophobic colloid tend to aggregate, in an effort to reduce the contact surface area between the particle and the media. Such colloids are, usually, thermodynamically unstable in the strict sense, but might actually be kinetically stable for a long period. In contrast, lyophilic colloids are, often, thermodynamically quite stable and consequently kinetics are not nearly as important.

It is possible to convert one colloid class to the other by applying appropriate surface modifications. We can coagulate a lyophilic dispersion even though it was initially thermodynamically stable, or we can stabilize a lyophobic dispersion even though initially unstable. It is simply a question of our overall objective. We will show, later (see Chapter 8), that ultrasound based techniques might be very helpful for optimizing these procedures. In order to explain the possibilities open to ultrasound characterization we first need to give a short description of the particles' interaction and coagulation.

There are several mechanisms of particle-particle interaction that are described in detail in the DLVO theory [34-37], and can be found in most handbooks on Colloid Science [1, 42, 50].

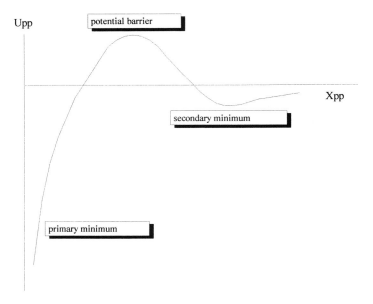

Figure 2.12. Typical DLVO interaction energy.

The presentation of the DLVO theory usually starts with the simplest case, in which the potential energy, U_{pp}, on bringing two surfaces close together, can be expressed as the sum of the Van der Waals attraction and DL repulsion. These two basic mechanisms exhibit a quite different dependence with respect to the separation distance, x_{pp}, between the surfaces. The Van der Waals attraction decays quite rapidly as an inverse power of the separation distance. The DL repulsion, on the other hand, typically decreases more slowly, more or less exponentially with this distance, and, typically, has a range on the order of the Debye length κ^{-1}. As a result, the Van der Waals attraction is dominant for small separations, whereas the DL repulsion may be dominant at larger distances. A general potential energy plot illustrating qualitatively the main features of this interaction is given in Figure 2.12.

The key feature of this potential energy plot is that there is often a well defined potential energy barrier at a distance comparable to the Debye length. A particle which approaches this barrier, perhaps due to its random Brownian motion, must have sufficient energy to overcome this barrier if it is to fall into the primary minimum and result in an aggregated particle. This barrier is therefore one important key to achieving stability in a colloid system.

This potential barrier exists when the DL repulsion exceeds the Van der Waals attraction. We can not do much to change the attractive forces, since these are fixed by the Hamaker constants of the material. We can only manipulate the repulsive forces by changing either the potential at the Stern Plane, or by changing the rate at which this potential decreases with distance, (as determined by the Debye length). The Stern potential can be affected strongly by the addition of potential-determining ions to the media. The Debye length can be decreased by increasing the ionic strength. Consequently, the height of this potential barrier depends on the Stern potential of the particles and electrolyte concentration or ionic strength. An increase in the ionic strength reduces this barrier height by shifting it closer to the surface where the Van der Waals attraction is stronger. There is a critical concentration of electrolyte, C_{ccc}, at which this barrier disappears altogether, and the particles might therefore coagulate.

$$C_{ccc}[mol\ l^{-1}] = \frac{9.85 \otimes 10^4 \varepsilon_m^3 R^5 T^5}{F^6 A^2 z^6} \frac{(\tanh \frac{z\overline{\psi}^d}{2} - 1)^4}{(\tanh \frac{z\overline{\psi}^d}{2} + 1)^4} \quad (2.32)$$

A more detailed and accurate calculation of this critical coagulation concentration, as well as other stability criteria, can be found in [1, 37].

In addition to ionic strength, aggregation stability is also sensitive to the concentration of DL potential-determining ions that control the value of the Stern potential. In principle, these two parameters (overall ionic strength and concentration of potential-determining ions) might be considered as independent parameters, which brings us to the conclusion that both aggregation stability and DL potentials should be considered in some multidimensional space. Contours of constant ς-potential would then appear as a "fingerprint" of the particular colloid. Indeed, this idea of "fingerprinting" is the logical way to characterize a colloidal system, and Marlow and Rowell have successfully applied this principle [38, 39] for various applications. We show, as one example, the three-dimensional plot of a 40% wt kaolin slurry that was titrated with sodium hexametaphosphate. From examination of this it is clear that the highest surface charge occurs at a pH value of 9.5 and at a hexametaphosphate to kaolin ratio of 0.6%wt. This, then, is the optimum condition for gaining aggregative stability.

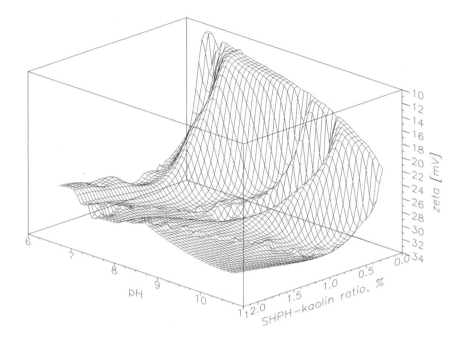

Figure 2. 13. Electrokinetic three-dimensional plot for kaolin slurry in the sodium hexametaphosphate solution.

The hexametaphosphate affects the stability of the kaolin particles through electrostatic interactions. There are different types of stabilizing agents which might enhance stability via other mechanisms, such as steric interactions. For instance, various polymers adsorb on the particle surface and change its affinity to the solvent from lyophobic to lyophilic. However, there is a danger, if the stabilizing agent is overdosed, of flocculating the colloid through the bridging effect. This danger is the reason why surface characterization is important. It allows one to determine the optimum dose of the stabilizing agent and thus eliminate the dangers of flocculation.

2.5.2. *Interaction in a hydrodynamic field. Cell and core-shell models. Rheology.*

A hydrodynamic field exists when a particle moves relative to the liquid media. There are a number of effects caused by this motion (see [46]). When a particles moves relative to a viscous liquid, it creates a sliding motion of the liquid in its vicinity. The volume over which this disturbance exists is usually referred to as the "hydrodynamic boundary layer". As the volume fraction of the sample is increased, the average distance between particles becomes smaller.

When this distance becomes comparable to the thickness of the hydrodynamic layer these layers overlap and, as a result, one moving particle affects its immediate neighbors. The particles interact through their respective hydrodynamic fields, and such interactions can be defined mathematically.

The motion of the liquid in the hydrodynamic layer surrounding a particle, is described in the terms of a local liquid velocity, u, and a hydrodynamic pressure, P. Liquids are usually assumed to be incompressible, described mathematically as:

$$Div\ u = 0 \tag{2.33}$$

We will see later that this assumption can be used even for ultrasound related phenomena in the long wavelength range. The assumption of fluid incompressibility then allows one to use a simplified version of the Navier-Stokes' equation [46], namely:

$$\rho_m \frac{du}{dt} = \eta\, rot\, rot\, u + grad P \tag{2.34}$$

Equation 2.34 neglects terms related to the volume viscosity, η_m^v, of the liquid.

These last two equations require a set of boundary conditions in order to calculate the liquid velocity and hydrodynamic pressure. The boundary conditions at the surface of the particle are rather obvious:

$$u_r(r = a) = u_p - u_m \tag{2.35}$$

$$u_\theta(r = a) = -(u_p - u_m) \tag{2.36}$$

where r and θ are spherical coordinates associated with the particle.

The other set of boundary conditions must specify the velocity of the liquid somewhere in the bulk. For a dilute system this presents no problem. Particles do not interact and this point can be selected infinitely far from the particles. This yields, for instance, a well-known Stokes' law [47] for the frictional force, F^h, exerted on particles moving relative to the liquid:

$$F^h = 6\pi \eta a \Omega (u_p - u_m) \tag{2.37}$$

where Ω is the drag coefficient.

In the case of a concentrated colloid this approach is not viable because the particles interact and the hydrodynamic field of one particle affects the hydrodynamic fields of its neighbors. The concept of a "cell model" [48, 49] allows one to take into account this interaction and resolve the hydrodynamic problem for a collection of interacting particles.

The basic feature of a "cell model" is that each particle, in the concentrated system, is considered separately inside of spherical cell of liquid that is associated only with the given individual particle (Figure 2.14). In the past, the cell model has been applied only to monodisperse systems. This restriction allows one to define the radius of the "cell". Equating the solid volume fraction of each cell, to the volume fraction, φ, of the entire system, yields the following expression for the cell radius, b:

$$b = \frac{a}{\sqrt{\varphi}} \qquad (2.38)$$

The cell boundary conditions formulated on the outer boundary, at r = b, of the cell reflect the particle-particle interaction.

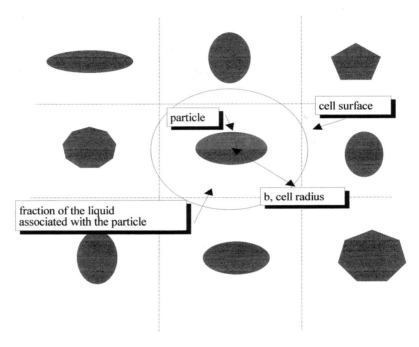

Figure 2.14. Illustration of the cell model

The two most widely used versions of these boundary conditions are named after their authors: the Happel cell model [48], and the Kuwabara cell model [49]. Both of them are formulated for an incompressible liquid.

For the Kuwabara cell model, the cell boundary conditions are given by the following equations:

$$rot\, u_{r=b} = 0 \qquad (2.39)$$
$$u_r(r = b) = 0 \qquad (2.40)$$

In the case of the Happel cell model they are:

$$\Pi_{r\theta}(r = b) = \frac{1}{r}\frac{\partial u_r}{\partial \theta} + r\frac{\partial \frac{u_\theta}{r}}{\partial r} = 0 \qquad (2.41)$$

$$u_r(r = b) = 0 \qquad (2.42)$$

The general solution for the velocity field contains three unknown constants C, C_1 and C_2

$$u_r(r) = C(1-\frac{b^3}{r^3}) + 1.5\int_r^b (1-\frac{x^3}{r^3})h(x)dx \qquad (2.43)$$

$$u_\theta(r) = -C(1+\frac{b^3}{2r^3}) - 1.5\int_r^b (1+\frac{x^3}{2r^3})h(x)dx \qquad (2.44)$$

$$C = -\frac{\tilde{b}}{3}(C_1 h_1(\tilde{b}) - C_2 h_2(\tilde{b})) \qquad (2.45)$$

where:

$$h(x) = C_1 h_1(x) + C_2 h_2(x) \qquad (2.46)$$

$$h_1(x) = \frac{\exp(-x)}{x}[\frac{x+1}{x}\sin x - \cos x + i(\frac{x+1}{x}\cos x + \sin x)]$$

$$h_2(x) = \frac{\exp(x)}{x}[\frac{x-1}{x}\sin x + \cos x + i(\frac{1-x}{x}\cos x + \sin x)]$$

The drag coefficient can be expressed, in a general form, for both Kuwabara and Happel cell models:

$$\Omega = -\frac{\tilde{a}^2}{3}(\frac{d(C_1 h_1 + C_2 h_2)}{dx} + \frac{C_1 h_1 + C_2 h_2}{\tilde{a}})_{x=\tilde{a}} - \frac{4j\tilde{a}^2}{9} \qquad (2.47)$$

where :

$\tilde{a}^2 = a^2 \omega \rho_m / 2\eta_m$

$\tilde{b} = b\tilde{a}/a$

x is normalized in the same manner as $\tilde{a}$.

Coefficients C_1 and C_2 are different in each of the two cell models, (see Table 2.2).

Table 2.2. Parameters of the cell model solutions.

	Kuvabara	Happel
C_1	$\dfrac{h_2(\tilde{b})}{I}$	$\dfrac{\tilde{b} h_2(\tilde{b}) - 2I_{23}}{\tilde{b} I + 2(I_2 I_{13} - I_1 I_{23})}$
C_2	$-\dfrac{h_1(\tilde{b})}{I}$	$-\dfrac{\tilde{b} h_1(\tilde{b}) - 2I_{13}}{\tilde{b} I + 2(I_2 I_{13} - I_1 I_{23})}$

There are several special functions used in this theory. They are defined as follows:

$I = I(\tilde{b}) - I(\tilde{a})$

$I_{1,2} = -j\dfrac{e^{\mp x(1+j)}}{x}$

$$I(x) = -h_1(\tilde{b})\ell^{x(1+j)}[\frac{3(1-x)}{2\tilde{b}^3} + j(\frac{x^2}{\tilde{b}^3} - \frac{3x}{2\tilde{b}^3} - \frac{1}{x})] - h_2(\tilde{b})\ell^{-x(1+j)}[\frac{3(1+x)}{2\tilde{b}^3} + j(\frac{x^2}{\tilde{b}^3} + \frac{3x}{2\tilde{b}^3} - \frac{1}{x})]$$

$$I_{13} = -\frac{\ell^{-x(1+j)}}{\tilde{b}^3}[\frac{3}{2}(1+x) + j(x^2 + \frac{3}{2}x)]$$

$$I_{13} = -\frac{\ell^{x(1+j)}}{\tilde{b}^3}[\frac{3}{2}(-1+x) + j(-x^2 + \frac{3}{2}x)]$$

The Happel cell model is better suited to acoustics, because it describes energy dissipation more adequately, whereas the Kuwabara cell model is better suited to electroacoustics because it automatically yields the Onsager relationship [22, 50].

In the case of a polydisperse system, the introduction of the cell is more complicated, because the total liquid can be distributed between size fractions in an infinite number of ways. However, the condition of mass conservation is still necessary.

Each fraction can be characterized by particles having radii a_i, cell radii b_i, thickness of the liquid shell in the spherical cell $l_i = b_i - a_i$, and volume fraction φ_i. The mass conservation law ties these parameters together as:

$$\sum_{i=1}^{N}(1+\frac{l_i}{a_i})^3 \varphi_i = 1 \tag{2.48}$$

This expression might be considered as an equation with N unknown parameters, l_i. An additional assumption is still necessary to determine the cell properties for the polydisperse system. This additional assumption should define for each fraction the relationship between particle radii and shell thickness. We have suggested the following simple relationship [50]:

$$l_i = la_i^n \tag{2.49}$$

This assumption reduces the number of unknown parameters to only two, related by:

$$\sum_{i=1}^{N}(1+la_i^{n-1})^3 \varphi_i = 1 \tag{2.50}$$

The parameter, n, is referred to as a "shell factor". Two specific values of the shell factor correspond to easily understood cases. A shell factor of 0 depicts the case in which the thickness of the liquid layer is independent of the particle size. A shell factor of 1 corresponds to the normal "superposition assumption", giving the same relationship between particles and cell radiuses as in the monodisperse case, i.e. each particle is surrounded by a liquid shell that provides each particle with the same volume concentration as the volume concentration of the overall system.

In general, the "shell factor" might be considered to be an adjustable parameter because it adjusts the dissipation of energy within the cells. However, our experience using this cell model with acoustics for particle sizing [51, 52] indicates that a shell factor of unity is almost always more suitable.

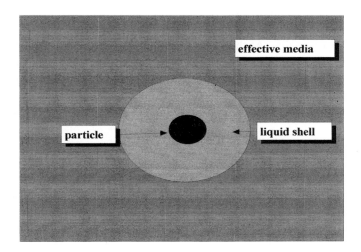

Figure 2.15. Illustration of the core-shell model.

There is another approach to describe the hydrodynamic particle interaction; it is the so-called "effective media" approach. It was used originally by Brinkman [53] for hydrodynamics, and later by Bruggeman for dielectric spectroscopy [33]. This approach and its relationship to the "cell model" are fully described in a recent review [54]. The effective media approach is very useful for interpreting acoustic spectroscopy results, and will be discussed later in Chapter 8.

Recently, a new "core-shell model" has become quite popular [54]. It attempts to combine the "cell model" and "effective media" approaches as illustrated in Figure 2.15. Here, each particle is surrounded with a spherical layer of liquid, as in the regular cell model. However, the colloid beyond this spherical layer is modeled as some "effective media". The properties of the effective media and the liquid shell thickness depend on the particular application. There are applications for this core-shell model in acoustics, in particular with the theory of the thermodynamic effects associated with sound propagation through the colloid (see Chapter 4).

Theoretical developments of the hydrodynamic particle-particle interaction are the backbone of rheological theories that describe reactions of colloids to mechanical stresses. Such reactions might be very complex because colloids possess features of both liquids and solids. They are able to flow and exhibit viscous behavior as liquids. At the same time they have a certain elasticity as with solids. This is why colloids are referred to as viscoelastic materials. In the simplest linear case, the total stress of the colloid, σ_{shear}, in a shear flow of shear rate, R_{shear}, can be presented as the combination of a Newtonian liquid, (stress is proportional to the rate of strain) and a Hookean solid, (stress is proportional to the strain) [42]:

$$\sigma_{shear} = \eta_s(\omega) R_{shear} \sin\omega t - \frac{M_s(\omega)}{\omega} R_{shea} \cos\omega t \qquad (2.51)$$

where ω is the frequency of the shear oscillation, and η_s and M_s are, respectively, the macroscopic viscosity and elastic bulk modulus of the colloid.

The macroscopic viscosity and bulk modulus depend on the microscopic properties of colloids, such as the particle size distribution and the nature of the bonds that bind the particles together. These links are usually modeled as Hookean springs. There are a variety of theories [40-45] that relate the macroscopic viscosity and bulk modulus to the microscopic and structural properties of particular colloids using the Hookean spring model. Later (Chapter 4), we will use this model to describe ultrasound attenuation in structured colloids. This is possible because of similarities between rheology and acoustics; both deal with responses to mechanical stress: shear stress in rheology, and longitudinal stress in acoustics. This parallelism is explored in more detail in the same chapter.

2.5.3 Linear interaction in an electric field. Electrokinetics and dielectric spectroscopy.

Applying an electric field to a colloid system generates a variety of effects. One group of such effects is called "electrokinetic phenomena". Electrophoresis is the best known member of this group, which also includes electroosmosis, streaming potential, and sedimentation potential [55-59]. These electrokinetic effects depend on the first power of the applied electric field strength, E. There are other effects, induced by an electric field, that are non-linear in terms of the electric field strength, but they are usually less well known and less important. They are not well known because the electric field at normal conditions in aqueous systems is rather weak. It becomes especially clear when the electric field is presented in the natural normalized form as follows:

$$\tilde{E} = \frac{EaF}{RT} \approx \frac{E[V/cm]}{250}\bigg|_{a=1micron} \qquad (2.52)$$

In aqueous colloids the electric field strength is limited by high conductivity and is typically no more than 100 V/cm. A stronger applied electric field creates thermal convection and this masks any electrokinetic effects. Accordingly, the dimensionless electric field strength is small, and linear effects are dominant. There are still some peculiar non-linear effects (described in the next section).

The most widely known and used electrokinetic theory was developed a century ago by Smoluchowski [60], to describe the electrophoretic mobility, μ, of colloidal particles. It is important because it yields a relationship between the ζ-potential of particles and the experimentally measured parameter, μ:

$$\mu = \frac{\varepsilon_m \varepsilon_0 \zeta}{\eta_m} \qquad (2.53)$$

There are two conditions restricting the applicability of Smoluchowski's equation. The first restriction is that the double layer thickness must be much smaller than particle radius a:

$$\kappa a \gg 1 \qquad (2.54)$$

where κ is the reciprocal of the Debye length.

In the situation when the DL is not thin compared to the particle radius, Henry [see in [62]] applied a correction to modify Smoluchowski's equation:

$$\mu = \frac{\varepsilon_0 \varepsilon_m \zeta}{\eta_m} f_H(\kappa a) \qquad (2.55)$$

where $f_H(\kappa a)$ is called the Henry function. A simple approximate expression has been derived by Ohshima [103]:

$$f_H(\kappa a) = \frac{2}{3} + \frac{1}{3[1 + \frac{2.5}{\kappa a[1 + 2EXP(-\kappa a)]}]^3} \qquad (2.56)$$

Figure 2.16 illustrates the form of the Henry function values as a function of κa.

The second restriction on the Smoluchowski equation is that the contribution of surface conductivity, κ^σ, to the tangential component of electric field near the particle surface is negligibly small. This condition is satisfied when the dimensionless Dukhin number, Du, is sufficiently small:

$$Du = \frac{\kappa^\sigma}{K_m a} \ll 1 \qquad (2.57)$$

where K_m is the conductivity of medium outside of the double layer.

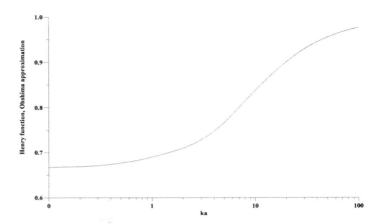

Figure 2.16. Henry function calculated according to the Ohshima expression.

Equation 2.53 is termed the Smoluchowski electrophoresis equation and is usually derived for a single particle in an infinite liquid. In this derivation the electrophoretic mobility, μ, is defined as the ratio of field induced particle velocity, V, to the homogeneous field strength, E_{ext}, in the liquid at a large distance from the particle:

$$\mu = \frac{V}{E_{ext}} \qquad (2.58)$$

Smoluchowski's equation is applicable to a particle with any geometrical form; a convenient property that makes it applicable to any arbitrary system of particles, including a cloud of particles, or a concentrated suspension. This property was experimentally tested by Zukoski and Saville [67].

However, in order to apply Smoluchowski's equation to a concentrated suspension, one should take into account the difference between the external electric field in the free liquid outside the suspension, E_{ext}, and the averaged electric field inside the dispersion, $<E>$. The static mobility μ in Equation 2.55 is determined with respect to the external electric field, i.e. the field strength in the free liquid outside the suspension.

For concentrated suspensions, an alternative electrophoretic mobility, $<\mu>$, (or electroosmotic velocity), is defined with respect to the field $<E>$:

$$<\mu> = \frac{\varepsilon \varepsilon_0 \zeta}{\eta} \frac{K_s}{K_m} \qquad (2.59)$$

where K_s is the macroscopic conductivity of the dispersed system, and

$$<\mu> = \frac{V}{<E>} \qquad (2.60)$$

An equation, similar to Eq.2.59 but with opposite sign, is termed Smoluchowski's electroosmosis equation, and is usually expressed in this form (Kruyt and Overbeek [50], Dukhin [22], and O'Brien [65]). It is valid under the same conditions as given above for electrophoresis, i.e. a thin DL and negligible surface conductivity. The same absolute values and opposite signs of electrophoretic and electroosmotic velocities reflect the difference in the frame of references between these two phenomena. In the case of static electrophoresis, the frame of references is the liquid, whereas in the case of electroosmosis it is the particle matrix.

These two definitions of the electrophoretic mobility are identical if we take into consideration the well known [33] relationship between E_{ext} and $<E>$:

$$\frac{<E>}{E_{ext}} = \frac{K_m}{K_s} \qquad (2.61)$$

Both expressions (Equation 2.53 and 2.59) specify an electrophoretic mobility (μ or $<\mu>$), that is independent of both the volume fraction and the system geometry. This happens because the hydrodynamic and electrodynamic interactions of particles have the same geometry [22] when conditions 2.54 and 2.57 are valid. From this viewpoint Smoluchowski's equation is unique.

Surface conductivity and the associated concentration polarization of the DL destroys this geometric similarity between the electric and hydrodynamic fields, and consequently the electrophoretic mobility becomes dependent on the particle size and shape; a correction is required and several theories have been proposed. One numerical solution for a single particle in an infinite liquid (for very dilute colloids) was derived by O'Brien and White [61]. However, experimental tests by Midmore and Hunter [63] indicate that the Dukhin-Semenikhin theory [62] is probably the most adequate among the several approximate formulas [64, 68], because it has the advantage of distinguishing between the ζ- potential and the Stern potential, making it more useful for characterizing surface conductivity. The Dukhin-Semenikhin theory yields the following expression for electrophoretic mobility:

$$\mu = \frac{\varepsilon \varepsilon_0 \zeta}{\eta}[1 - \frac{4M sh^2\overline{\zeta} + 2G1 + \frac{\ln ch\overline{\zeta}}{\overline{\zeta}}(2M sh2\overline{\zeta} - 12m\overline{\zeta} + 2G2)}{\kappa a + 8M sh^2\overline{\zeta} - \frac{24m\ln ch\overline{\zeta}}{z^2} + 4G1}] \qquad (2.62)$$

where

$$G1 = \frac{D_s^\pm}{D^\pm}(ch\frac{\overline{\psi}^d}{2} - ch\frac{\overline{\varsigma}}{2}); \quad G2 = \frac{D_s^\pm}{D^\pm}(sh\frac{\overline{\psi}^d}{2} - sh\frac{\overline{\varsigma}}{2}); \quad M = 1 + \frac{3m}{z^2};$$

$$m = \frac{0.34(z^+D^+ - z^-D^-)}{D^+D^-(z^+ - z^-)} \quad \overline{\psi}^d = \frac{zF\psi^d}{RT}; \quad \overline{\varsigma} = \frac{zF\varsigma}{RT};$$

In general, there is a great difference between the theory in dilute systems and that in concentrated systems. Particle interaction makes the theory for concentrated systems much more complicated. Unfortunately, the simplicity and wide applicability of Smoluchowski's equation is a rather unique exception. The simplest and the most traditional way to resolve the complexity of the electrokinetic theory for concentrated systems is to employ a cell model approach, in a manner similar to that used in the case of hydrodynamics.

The first electrodynamic cell model was applied for pure electrodynamic problem by Maxwell, Wagner and later O'Konski [28-30] for calculating conductivity and dielectric permittivity of concentrated colloids comprised of particles having a given surface conductivity. This theory gives the following relationship between complex conductivity and dielectric permittivity:

$$\frac{\varepsilon_s^* - \varepsilon_m^*}{\varepsilon_s^* + 2\varepsilon_m^*} = \frac{\varepsilon_p^* - \varepsilon_m^*}{\varepsilon_p^* + 2\varepsilon_m^*} \varphi \qquad (2.63)$$

$$\frac{K_s^* - K_m^*}{K_s^* + 2K_m^*} = \frac{K_p^* - K_m^*}{K_p^* + 2K_m^*} \varphi \qquad (2.64)$$

where the index, s, corresponds to the colloid, m corresponds to the media, and p to the particle. Complex parameters are related to the real parameters as:

$$K_m^* = K_m - j\omega\varepsilon_o\varepsilon_m \qquad (2.65)$$

$$\varepsilon_m^* = \varepsilon_m(1 + j\frac{K_m}{\omega\varepsilon_m\varepsilon_0})$$

$$\varepsilon_p^* = \varepsilon_p + j\frac{2K_m Du}{\omega\varepsilon_0}$$

According to the Maxwell-Wagner-O'Konski theory, there is a dispersion region where the dielectric permittivity and the conductivity of the colloid are frequency dependent. Two simple interpretations exist for this Maxwell-Wagner frequency, ω_{MW}. From DL theory it is the frequency of the DL relaxation to the external field disturbance. From general electrodynamics it is the frequency at which active and passive currents are equal. Thus ω_{MW} can be defined as two expressions:

$$\kappa^2 D_{eff} = \omega_{MW} = \frac{K_m}{\varepsilon_0\varepsilon_m} \qquad (2.66)$$

The Maxwell-Wagner-O'Konski cell model is applied to the pure electrodynamic problem only. The other cell model given above (Happel or Kuvabara cell model, Section 2.5.2) covers only hydrodynamic effects. An electrokinetic cell model must be valid for both electrodynamic and hydrodynamic effects. It must specify the relationship between macroscopic experimentally measured electric properties and the local electric properties calculated using the cell concept. There are multiple ways to do this. For instance, the Levine-Neale cell model [69] specifies this relationship using one of the many possible analogies between local and macroscopic properties. The macroscopic properties are current density, $<I>$, and electric field strength, $<E>$. According to the Levine-Neale cell model, they are related to the local electric current density, I, and electric field, $\nabla\phi$ as:

$$<I> = \frac{I_r}{b\cos\theta}\bigg|_{r=b} \qquad (2.67)$$

$$<E> = -\frac{1}{\cos\theta}\frac{\partial\phi}{\partial r}\bigg|_{r=b} \qquad (2.68)$$

Relationships (2.67-68) are not unique. There are many other ways to relate macroscopic and local fields. This means that we need a set of criteria to select a proper cell model. One set has been suggested in the electrokinetic cell model created by Shilov and Zharkikh [2]. Their criteria determine a proper choice of expressions for macroscopic "fields" and "flows".

The first criterion is a well-known Onsager relationship which constrains the values of the macroscopic velocity of the particles relative to the liquid, $<V>$, the macroscopic pressure, $<P>$, the electric current, $<I>$ and the field $<E>$:

$$\frac{<V>}{<I>_{<\nabla P>=0}} = \frac{<E>}{<\nabla P>_{<I>=0}} \qquad (2.69)$$

This relationship requires a certain expression for entropy production, Σ:

$$\Sigma = \frac{1}{T}(<I>\times<E> + <V>\times<\nabla P>) \qquad (2.70)$$

It turns out that the Shilov and Zharkikh derived expression for the macroscopic field strength is different as compared to the Levine-Neale model. It is:

$$<E> = \frac{\phi}{b\cos\theta}\bigg|_{r=b} \qquad (2.71)$$

However, the expression for the macroscopic current is the same in both models. It is important to mention that the Shilov-Zharkikh electrokinetic cell model condition 2.71 fully coincides with the relationship between macroscopic and local electric fields of the Maxwell-Wagner-O'Konski electrodynamic cell model.

The Shilov-Zharkikh cell model, allowing for some finite surface conductivity but assuming no concentration polarization, yields the following expressions for the electrophoretic mobility and conductivity in a concentrated dispersion:

$$<\mu> = \frac{\varepsilon_0 \varepsilon_m \varsigma}{\eta_m} \frac{1-\varphi}{1+Du+0.5\varphi(1-2Du)} \qquad (2.72)$$

$$\frac{K_s}{K_m} = \frac{1+Du-\varphi(1-2Du)}{1+Du+0.5\varphi(1-2Du)} \qquad (2.73)$$

Identical electrodynamic conditions in the Shilov-Zharkikh and Maxwell-Wagner-O'Konski cell models yields the same low-frequency limit for the conductivity, Eq.2.73.

In addition, for the extreme case of $Du \rightarrow 0$, Eqs.2.72 and 2.73 lead to Eq.2.59 for electrophoretic mobility. Consequently, the Shilov-Zharkikh cell model reduces to Smoluchowski's equation in the case of negligible surface conductivity. This is an important test for any electrokinetic theory because Smoluchowski's law is known to be valid for any geometry and volume fraction.

In the case of dielectric spectroscopy there is a method that is similar to the "effective media" approach used by Brinkman for hydrodynamic effects. It was suggested first by Bruggeman [31]. We will use both these approaches for describing particle interaction in Acoustic and Electroacoustic phenomena.

2.5.4. Non-linear interaction in the electric field. Electrocoagulation and electro-rheology.

The contribution of the electric field to particle-particle interaction is non-linear with the electric field strength, E. This follows from the simple symmetry considerations. If such a force proportional to E would exist it would depend on the direction of the electric field, or on the sign of E. This implies that in changing the direction of E, we would convert attraction to repulsion and vice versa. This is clearly impossible. Any interaction force must be independent of the direction of the electric field. This requirement becomes valid if the interaction force is proportional, for example, to the square of the electric field strength, E^2.

A linear interaction between particles affects their translation motion in an electric field. A mechanism exists for a linear particle interaction with an electrode in a DC field [80] which leads to a particle motion either towards the electrode, or away from it [81]. In contrast, a non-linear interaction induces a mutual or relative motion on the particles. We now give a short description of the theory and experiments related to this non-linear particle-particle interaction.

This non-linear particle-particle interaction is the cause of two readily observed phenomena that occur under the influence of an electric field. The first phenomenon is the formation of particle chains oriented parallel to the electric field, as shown in Figure 2.17. These chains affect the rheological properties of the colloid [72, 104].

Figure 2.17. Chains of particles in parallel with the electric field strength. Obtained from Murtsovkin [77].

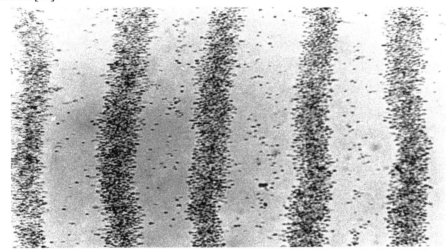

Figure 2.18. Structure of particles perpendicular to the electric field strength. Obtained from Murtsovkin [77].

The second effect is less well known but first was reported by Murtsovkin [77]. He observed that under certain conditions the particles build structures perpendicular, or at some other angle, to the electric field, as shown in Figure 2.17.

These experiments clearly indicate that the application of an electric field can affect the stability of colloids, and that is why this condition is of general importance for colloid science.

We have already mentioned that the volume of the dispersion medium adjacent to a colloid particle in the external electric field differs in its characteristics from the bulk dispersion medium. Such intensive parameters as the electric field, the chemical potential, and the pressure vary from the immediate vicinity of a particle to distances on the order of the dimension of the particle itself. If the volume fraction of the particles is sufficiently large, the particles will frequently fall into these regions of inhomogeneity of the intensive parameters. Naturally, the particles' behavior in these regions will differ from that in the free bulk. Historically, it happened that the influence of each intensive parameter on the relative particle motion was considered separately from the others. As a result, three different mechanisms of particle interactions in an electric field are known. A short description of each follows.

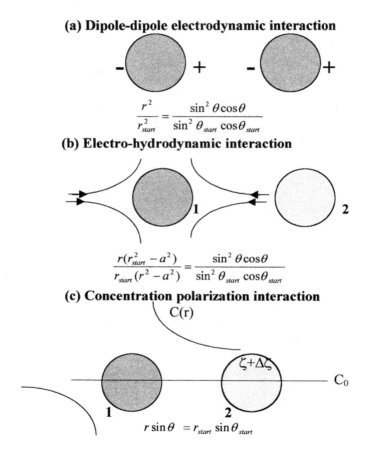

Figure 2.19. Three mechanisms for the interaction of particles in an electric field.

The oldest, and best known, is the mechanism related to the variation of the electric field, itself, in the vicinity of the particles. The particles gain dipole moments, p_{ind}, in the electric field, and these dipole moments change the electric field strength in the vicinity of the particles. This, in turn, creates a particle-particle attraction. This mechanism is usually referred to as a "dipole-dipole electrodynamic interaction" [78, 79].

This dipole-dipole interaction is responsible for the formation of particle chains, and the phenomena of electro-rheology [72].

The dipole-dipole interaction is potential, which means that the energy of interaction, normally characterized with a potential energy, U_{dd}, is independent of the trajectory of the relative motion:

$$U_{dd} = \frac{p_{ind}^2 (1 - 3\cos^2 \theta)}{4\pi \varepsilon_0 \varepsilon_m r^3} \qquad (2.74)$$

where r and θ are spherical coordinates associated with one of the particles. The trajectory of the relative particle motion is shown as (a) in Figure 2.19, where r_{start} and θ_{start} are the initial particle coordinates.

Dipole-dipole interaction is the only potential mechanism of particle interaction in an electric field. The two others are not potential, and this means that the energy spent on the relative motion depends on the trajectory of the motion.

In order to compare intensities of these different mechanisms, it is convenient to define them in terms of the radial component of the relative motion velocity, V^r. For the dipole-dipole interaction V^r is given by:

$$V_{dd}^r = \frac{\varepsilon_0 \varepsilon_m a^5 E^2 (1 - Du)^2 (1 - 3\cos^2 \theta)}{2\eta(1 + 2Du)r^4} \qquad (2.75)$$

where the value of the dipole moment is substituted according to Equation 2.29.

The second, "electro-hydrodynamic", mechanism was suggested by Murtsovkin [72, 73, 74, 76, 77]. He observed that an electric field induces an electroosmotic flow that has quadrupole symmetry around the particle. The radial component of the relative (non-conducting) particle motion velocity is:

$$V_h^r = \frac{3\varepsilon_0 \varepsilon_m a^3 E^2 Du |\zeta|(1 - 3\cos^2 \theta)}{2\eta(1 + 2Du)r^2} \qquad (2.76)$$

The trajectory of this relative motion is shown in Figure 2.19(b).

The third, "concentration" mechanism [75] is related to the concentration polarization of the particles. It is defined as:

$$V_c^r = \frac{3a^3 C_s F E^2 Du (1 - \cos^2 \theta)}{2(1 + 2Du)RTr^2} \frac{d\mu}{dC_s} \qquad (2.77)$$

where C_s is the concentration of the electrolyte in the bulk of the solution. Again, the corresponding trajectory is given in Figure 2.19 (c).

Comparison of V^r, for various mechanisms, indicates that the "electro-hydrodynamic" and "concentration" mechanisms decay with distance as r_{pp}^2 and are longer ranged than the "dipole-dipole" which decays as r_{pp}^4. This explains why, in some cases, particles build structures perpendicular to the electric field.

These effects are especially pronounced for systems of biological cells and for metal particles [16].

The "electro-hydrodynamic" and "concentration" mechanisms for non-conducting colloidal particles, are only appropriate for static or low frequency electric fields. They do not operate above the frequency of concentration polarization. This means that, at the high frequency of ultrasound, only the dipole-dipole interaction can affect the stability of the colloid.

2.6. Traditional particle sizing

There are three groups of traditional particle sizing methods that can be applied to characterize the size of colloidal particles [88, 89]. These groups are: (1) counters, (2) fractionation techniques, and (3) macroscopic fitting techniques.

Particle counters yield a full particle size distribution weighted by number. This group includes various microscopic image analyzers [87], electrozone counters (Coulter counters), and optical zone counters.

Fractionation techniques also yield a full particle size distribution, but weighted by mass. They achieve this by separating particles into fractions with different sizes. This group includes sieving, sedimentation, and centrifugation.

The last group, the "macroscopic technique", includes methods, related to the measurement of various macroscopic properties that are particle size dependent. All of these methods require an additional step in order to deduce a particle size distribution from the measured macroscopic data. This is the so-called "ill-defined" problem [85, 86]. We further discuss this problem later with regard to Acoustics in Chapter 7.

Each of these three groups has advantages and disadvantages over the others. There is no universal method available that is able to solve all particle sizing applications. For example, the "counters group" has an advantage over "macroscopic techniques" in that it yields a full particle size distribution. However, statistically representative analysis requires the counting of an enormous number of particles, especially for polydisperse systems. The technique can be very slow. Sample preparation is rather complicated and might affect the results. Transformation of a number weighted PSD to a volume based PSD is not accurate in many practical systems.

Fractionation methods are, statistically, more representative than counters, but they are not suitable for small particles. It can take hours to perform a single particle size analysis, particularly of submicron particles, even using an ultra-centrifuge. In these methods hydrodynamic irregularities can dramatically disturb the samples, and consequently affect the measured particle size distribution.

Macroscopic methods are fast, statistically representative, and easy to use. However, the price to be paid is the necessity to solve the "ill-defined" problem, and that limits particle size information.

Fortunately, in many cases there is no need to know the exact details of the full particle size distribution. Median size, PSD width, or bimodality are sufficient for an adequate description of many practical colloids and colloid related technologies. This is why "macroscopic methods" are very popular, and widely used in many laboratories and plants for routine analysis.

Acoustics, as a particle sizing technique, belongs to the third group of macroscopic methods. It competes with the most widely used macroscopic method - light scattering [82, 84]. It can be considered, also, as an alternative to neutron scattering for characterizing microemulsions and polymer solutions. The advantages of Acoustics over light scattering were given in the Introduction.

There are a number of points of similarity between light and ultrasound that are important in the discussion of the application of acoustics to particle sizing analysis. To this end, a short description of light scattering follows. A full review can be found in the book by Bohren and Huffman [82].

2.6.1. Light Scattering. Extinction=scattering + absorption.

This section is a selection of various important statements and notions presented in the book by Bohren and Huffman [82].

The presence of the particles results in an *extinction* of the incident beam. The extinction energy rate, W_{ext}, is the sum of the energy scattering rate, W_{sca}, plus the energy absorption rate, W_{abs}, through an imaginary sphere around a particle.

$$W_{ext} = W_{abs} + W_{sca} \tag{2.78}$$

In addition to re-irradiating electromagnetic energy (*scattering*), the excited elementary charges may transform part of the incident electromagnetic energy into other forms (for example thermal energy), a phenomenon called *absorption*. Scattering and absorption are not mutually independent processes, and although, for brevity, we often refer only to scattering, we shall always as well imply absorption. Also, we will restrict our treatment to elastic scattering: i.e. the frequency of the scattered light is the same as that of the incident light.

If the particle is small compared with the wavelength, all the secondary wavelets are approximately in phase; for such a particle we do not expect much variation of scattering with direction.

Single scattering occurs when the number of particles is sufficiently small and their separation sufficiently large that, in the neighborhood of any particle, the total field scattered by all particles is small compared with the external field.

Incoherent scattering occurs when there is no systematic relation between the phases of the waves scattered by the individual particles. However, even in a collection of randomly separated particles, the scattering is *coherent* in the forward direction.

The *extinction cross section* C_{ext} may be written as the sum of the absorption cross section, C_{abs}, and the scattering cross section, C_{sca}:

$$C_{ext} = \frac{W_{ext}}{I_i} = C_{abs} + C_{sca} \qquad (2.79)$$

where I_i is the incident irradiance.

There is an important Optical theorem [82] that defines *Extinction as depending only on the scattering amplitude in the forward direction.*

This theorem, common to all kinds of seemingly disparate phenomena, involves acoustic waves, electromagnetic waves, and elementary particles. It is anomalous because extinction is the combined effect of absorption and scattering in all directions by the particle.

Extinction cross-section is a well-defined, observable, quantity. We measure the power, incident at the detector, with and without a particle interposed between the source and the detector. The effect of the particle is to reduce the detector area by C_{ext}. Hence the description of C_{ext} as an area. In the language of geometrical optics we would say that the particle "casts a shadow" of area C_{ext}. However, this "shadow" can be considerably larger or smaller than the particle's geometric shadow.

Importantly, for a particle that is much larger than the wavelength of the incident light, the light scattered tends to be concentrated around the forward direction. Therefore, the larger the particle, the more difficult it is to exclude scattered light from the detector. The *Optical theorem neglects receiving the scattered light, but it creates a problem for large particles [82].*

By observing only the transmitted light it is not possible to determine the relative contribution of absorption and scattering to extinction; to do so requires an additional, independent, observation.

Multiple scattering aside, the underlying assumption is that all light scattered by the particles is excluded from the detector. The larger the particle the more the scattering envelope is peaked in the forward direction and hence the greater the discrepancy between the measured and calculated extinction. Thus, a detection system has to be carefully designed in order to measure an extinction that can be legitimately compared with the theoretical extinction. This is particularly important for large particles.

An Extinction efficiency, Q_{ext}, can be defined as:

$Q_{ext} = C_{ext}/\pi a^2$

As the particle radius, a, increases, Q_{ext} approaches a limiting value of 2. This is twice as large as that predicted from geometrical optics. This puzzling result is called the *extinction paradox,* since it seemingly contradicts geometrical optics. Yet geometrical optics is considered to be a good approximation, especially when all particle dimensions are much larger than the incident wavelength. Moreover, the result contradicts "common sense"; we do not expect a large object to dissipate twice the amount of energy that is incident upon it!

Although for large objects, geometrical optics is a good approximation for the exact wave theory, even for very large objects, geometrical optics is still not exact. This arises because all practical materials are anisotropic; they posses "edges". The edge deflects rays in its neighborhood, rays that, from the viewpoint of geometrical optics, would have passed unimpeded. Irrespective of how small the angle is through which they are deflected, rays are counted as having been removed from the incident beam and this contributes, therefore, to the total extinction (Roughly speaking, we may say that the incident wave is influenced beyond the physical boundaries of the obstacle).

These edge deflected rays cause diffraction that can be described in terms of Fresnel zones [82, 83]. It is convenient to describe diffraction by considering light transmission through an aperture instead of the scattering by the particle. For instance, we may ask how the intensity changes when the point of observation, P_{obs}, moves along the axis of aperture of fixed dimensions which models the particle. Since the radii of the Fresnel zones depend on the position of P_{obs}, we find that the intensity goes through a series of maxima and minima, occurring, respectively, when the aperture includes an odd, or an even, number of Fresnel zones.

A special case is when the source is very far away, so that the incident wave may be regarded as a plane wave. The radius of the first Fresnel zone, R_1, increases indefinitely as the distance from the aperture R_{obs} of P_{obs} goes to infinity. Thus, if the point of observation is sufficiently far away, the radius of the aperture is certainly smaller than that of the first Fresnel zone. As the distance, R_{obs}, gradually decreases, i.e. P_{obs} approaches the aperture, the radius, R_1, decreases correspondingly. Hence an increasingly large fraction of the first Fresnel zone appears through the aperture.

Again, as P_{obs} moves closer and closer to the aperture, maxima and minima follow each other at decreasing distances. When the distance R_{obs} of P_{obs} from the aperture is not much larger than the radius of the aperture itself, R, then the distances between successive maxima and minima approach the magnitude of the wavelength. Under these conditions, of course, the maxima and minima can no longer be practically observed.

From our previous discussion, when the distance of the source and the point of observation from the aperture are large, compared with the square of the radius of the aperture divided by the wavelength (Rayleigh distance, R_o), it follows that only a small fraction of a Fresnel zone appears through the aperture. The diffraction phenomenon observed at these circumstances is called Fraunhofer diffraction. When, however, the distance from the aperture of either the point of observation or the source is R_0 or smaller, then the aperture uncovers one or more Fresnel zones. The diffraction phenomenon observed is then termed Fresnel diffraction.

The exact solutions for scattering and absorption cross-sections of spheres are given in the Mie theory [91]. It might appear that we are faced with a straightforward task to obtain quantitative results from the Mie theory. However, the number of terms required for convergence can be quite large. For example, if we were interested in investigating the rainbow created by 1 mm water droplets, we would need to sum about 12,000 terms. Even for smaller particles the number of calculations can be exceedingly large. Computers can greatly reduce this computation time, but problems still remain related to the unavoidable need to represent a number having an infinite number of digits by a number with a finite number. The implicit round-off error accumulates as the number of terms required to be counted increases. Unfortunately, there is no unanimity of opinion defining the condition under which round-off error accumulation becomes a problem.

If we were interested only in scattering and absorption by spheres, we would need to go no further than the Mie theory. But physics is, or should be, more than just a semi-infinite strip of computer output. In fact, great realms of calculations often serve only to obscure from view the basic physics that can be quite simple. Therefore, it is worthwhile to consider approximate expressions, valid only in certain limiting cases, that point the way toward approximate methods to be used to tackle problems for which there is no exact theory.

If a particle has other than a regular geometrical shape, then it is difficult, if not impossible, to solve the scattering problem in its most general form. There are, however, frequently encountered situations when the particles and the medium have similar optical properties. If the particles, which are sometimes referred to as "soft", are not too large (but they might exceed the Rayleigh limit), it is possible to obtain relatively simple approximate expressions for the scattering matrix. This is expressed as the Rayleigh-Gans theory.

Geometrical optics is a simple and intuitively appealing approximate theory that need not be abandoned because an exact theory is at hand. In addition to its role in guiding intuition, geometrical optics can often provide quantitative answers to small-particle problems; solutions that are sufficiently accurate for many applications, particularly in light of the precision, accuracy and reproducibility of actual measurements.

Transition from single particle scattering to scattering by a collection of particles can be simplified using the concept of an *effective refractive index*. This concept is meaningful for a collection of particles that are small compared with wavelength, at least as far as transmission and reflection are concerned. However, even when the particles are small compared with the wavelength, this effective refractive index should not be interpreted too literally as a true refractive index on the same footing as the refractive index of, say, a homogeneous medium. For example, attenuation, in a strictly homogeneous medium, is a result of absorption and is accounted for quantitatively by the imaginary part of the refractive index. In a particulate medium, however, attenuation may be wholly or in part the result of scattering. Even if the particles are non-absorbing, the imaginary part of the effective refractive index can be non-zero [82].

When an incident beam traverses a distance, x, through an array of particles the irradiance attenuates according to

$$I_t = I_i \exp(-\alpha x) \tag{2.80}$$

where

$$\alpha = N_p C_{ext} = N_p C_{abs} + N_p C_{sca} \tag{2.81}$$

Now, the exponential attenuation of any irradiance in particulate media requires that:

$$\alpha x \ll 1 \tag{2.82}$$

This condition might be relaxed somewhat if the scattering contribution to total attenuation is small:

$$N_p C_{sca} x \ll 1 \tag{2.83}$$

An amount of light, dI, is removed from a beam propagating in the x direction through an infinitesimal distance between x and $x+dx$ in an array of particles

$$dI = -\alpha I dx \tag{2.84}$$

where I is the beam irradiance at x.

However, light can get back into the beam through multiple scattering; i.e. light scattered at any other position in the array may ultimately contribute to the irradiance at x. Scattered light, in contradiction to absorbed light, is not irretrievably lost from the system. It merely changes direction, is lost from a beam propagating in a particular direction, but then contributes in other directions. Clearly, the greater the scattering cross-section, the number density of particles, and thickness of the array, the greater will be the multiple scattering contributions to the irradiance at x. Thus, if $N_p C_{sca}$ is sufficiently small, we may ignore multiple scattering and can integrate Eq.2.84 to yield the exponential attenuation Eq.2.80.

The expression for the field, or envelope, of light scattered by a sphere is normally obtained under the assumption that the beam is infinite in lateral extent; such a beam, however, is difficult to produce practically in a laboratory. Nevertheless, it is physically plausible that scattering and absorption, by any particle, will be independent of the extent of the beam provided that the beam is large compared with the particle size; that is, the particle is completely bathed in the incident light. This is supported by the analysis of Tsai and Pogorzelski [92], who obtained an exact expression for the field scattered by a conducting sphere when the incident beam is cylindrically symmetric with a finite cross-section. Their calculations show no difference in the angular dependency of the light scattered by a conducting sphere between infinite and finite beams provided that the beam radius is about 10 times larger than the sphere radius.

REFERENCES.

1. Lyklema, J. "Fundamentals of Interface and Colloid Science", Volumes 1, Academic Press, (1993)
2. Shilov, V.N., Zharkih, N.I. and Borkovskaya, Yu.B. "Theory of Nonequilibrium Electrosurface Phenomena in Concentrated Disperse System.1.Application of Nonequilibrium Thermodynamics to Cell Model.", Colloid J., 43,3, 434-438 (1981)
3. Handbook of Chemistry and Physics, Ed. R.Weast, 70th addition CRC Press, Florida, (1989)
4. Temkin S. "Elements of Acoustics", 1st sd., John Wiley & Sons, NY (1981)
5. Morse, P.M. and Uno Ingard, K., "Theoretical Acoustics", 1968 McGraw-Hill, NY, 1968, Princeton University Press, NJ, 925 p. (1986)
6. Litovitz T.A. and Lyon, "Ultrasonic Hysteresis in Viscous Liquids", J. Acoust. Soc. Amer., vol.26, 4, pp. 577- 580, (1954)
7. Stokes, "On a difficulty in the Theory of Sound", Phil. Mag., Nov. (1848)
8. Cannon D.W. "New developments in electroacoustic method and instrumentation", in S. B. Malghan (Ed.) Electroacoustics for Characterization of Particulates and Suspensions, NIST, 40-66 (1993)
9. Hunter, R.J. "Review. Recent developments in the electroacoustic characterization of colloidal suspensions and emulsions", Colloids and Surfaces, 141, 37-65 (1998)
10. Treffers, R. and Cohen, M., "High resolution spectra of cool stars in the 10- and 20-micron region", Astrophys. J., 188, 545-552 (1974)
11. Fuchs, R., "Theory of the optical properties of ionic crystal cubes", Phys. Rev., B11, 1732-1740 (1975)
12. Pendse, H.P., Bliss T.C. and Wei Han "Particle Shape Effects and Active Ultrasound Spectroscopy", Ultrasonic and Dielectric Characterization Techniques for Suspended Particulates, Edited by V.A. Hackley and J. Texter, American Ceramic Society, Westerville, OH, (1998)
13. Zinin, P.V. "Theoretical Analysis of Sound Attenuation Mechanisms in Blood and Erythrocyte Suspensions", Ultrasonics, 30, 26-32, (1992)
14. Leschonski, K. "Representation and Evaluation of Particle Size Analysis data", Part. Charact., 1, 89-95 (1984)

15. Irani, R.R. and Callis, C.F., "Particle Size: Measurement, Interpretation and Application", John Wiley & Sons, NY-London, (1971)
16. Dukhin, A. S. "Biospecifical Mechanism of Double Layer Formation in Living Biological Cells and Peculiarities of Cell Electrophoresis". Colloids and Surfaces, 73, 29-48, (1993)
17. Karamushka, V.I., Ulberg, Z.R., Gruzina, T.G. and Dukhin, A.S. "ATP-Dependent gold accumulation by living Chlorella Cells", Acta Biotechnologica, 11, 3, 197-203 (1991)
18. Loeb, A.L., Overbeek, J.Th.G. and Wiersema, P.H. "The Electrical Double Layer around a Spherical Colloid Particle", MIT Press (1961)
19. Overbeek, J.Th.G., Verhiekx, G.J., de Bruyn, P.L and Lakkerkerker, J. Colloid Interface Sci., 119, 422 (1987)
20. Dukhin, S.S., Semenikhin, N.M., Shapinskaya, L.M. transl. Dokl. Phys. Chem., 193, 540 (1970)
21. Zholkovskij, E., Dukhin, S.S., Mischuk, N.A., Masliyah, J.H., Charnecki, J., "Poisson-Bolzmann Equation for Spherical Cell Model: Approximate Analytical Solution and Applications", Colloids and Surfaces, accepted
22. Dukhin S.S. and Derjaguin B.V. "Electrokinetic Phenomena", Surface and Colloid Science, Ed. E. Matijevic, John Willey & Sons, NY, (1974)
23. Hidalgo-Alvarez, R., Moleon, J.A., de las Nieves, F.J. and Bijsterbosch, B.H., "effect of Anomalous Surface Conductance on ζ-potential Determination of Positively Charged Polystyrene Microspheres", J.Colloid and Interface Sci., 149, 1, 23-27 (1991)
24. Kijlstra, J., van Leeuwen, H.P. and Lyklema, J. "Effects of Surface Conduction on the Electrokinetic Properties of Colloid", Chem. Soc., Faraday Trans., 88, 23, 3441-3449 (1992)
25. Kijlstra, J., van Leeuwen, H.P. and Lyklema, J. "Low-Frequency Dielectric Relaxation of Hematite and Silica Sols", Langmuir, (1993)
26. Myers, D.F. and Saville, D.A. "Dielectric Spectroscopy of Colloidal Suspensions", J. Colloid and Interface Sci., 131, 2, 448-460 (1988)
27. Rose, L.A., Baygents, J.C. and Saville, D.A. "The Interpretation of Dielectric Response Measurements on Colloidal Dispersions using dynamic Stern Layer Model", J. Chem. Phys., 98, 5, 4183-4187 (1992)
28. Maxwell, J.C. "Electricity and Magnetism", Vol.1, Clarendon Press, Oxford (1892)
29. Wagner, K.W., Arch. Elektrotech., 2, 371 (1914)
30. O'Konski, C.T., "Electric Properties of Macromolecules v. Theory of Ionic Polarization in Polyelectrolytes", J. Phys. Chem, 64, 5, 605-612 (1960)
31. Bruggeman, D.A.G., Annln Phys., 24, 636 (1935)
32. DeLacey, E.H.B. and White, L.R., "Dielectric Response and Conductivity of Dilute Suspensions of Colloidal Particles", J. Chem. Soc. Faraday Trans., 77, 2, 2007 (1982)
33. Dukhin, S.S. and Shilov V.N. "Dielectric phenomena and the double layer in dispersed systems and polyelectrolytes", John Wiley and Sons, NY, (1974)
34. Derjaguin, B.V. and Landau, L., "Theory of the stability of strongly charged lyophobic sols and the adhesion of strongly charged particles in solution of electrolytes", Acta Phys. Chim, USSR, 14, 733 (1941)
35. Verwey, E.J.W. and Overbeek, J.Th.G., "Theory of the Stability of Lyophobic Colloids", Elsevier (1948)

36. Liftshitz, E.M., "Theory of molecular attractive forces", Soviet Phys. JETP 2, 73 (1956)
37. Derjaguin, B.V. and Muller, V.M. "Slow coagulation of hydrophobic colloids", Dokl. Akad. Nauk SSSR, 176, 738-741 (1967)
38. Marlow, B.J. and Rowell, R.L. "Electrophoretic Fingerprinting of a Single Acid Site Polymer Colloid Latex", Langmuir, 7, 2970-2980 (1991)
39. Marlow, B.J, Fairhurst, D. and Schutt, W. "Electrophoretic Fingerprinting an the Biological Activity of Colloidal Indicators", Langmuir, 4, 776 (1988)
40. Goodwin, J.W., Gregory, T., and Stile, J.A. "A Study of Some of the Rheological Properties of Concentrated Polystyrene Latices", Adv. In Colloid and Interface Sci., 17, 185-195, (1982)
41. Probstein, R,F., Sengun, M.R., and Tseng, T.C., "Bimodal model of concentrated suspension viscosity for distributed particle sizes", J. Rheology, 38, 4, 811-828, (1994)
42. Russel, W.B., Saville, D.A. and Schowalter, W.R. "Colloidal Dispersions", Cambridge University Press, (1989)
43. Kamphuis, H., R.J.J.Jongschaap and P.F.Mijnlieff, "A transient-network model describing the rheological behavior of concentrated dispersions", Rheological Acta, 23, 329-344, (1984)
44. Ferry, J.D., Sawyer, W.M., and Ashworth, J.N., "Behavior of Concentrated polymer Solutions under Periodic Stresses", J. of Polymer Sci., 2, 6, 593-611, (1947)
45. T.G.M. van de Ven and Mason, S.G., "The microrheology of colloidal dispersions", J. Colloid Interface Sci., 57, 505 (1976)
46. T.G.M. van de Ven, "Colloidal Hydrodynamics", Academic Press, (1989)
47. Happel J. and Brenner, H, "Low Reynolds Number Hydrodynamics", Martinus Nijhoff Publishers, Dordrecht, The Netherlands, (1973)
48. Happel J., "Viscous flow in multiparticle systems: Slow motion of fluids relative to beds of spherical particles", AICHE J., 4, 197-201 (1958)
49. Kuwabara, S. "The forces experienced by randomly distributed parallel circular cylinders or spheres in a viscous flow at small Reynolds numbers", J. Phys. Soc. Japan, 14, 527-532 (1959)
50. Kruyt, H.R. "Colloid Science", Elsevier: Volume 1, Irreversible systems, (1952)
51. Dukhin, A.S. and Goetz, P.J. "Acoustic Spectroscopy for Concentrated Polydisperse Colloids with High Density Contrast", Langmuir, 12 [21] 4987-4997 (1996)
52. Dukhin, A.S. and Goetz, P.J. "New Developments in Acoustic and Electroacoustic Spectroscopy for Characterizing Concentrated Dispersions", Colloids and Surfaces, 192, 267-306 (2001)
53. Brinkman, H.C. "A calculation of viscous force exerting by a flowing fluid on a dense swarm of particles", Appl. Sci. Res., A1, 27 (1947)
54. Li, Yongcheng., Park, C.W. "Effective medium approximation and deposition of colloidal particles in fibrous and granular media", Adv. In Colloid and Interface Sci., 87, 1-74 (2000)
55. Christoforou, C.C., Westermann-Clark, G.B. and J.L.Anderson, "The Streaming Potential and Inadequacies of the Helmholtz Equation", J. Colloid and Interface Science, 106, 1, 1-11 (1985)
56. Gonzalez-Fernandez, C.F., Espinosa-Jimenez, M., Gonzalez-Caballero, F. "The effect of packing density cellulose plugs on streaming potential phenomena", Colloid and Polymer Sci., 261, 688-693 (1983)

57. Hidalgo-Alvarez, R., de las Nieves, F.J., Pardo, G., "Comparative Sedimentation and Streaming Potential Studies for ζ-potential Determination", J. Colloid and Interface Sci., 107, 2, 295-300 (1985)
58. Groves, J. and Sears, A. "Alternating Streaming Current Measurements", J. Colloid and Interface Sci., 53, 1, 83-89 (1975)
59. Chowdian, P., Wassan, D.T., Gidaspow, D., "On the interpretation of streaming potential data in nonaqueous media", Colloids and Surfaces, 7, 291-299 (1983)
60. Smoluchowski, M., in Handbuch der Electrizitat und des Magnetismus", vol.2, Barth, Leipzig (1921).
61. O'Brien, R.W. and White, L.R. "Electrophoretic mobility of a spherical colloidal particle", J. Chem. Soc. Faraday Trans., II, 74, 1607-1624 (1978)
62. Dukhin, S.S. and Semenikhin, N.M. "Theory of double layer polarization and its effect on the electrokinetic and electrooptical phenomena and the dielectric constant of dispersed systems", Kolloid. Zh., 32, 360-368 (1970)
63. Midmore, B.R. and Hunter, R.J., J. Colloid Interface Science, 122, 521 (1988)
64. O'Brien, R.W., "The solution of electrokinetic equations for colloidal particles with thin double layers", J. Colloid Interface Sci, 92, 204-216 (1983)
65. O'Brien, R.W. "Electroosmosis in Porous Materials", Journal of Colloid and Interface Science", 110, 2, 477-487 (1986)
66. Dukhin, A. S., Shilov, V.N. and Borkovskaya Yu. "Dynamic Electrophoretic Mobility in Concentrated Dispersed Systems. Cell Model.", Langmuir, 15, 10, 3452-3457 (1999)
67. Zukoski, C.F. and Saville, D.A. "Electrokinetic properties of particles in concentrated suspensions", J. Colloid Interface Sci., 115, 422-436 (1987)
68. Hunter, R.J. "Zeta potential in Colloid Science", Academic Press, NY (1981)
69. Levine, S. and Neale, G.H. "The Prediction of Electrokinetic Phenomena within Multiparticle Systems.1.Electrophoresis and Electroosmosis.", J. of Colloid and Interface Sci., 47, 520-532 (1974)
70. Kozak M.W. and Davis, J.E. "Electrokinetic phenomena in Fibrous Porous Media", Journal of Colloid and Interface Science, 112, 2, 403-411 (1986)
71. Smith, K.L., and Fuller, G.G., "Electric Field Induced Structure in Dense Suspensions", J. of Colloid and Interface Sci., 155, 1, 183-190, (1993)
72. Marshall, L., Goodwin, J.W., and Zukoski, C.F., "The effect of Electric Fields on the Rheology of Concentrated Suspensions", J.Chem.Soc., Faraday Trans., (1988)
73. Murtsovkin, V.A. and Muller, V.M. "Steady-State Flows Induced by Oscillations of a Drop with an Adsorption Layer", J. of Colloid and Interface Sci., 151, 1, 150-156, (1992)
74. Murtsovkin, V.A. and Muller, V.M. "Inertial Hydrodynamic Effects in Electrophoresis of Particles in an Alternating Electric Field", J. of Colloid and Interface Sci., 160, 2, 338-346 (1993)
75. Malkin, E. S. ; Dukhin, A. S. "Interaction of Dispersed Particles in an Electric Fields and Linear Concentration Polarization of the Double Layer", Kolloid. Zh., 5, 801-810, (1982)
76. Dukhin, A. S. ; Murtsovkin, V. A. "Pair Interaction of Particles in Electric Field. 2. influence of Polarization of Double Layer of Dielectric Particles on their Hydrodynamic Interaction in Stationary Electric Field", Kolloid. Zh., 2, 203-209, (1986)

77. Gamayunov, N.I., Murtsovkin, V.A. and Dukhin, A.S. "Pair interaction of particles in electric field. Features of hydrodynamic interaction of polarized particles", Kolloid. Zh., 48, 2, 197-209 (1986)
78. Estrela-L'opis, V.R, Dukhin, S.S. and Shilov, V.N., Kolloid. Zh., 36, 6, 1140 (1974)
79. Shilov, V.N. and Estrela-L'opis, V.R., in "Surface forces in Thin Films", ed.B.V.Derjaguin, Nauka, Moskow, p.39 (1979)
80. Andreson, J. and oth. Langmuir, 16, 9208-9216 (2000)
81. Ulberg, Z. R. ; Dukhin, A. S. "Electrodiffusiophoresis- Film Formation in AC and DC Electrical Fields and Its Application for Bactericidal Coatings", Progress in Organic Coatings, 1, 1-41, (1990)
82. Bohren, C. and Huffman, D. "absorption and Scattering of Light by Small Particles", J. Wiley & Sons, 530 p., (1983)
83. Bruno Rossi, "Optics", Addison-Wesley, Reading, MA, 510 p, (1957).
84. Chu, B., and Liu, T. "Characterization of nanoparticles by scattering techniques", J. of Nanoparticle Research, 2, 29-41, (2000)
85. Phillips, D.L. "A Technique for the Numerical Solution of Certain Integral Equations of the First Kind", J. Assoc. Comput. Mach., 9, 1, 84-97 (1962)
86. Twomey, S. "On the Numerical Solution of Fredholm Integral Equation of the First Kind by the Inversion of the Linear System Produced by Quadrature", J. Assoc. Comput. Mach., 10, 1, 97-101 (1963)
87. Fisker, R., Carstensen, J.M., Hansen, M.F., Bodker, F. and Morup, S., "Estimation of nanoparticle size distribution by image analysis", J. of Nanoparticle Research, 2, 267-277 (2000)
88. Heywood, H, "The origins and development of Particle Size Analysis", in "Particle size analysis", ed. Groves, M.J and oth., The Society for analytical chemistry, London, (1970)
89. Allen, T., "Particle size measurement", 4th edition, Chapman and Hall, NY, (1990)
90. Allen, T. "Sedimentation methods of particle size measurement", Plenary lecture presented at PSA'85 (ed. P.J. Llloyd), Wiley, NY (1985)
91. Mie, G., "Beitrage zur Optik truber Medien speziell kolloidaler Metallosungen", Ann Phys., 25, 377-445 (1908)
92. Tsai, W.C. and Pogorzelski, R.J. "Eigenfunction solution of the scattering of beam radiation fields by spherical objects", J. Opt. Soc. Am, 65, 1457-1463 (1975)
93. Hackley, V.A. and Ferraris, C.F. "The Use of Nomenclature in Dispersion Science and Technology", NIST, special publication 960-3 (2001)
94. Anson, L.W. and Chivers, R.C. "Thermal effects in the attenuation of ultrasound in dilute suspensions for low values of acoustic radius", Ultrasonic, 28, 16-25 (1990)
95. Kosmulski, M. and Rosenholm, J.B. "Electroacoustic study of adsorption of ions on anatase and zirconia from very concentrated electrolytes", J. Phys. Chem., 100, 28, 11681-11687 (1996)
96. Kosmulski, M. "Positive electrokinetic charge on silica in the presence of chlorides", JCIS, 208, 543-545 (1998)
97. Kosmulski, M., Gustafsson, J. and Rosenholm, J.B. "Correlation between the Zeta potential and Rheological properties of Anatase dispersions", JCIS, 209, 200-206 (1999)

98. Kosmulski, M., Durand-Vidal, S., Gustafsson, J. and Rosenholm, J.B. "Charge interactions in semi-concentrated Titania suspensions at very high ionic strength", Colloids and Surfaces A, 157, 245-259 (1999)
99. Deinega, Yu.F., Polyakova, V.M., Alexandrove, L.N. "Electrophoretic mobility of particles in concentrated electrolyte solutions", Kolloid. Zh., 48, 3, 546-548 (1982)
100. Dukhin, S.S., Churaev, N.V., Shilov, V.N. and Starov, V.M. "Problems of the reverse osmosis modeling", Uspehi Himii, (Russian) 43, 6 , 1010-1023 (1988), English, 43, 6 (1988).
101. Alekseev, O.L., Boiko, Yu.P., Ovcharenko, F.D., Shilov, V.N., Chubirka, L.A. "Electroosmosis and some properties of the boundary layers of bounded water", Kolloid. Zh., 50, 2, 211-216 (1988)
102. Rowlands, W.N., O'Brien, R.W., Hunter, R.J., Patrick, V. "Surface properties of aluminum hydroxide at high salt concentration", JCIS, 188, 325-335 (1997).
103. Ohshima, H. "A simple expression for Henry's function for retardation effect in electrophoresis of spherical colloidal particles", JCIS, 168, 269-271 (1994)
104. Tian Hao, "Electrorheological fluids", Adv. Materials, 13, 24, 1847-1857 (2001)

Chapter 3. FUNDAMENTALS OF ACOUSTICS IN LIQUIDS

The purpose of this chapter is to provide a general overview characterizing sound wave propagation in homogeneous liquids. A large body of existing literature describes various theories that relate the sound speed and attenuation of such liquids to rheological and thermodynamic parameters. Fortunately, the attenuation of pure liquids is usually negligibly small compared to the incremental attenuation resulting from the addition of a large number of particles in a concentrated colloid. Even in the more dilute case, we can readily measure the sound speed and attenuation of the homogeneous liquid used as the dispersing media, and then use this background information as an input parameter to our colloid measurement.

Since the properties of the media are of less significance, we give here just a short overview of ultrasound in the pure liquids, stressing those points that will be important in later chapters describing heterogeneous systems. For this review we draw on two handbooks: *Theoretical Acoustics* by Morse and Ingard [4], and *Elements of Acoustics* by Temkin [5].

3.1 Longitudinal waves and the wave equation

We call traveling compression waves in liquids "longitudinal waves", in contrast to "transverse waves" typified by a vibrating string. The direction that the material moves, relative to the direction of wave propagation, makes the difference. For a longitudinal wave, the liquid molecules move back and forth in the direction of the wave propagation, whereas for a transverse wave this motion is perpendicular to the direction of propagation. Longitudinal waves, in the most general case, are three dimensional with a potentially very complex geometry. Here, we will consider only two dimensional plane waves, which have the same direction of propagation everywhere.

Longitudinal waves can propagate in a fluid because the fluid has a finite compressibility that allows the energy to be transported across space. In addition to these compression waves, the sound wave itself may induce shear and thermal waves at the boundaries with other surfaces. Compression waves are able to propagate long distances in the liquid, whereas shear and thermal waves exist only in the close vicinity of phase boundaries. This chapter covers only compression longitudinal waves. Shear and thermal waves play a major role in colloidal systems because of the extended surface area. This aspect is dealt with in detail later, in Chapter 4.

We consider the liquid itself to be in equilibrium, and having a certain density, ρ_m, at a given static pressure, P, and temperature, T. The relationship between these three parameters can be described by two partial derivatives, one at constant temperature, and one at constant pressure. Together, these equations constitute the so-called "equations of state", namely:

$$\beta^P = \frac{1}{\rho_m}(\frac{\partial \rho_m}{\partial P})_T \qquad (3.1)$$

$$\beta = \frac{1}{\rho_m}(\frac{\partial \rho_m}{\partial T})_P \qquad (3.2)$$

We commonly refer to the coefficient β^P as the "isothermal compressibility" and the coefficient β as the "thermal expansion coefficient".

The presence of a sound wave at a particular instant of time, and point in space, causes an incremental variation of the density, pressure, and temperature about their equilibrium values ρ_m, P_{eq}, and T. We designate the instantaneous values of these incremental changes as $\delta\rho_m$, δP, and δT respectively. The pressure variation is the easiest to detect, and is in fact a typical output of an acoustic experiment. The sound wave pressure is usually expressed in decibels, abbreviated as dB, and defined as 20 times the logarithm to the base 10 of the ratio of the pressure amplitude, in dynes per centimeter squared, and a reference level of 1 dyne per centimeter squared.

The variation of the temperature in the liquid, δT, due to the presence of the sound wave is usually very small, because typically the high thermal conductivity of the liquid rapidly smoothes out any local perturbations. Consequently, the density variation, $\delta\rho_m$, is adequately described considering only the isothermal compressibility β^P:

$$\rho_m + \delta\rho_m = \rho_m + (\frac{\partial \rho_m}{\partial P})_T P = \rho_m + \rho_m \beta^P P \qquad (3.3)$$

The variation of the density is independent of the equilibrium pressure. The variation of the pressure in the plane sound wave is described with a wave equation:

$$\frac{\partial^2 P}{\partial x^2} = \frac{1}{c_m^2}\frac{\partial^2 P}{\partial t^2} \qquad (3.4)$$

where the sound speed c_m is found from:

$$c_m^2 = \frac{1}{\beta^P \rho_m} \qquad (3.5)$$

The values for sound speed and density of several fluids are given in Table 3.1. The sound speed through most solid materials is much higher than that of fluids, and is typically between 5000 and 6000 m/sec.

The sound speed is temperature dependent. The magnitude of this dependence is usually several meters per second per degree, as shown in Table 3.1.

Table 3.1. Sound speed of several fluids.

Fluid	Density [g/cm^3]	Sound Speed at 25^0 C [m/sec]	Sound Speed variation per ^{0}C [m/sec]
Water	0.998	1497	2.4
Acetone	0.79	1174	4.5
Ethanol	0.79	1207	4.0
Methanol	0.79	1103	3.2
Glycerol	1.26	1904	2.2
Castor oil	0.969	1477	3.6
Chloroform	1.49	987	3.4
Ethanolamide	1.018	1724	3.4
Ethylene glycol	1.113	1658	2.1
Mercury	13.5	1450	-

The simple wave equation (3.4) is valid for uniform homogeneous liquids having negligible attenuation due to viscous and thermal effects and, furthermore, only if the acoustic pressure is small compared to the equilibrium pressure. This simplified model of acoustic phenomena turns out to be quite adequate in practice, since the typical ultrasound pressure levels used for acoustic characterization are thousands of times weaker than the normal equilibrium pressure of 1 atm. We do not consider any nonlinear effects that might occur with the very high intensity ultrasound sometimes used for various modifications of liquids and colloids.

The general solution of the above wave equation describes a pressure wave traveling in the positive, x, direction as a function $P(x-c_m t)$:

$$P(x,t) = P_0 EXP(-\alpha_m x) EXP[j(\omega t - \Phi)] \qquad (3.6)$$

where α_m is the attenuation of the media, ω is a frequency of ultrasound, and P_0 is the initial amplitude of the pressure.

A similar equation can be written for the fluid velocity, v. The ratio between the acoustic pressure and the fluid velocity is called the acoustic impedance, which will be discussed later.

3.2 Acoustics and its relation to Rheology

There are dozens of papers [38-73] and patents [73-76] devoted to the application of acoustics for characterizing the rheological properties of pure liquids, polymer solutions, or colloids. Despite this, acoustics remains practically an unknown tool for characterizing the rheological properties of colloids. We give here a short overview of acoustics from a rheological point of view.

Viscous drag and thermal conduction are two mechanisms, intrinsic to homogeneous fluids, which transform the organized motion of the sound wave into the disorganized motion of heat [1-6]. The potential contribution of various ions, molecules or colloids, which might be present in the fluid, will be considered later in this Chapter and in Chapter 4. Both viscous drag and thermal conductivity affect only slightly the propagation of the sound wave through a homogenous liquid; they cause only an exponential decay in the amplitude with increasing distance. In characterizing concentrated colloids, it is often possible to neglect this intrinsic attenuation effect because the incremental attenuation, upon adding the particles, is so large in comparison. However, both viscous and thermal effects become much more significant near the boundary between phases, and consequently play an important role in colloids with high surface area (Chapter 4). Here we give only a short description of how these two mechanisms affect ultrasound in a homogeneous liquid.

Viscosity dissipates acoustic energy through the diffusion of momentum. There are two components of this dissipation, commonly referred to as the "shear viscosity", η_m, and the "volume viscosity", $\eta_m^{\,v}$.

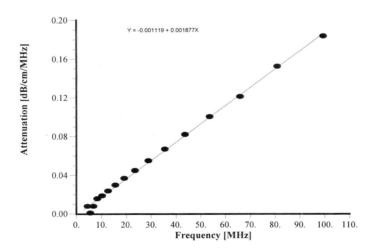

Figure 3.1. Attenuation of distilled water.

There is a large volume of literature describing the viscosity contributions to ultrasound attenuation in the liquid, α_m. The first theory relating attenuation and viscosity was derived in 1845 by Stokes [1-6]. He took into account only the shear viscosity contribution, and completely neglected volume viscosity. For a plane wave he derived the following expression for attenuation:

$$\alpha_m [np/cm] = \frac{2\omega^2 \eta_m}{3\rho_m c_m^3} \quad (3.7)$$

A salient feature of equation (3.7) is that the attenuation, expressed in *neper/cm*, increases with the square of frequency. It is convenient to normalize this function by dividing by frequency, and, furthermore, to convert from nepers to dB, and from radian frequency to MHz. Thus, in all of the following we present attenuation data in the resulting units of dB/cm/MHz. From equation (3.7) we would expect the attenuation for homogeneous liquids, expressed in these units, to be a linear function of frequency. This is, indeed, the case for water (Figure 3.1) and for other liquids (Figure 3.2). This linear dependence breaks down only for extremely viscous liquids having very high attenuation (Figure 3.3).

Further theoretical developments [5] showed that Stokes' theory yields only the frequency asymptotic limit of the viscous attenuation. The complete frequency dependence of the viscous attenuation is:

$$\alpha_m = \frac{\omega}{\sqrt{2} c_m} [\frac{\sqrt{1+(\omega \tau_\eta)^2} - 1}{1+(\omega \tau_\eta)^2}]^{1/2} \quad (3.8)$$

where τ_η is a viscous relaxation time given by:

$$\tau_\eta = (\frac{4}{3}\eta_m + \eta_m^v)/\rho_m c_m^2 \quad (3.9)$$

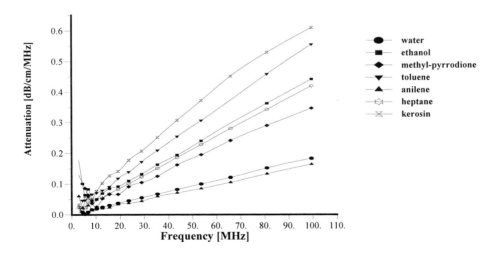

Figure 3.2 Attenuation frequency spectra for liquids with low viscosity.

The typical value of the viscous relaxation time is similar to the mean time between collisions of the fluid molecules, i.e. about 10^{-12} sec. This corresponds to high frequencies far beyond the typical ultrasound range used for colloid characterization (1 – 100 MHz). That is why, experimentally, we normally observe the frequency dependence predicted by the simple Stokes' theory.

Having addressed the attenuation due to the viscous drag mechanism, let us now turn our attention the thermal conductivity contribution. The energy loss due to the heat conduction arises because of thermodynamic coupling between pressure and temperature. The pressure variation in the sound wave generates small temperature gradients. In regions of higher temperature the molecules move about with a higher speed. These faster molecules will diffuse out into the cooler regions, in an attempt to equilibrate the temperature. The flux of heat J_h is proportional to the temperature gradient:

$$J_h = -\tau_m \, gradT \qquad (3.10)$$

where τ_m is the thermal conductivity of the liquid.

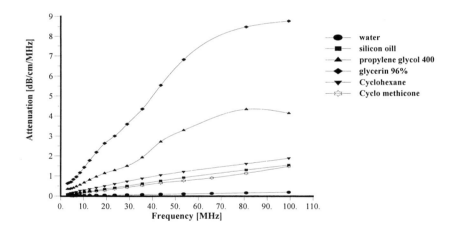

Figure 3.3. Attenuation frequency spectra for highly viscous liquids.

The net gain of heat in a unit volume is $-div \, J_h$. The rate of the temperature change equals the net heat gain divided by the heat capacity, which yields [3, 5] the following equation:

$$\frac{\partial T}{\partial t} = \frac{\tau_m}{\rho_m C_p^m} div \, gradT \qquad (3.11)$$

From equation (3.11) we can determine the loss of energy in the sound wave due to the heat conduction.

The attenuation due to heat conduction has the same frequency dependence as the viscous attenuation (Eq.3.8), but with a different relaxation time:

$$\tau_{h\eta} = \frac{\tau_m}{C_p^m \rho_m c_m^2} \tag{3.12}$$

The low frequency limit is thus given by:

$$\alpha_m [np/cm] = \frac{2\omega^2 \tau_m}{3\rho_m c_m^{\ 3} C_p^m}(\gamma - 1) \tag{3.13}$$

where $\gamma = C_p/C_v$, the ratio of specific heat at constant pressure, to the specific heat at constant volume.

A combination of viscous relaxation theory and thermal relaxation theory has been applied successfully in many non-associated non-polar liquids. However, it failed to explain the sound attenuation in associated polar liquids like alcohols and water. It appears that in these liquids the intermolecular forces are so strong that the thermal relaxation time is very short and, consequently, the contribution to the total attenuation is small.

To explain sound attenuation in polar liquids, the coefficient of the volume viscosity must be taken into account. It is associated with so-called structural relaxation. The more general theory, which takes into account volume viscosity together with thermal conduction and shear viscosity, yields the following expression for the attenuation:

$$\alpha_m [np/cm] = \frac{\omega^2}{2\rho_m c_m^{\ 3}}[(\frac{4}{3}\eta_m + \eta_m^v) + \tau_m(\frac{1}{C_v^m} - \frac{1}{C_p^m})] \tag{3.14}$$

This expression allows one to calculate the volume viscosity from the attenuation coefficient. For illustration we show this value for several liquids (Table 3.2), as a ratio of the volume viscosity to the shear viscosity. It is seen that volume viscosity is comparable in value with the shear viscosity. The volume viscosity contains useful information about liquid structure. A number of theories have been developed for liquids having different structure [29, 34, 35, 36].

In addition to viscosity, acoustics can also provide information about the elastic properties. The general characteristic of the elasticity is a bulk modulus. It is defined as the ratio of stress to strain, where the stress is the pressure increase, dP, and the accompanying strain is the decrease in volume per unit volume, $-dV/V$. Hence:

$$M_B = dP/(-dV/V) = -V\, dP/dV \tag{3.15}$$

In the case of the extended solid, the speed of *longitudinal* waves is given by:

$$c^l = \sqrt{\frac{M_B + 4/3 M_s}{\rho}} \tag{3.16}$$

which involves the bulk modulus, M_B, the shear modulus, M_s, and the density ρ.

Table 3.2 Volume and shear viscosities according to reference [35].

Liquid	Temperature (0^0 C)	Shear viscosity [poise]	η_m^v/η_m
Water	15	0.011	2.81
Methanol	2	0.0079	2.66
Propanol	-130	450	1.02
Glycerol	-14	616	1.03
Isobutyl bromide	0	0.83	0.44
Polyisobutylene	30	22	0.95
Hydrocarbon oil	30	2.1	1.33

In the case of a *transverse* wave in an extended solid, the shear modulus alone determines the speed, given by:

$$c^t = \sqrt{\frac{M_s}{\rho}} \quad (3.17)$$

It follows from these two equations (3.16 and 3.17), that the speed of a longitudinal wave in an extended solid is always greater than that of a transverse wave. A fluid cannot sustain a shear at substantial traveling distances ($M_S = 0$), so no transverse wave can be transmitted. In this case the speed of a longitudinal wave is given by:

$$c_m^l = \sqrt{\frac{M_B}{\rho}} \quad (3.18)$$

Compressions and rarefactions, caused by sound at high frequencies, must be regarded as *adiabatic* rather than *isothermal*; hence the M_B to be used in the above formulas is the adiabatic modulus. The distinction is unimportant in the case of those solids and liquids that are relatively incompressible, and whose properties vary little with small changes in temperature.

3.3 Acoustic Impedance

There is a convenient analogy between electric and acoustic phenomena. It is based on the linear relationship between driving force and corresponding flow. In the case of an alternating electric field it is electric voltage and electric current. According to Ohm's law, the current is proportional to the voltage; the coefficient of proportionality is called "electric impedance".

The acoustic impedance, Z, is introduced in a similar way, as a coefficient in the proportion between pressure, P, and velocity of the particles, v, in the sound wave:

$$Z = \frac{P}{v} \qquad (3.19)$$

As with the electric impedance, the acoustic impedance is a complex number. It depends on the other acoustic properties of the media. To derive this relationship, we consider the propagation of a plane longitudinal wave through the media along the x-axis. This media is characterized with a given density, ρ_m, sound speed, c_m, and longitudinal wave attenuation, α_m. The particle's displacement, u, in the plane sound wave is a simple harmonic function:

$$u(x,t) = u_0 EXP(-\alpha_m x) EXP[j(\omega t - \Phi)] \qquad (3.20)$$

where Φ is the phase, $\Phi=kx$, $k=2\pi/\lambda$, λ is a wavelength in meters, $\omega=2\pi f$, where f is frequency in Hertz. The attenuation coefficient, α_m, has dimension of neper/meter. Attenuation expressed in decibels, dB, is related to the attenuation expressed in nepers, np, through the following: dB/m =-8.686 np/m.

These equations allow us to calculate the wavelength for different fluids. For instance, the wavelength in water varies roughly from 15 microns at 100 MHz up to 1.5 millimeter at 1 MHz. It is important to note that the wavelength of ultrasound in this MHz range is much larger than the wavelength of light; for example the wavelength of green light is about 0.5 micron. This difference in wavelength, between light and ultrasound, results in a substantial variance between these two wave phenomena that otherwise have many similarities.

Combining together the two exponential functions and expressing the frequency and phase through the wavelength we can rewrite Eq.3.20 as:

$$u(x,t) = u_0 EXP[\frac{2\pi j}{\lambda}(c_m t - x(1 - j\frac{\alpha_m \lambda}{2\pi}))] \qquad (3.21)$$

To derive an expression for the pressure we use Hooke's law, which relates a force exerted on the element of the media, F_{hook}, and a strain caused by the sound wave, $\partial u/\partial x$:

$$F_{hook} = MS\frac{\partial u}{\partial x} \qquad (3.22)$$

where S is the cross-section area.

The pressure is simply the opposite of the force per unit of the surface:

$$P = -M\frac{\partial u}{\partial x} = -\rho_m c^2 \frac{\partial u}{\partial x} = \rho_m c_m^2 \frac{2\pi j}{\lambda}(1 - j\frac{\alpha_m \lambda}{2\pi})u(x,t) \qquad (3.23)$$

We express the velocity of the particles motion as a time derivative of the displacement:

$$v(x,t) = \frac{\partial u(x,t)}{\partial t} = \frac{2\pi j c_m}{\lambda} u(x,t)$$

Replacing displacement with the particle velocity in the expression for the pressure gives:

$$P = \rho_m c_m (1 - j\frac{\alpha_m \lambda}{2\pi})v \tag{3.24}$$

It directly follows by analogy with this expression that the acoustic impedance, Z, is:

$$Z_m = \rho_m c_m (1 - j\frac{\alpha_m \lambda}{2\pi}) \tag{3.25}$$

where attenuation is in np/m, wavelength in m, sound speed of the longitudinal wave in m/sec, and density is in kg/m^3.

The attenuation does not contribute significantly to the acoustic impedance. Even in highly concentrated colloids at high frequency, when the attenuation becomes about 1000 dB/cm, it still contributes only 2% to the acoustic impedance. This allows us to approximate the acoustic impedance as a real number equal to the product of the density and sound speed.

3.4 Propagation through phase boundaries - Reflection

Acoustic impedance is a very convenient property for characterizing effects that occur when the sound wave meets the boundary between two phases. There are certain similarities between longitudinal ultrasound and light reflection and transmission through the phase boundaries. For instance, the ultrasound reflection angle from a plane surface is equal to the incident angle; the same as for light (see Figure 3.4)

$$\theta_i = \theta_r \tag{3.26}$$

where index i corresponds to the incident wave, index r corresponds to the reflected wave.

The transmitted wave angle must satisfy a wave-front coherence at the border. Again, this yields the same relationship as with light transmission:

$$\frac{\sin\theta_i}{\sin\theta_t} = \frac{c_1}{c_2} \tag{3.27}$$

where indexes 1 and 2 correspond to the different phases.

Propagation of the sound wave through the phase border should not create any discontinuities in pressure or the particle's velocity. This condition yields the following relationships for the pressure in the reflected and transmitted waves:

$$\frac{P_r}{P_i} = \frac{Z_2 \cos\theta_i - Z_1 \cos\theta_t}{Z_2 \cos\theta_i + Z_1 \cos\theta_t} \tag{3.28}$$

$$\frac{P_t}{P_i} = \frac{2Z_2 \cos\theta_i}{Z_2 \cos\theta_i + Z_1 \cos\theta_t} \tag{3.29}$$

In the case of normal incidence, when $\theta_i = \theta_t = 0$, these equations simplify to:

$$\frac{P_r}{P_i} = \frac{Z_2 - Z_1}{Z_2 + Z_1} \tag{3.30}$$

and,

$$\frac{P_t}{P_i} = \frac{2Z_2}{Z_2 + Z_1} \qquad (3.31)$$

From these equations an important relationship is derived between the phase of the reflected and incident waves. *If $Z_2 > Z_1$, then the reflected pressure wave is in phase with the incident wave, otherwise it is 180 degrees out of phase.*

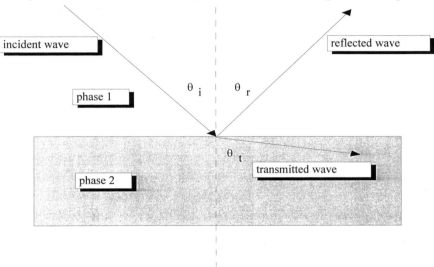

Figure 3.4 Illustration of the sound propagation through a phase boundary.

The pressure value determines the intensity of the ultrasound, *I*:

$$I = \frac{P^2}{2\rho c} \qquad (3.32)$$

For normal incidence, we can use equations (3.30 and 3.31) to obtain the ultrasound intensity of the reflected and transmitted waves:

$$\frac{I_r}{I_i} = \frac{(Z_2 - Z_1)^2}{(Z_2 + Z_1)^2} \qquad (3.33)$$

and,

$$\frac{I_t}{I_i} = \frac{4Z_2 Z_1}{(Z_2 + Z_1)^2} \qquad (3.34)$$

At normal incidence, the reflected wave interferes with the incident wave. This leads to the build-up of standing waves. For a perfect reflector the particle displacement in reflected and incident waves compensate each other completely when they are out of phase. They add together when they are in phase. This leads to a repeating pattern of nodes and maxima. Standing waves do not transmit any power, since the power coming back equals the power going out.

Standing waves can superimpose with the traveling waves when the reflection is not perfect. This effect occurs when ultrasound propagates through a multi-layer media. The case of three phases is important and well characterized. A standing wave appears in the first and the second layers. They superimpose here with traveling waves if reflection at the phase boundaries is not perfect.

In the case of normal incidence, it is possible to derive [5] an analytical expression for the intensities of the incident and transmitted waves:

$$\frac{I_{t3}}{I_{i1}} = \frac{4Z_3 Z_1}{(Z_3 + Z_1)^2 \cos^2 \frac{2\pi l_2}{\lambda_2} + (Z_2 + \frac{Z_3 Z_1}{Z_2})^2 \sin^2 \frac{2\pi l_2}{\lambda_2}} \quad (3.35)$$

where l_2 is a thickness of the second layer.

One important conclusion follows from Eq.3.35. There are two cases when the second layer becomes transparent for ultrasound propagation. The first one is rather obvious; it happens when the second layer thickness is much less than the wavelength ($l_2 << \lambda_2/4$). The second case is related to the standing waves built up in the intermediate layer, when $l_2 = n\lambda_2/2$. These conditions are important for designing acoustic and electroacoustic devices.

3.5 Propagation in porous media

Ultrasound provides an opportunity to characterize porous solids that are saturated with liquid. There is a large body of both theoretical [7-11] and experimental [12-21] literature on ultrasound propagation through various porous systems. Although this, for the most part, lies outside the scope of this book, we consider it to be an important application for acoustics for another type of dispersed heterogeneous system. This encourages us to give here a short overview of the major achievements in this field.

The simplest approach uses empirical relationships to define certain acoustic properties. Investigators found that the acoustic impedance, Z_{emp}, of many soils and sands can be described empirically according to the following expressions by a single parameter, namely flow resistivity, Φ_e:

$$Z_{emp} = R_{emp} + jX_{emp} \quad (3.36)$$

$$R_{emp} = \rho_m c_m (1 + 0.057 C_{emp}^{-0.754}) \quad (3.37)$$

$$X_{emp} = \rho_m c_m 0.057 C_{emp}^{-0.732} \quad (3.38)$$

$$C_{emp} = \frac{\omega \rho_m}{2\pi \Phi_e} \quad (3.39)$$

These empirical relationships work satisfactorily for many systems as long as this flow resistivity is greater that 0.01, and less than 1.0. This markedly limits the range of applicability. There are also many instances when these relationships do not work at all. A more elaborate theoretical analysis is possible. There are two different approaches to such a theory: phenomenological and microstructural [11].

The phenomenological approach, first developed by Morse and Ingard [4], treats a porous media, saturated with a liquid, as a new "effective media" (We will use a similar idea later for characterizing complex dispersions). They assumed that the solid part of the porous body is completely rigid and incompressible. This system of interconnected pores is a random network having a total volume fraction, Ω_e, called the *porosity*. The propagation of sound causes a motion of the liquid with a mean velocity, $v(x,t)$. The equation of continuity for v is:

$$\Omega_e \frac{\partial \delta\rho_s}{\partial t} - \rho_s \, div \, v = 0 \qquad (3.40)$$

This equation is still valid for the fluid flow for sufficiently large volume when the irregularities of the pores average out. The density, ρ_s, is the mass of the fluid occupying the fraction Ω_e of the space available to the fluid. The variation of the density, $\delta\rho_s$, is related to the acoustic pressure, P, and the effective compressibility, $\beta^p_{ef}(\omega)$ (which is generally frequency dependent):

$$\delta\rho_s = \beta^p_{ef}(\omega)\rho_s P \qquad (3.41)$$

The equation for dynamic force balance must include both inertia and drag effects arising from fluid motion in the pores. The effective density, ρ_s, of the moving mass might be larger than the actual density of the liquid because part of the structure might be involved into the motion. The frictional retardation of the flow is expressed in terms of a flow resistivity, $\Phi_e(\omega)$, which might also be frequency dependent. The force balance equation is thus:

$$\rho_s \frac{\partial v}{\partial t} + \Phi_e(\omega) v + grad P = 0 \qquad (3.42)$$

The equation of state contains the ratio of specific heats $\gamma^h(\omega)$ that also might be frequency dependent.

We stress here the frequency dependencies of these three parameters, ($\beta^p_{ef}(\omega)$, $\Phi_e(\omega)$, and $\gamma^h(\omega)$), because this represents the weakest point of the phenomenological approach. There is no independent way to measure these parameters. They must be determined from the acoustic measurements, and this takes away much of the predictive power of the phenomenological approach.

In contrast, the microstructural approach allows one to eliminate ill-defined parameters, while paying the high price of losing a certain level of abstraction. This approach requires a specifically defined shape and structure for the pores. It goes back to Lord Rayleigh [1, 2], who suggested a model consisting of a rigid matrix of identical parallel cylindrical pores. The most widely known theory of this kind is associated with Biot [7-9] who considered a flexible porous system. These theories introduce a total of five parameters: porosity, flow resistivity, dynamic and static shape factors, and tortuosity. All of these parameters can be either calculated or measured using non-acoustic means.

There is one interesting version of these theories that links acoustics to dielectric spectroscopy. For this reason it might be especially interesting for people involved in Colloid Science. This theory was presented by Johnson, Koplik, and Dashen [10]. They operate with two frequency dependent parameters: tortuosity $\alpha_{tor}(\omega)$, and permeability $k_{per}(\omega)$. These parameters are related to the each other by the following equation:

$$\alpha_{tor}(\omega) = \frac{j\eta\Omega_e}{k_{per}(\omega)\omega\rho_m} \qquad (3.43)$$

They consider equation 3.43 as an analog of the relationship between complex dielectric permittivity, $\varepsilon_s^*(\omega)$, and the conductivity, $K_s^*(\omega)$, of an effective continuous media:

$$\varepsilon_s^* = j\frac{K_s^*}{\omega\varepsilon_0} \qquad (3.44)$$

This analogy allows them to apply the methods developed for dielectric spectroscopy, and to obtain rather general and interesting results for acoustics in porous media.

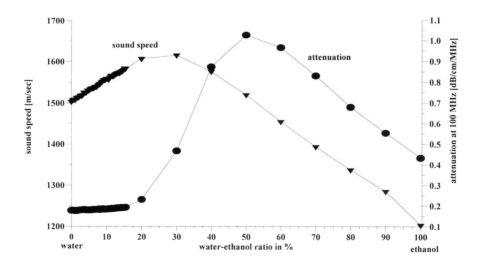

Figure 3.5 Attenuation and sound speed of water-ethanol mixture.

The similarity between acoustics and dielectric spectroscopy is very marked. It even justified combining them together under the one name "relaxation spectroscopy", following Eigen's Nobel Prize work for ions in liquids [22-24], described in the following section.

3.6 Chemical composition influence

It has been known for a long time that the acoustic properties of a homogeneous liquid depend on its chemical composition, and there have been successful attempts to use ultrasound for characterizing such chemical properties [22-32]. The homogeneous liquid, by itself, plays only the role of a background, a supporting character, nothing more. The background properties that are important in the present context include the intrinsic attenuation and sound speed (and, later when we speak of electroacoustics, an ion vibration current/potential).

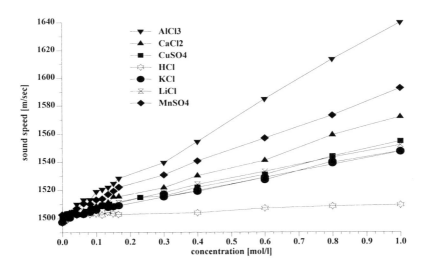

Figure 3.6 Influence of the various ions on the sound speed of water.

It is desirable that the intrinsic attenuation of the media be small in relationship to the incremental attenuation produced by adding particles, and that this intrinsic attenuation, as well as the sound speed, be relatively insensitive to the chemistry. In this case we would be able to use the same value for intrinsic attenuation and sound speed for all dispersions that use this particular liquid as the dispersion medium.

It turns out that the attenuation coefficient, as the measured acoustic parameter, satisfies these two requirements in the vast majority of practical dispersions. The effect of the chemical composition on the attenuation is measurable only for high concentrations of a very few chemicals. In contrast, sound speed is very sensitive to the chemical composition. Corresponding variations are easily measurable, even at relatively low concentrations of many ordinary chemicals, and are quite comparable with the contribution made by colloidal particles.

The intrinsic attenuation of many pure liquids is, indeed, small compared to the contribution of the colloid particles added to that media. Figures 3.2 and 3.3 illustrate the attenuation for several liquids. The intrinsic attenuation of water is about 0.2 dB/cm/MHz at 100 MHz. It is the most transparent liquid for ultrasound. The attenuation of alcohols, acetone, toluene, and heptanes is somewhat higher than water, but still much less than 1 db/cm/MHz at 100 MHz. We will show in Chapters 4 and 6 that the incremental attenuation resulting from adding colloidal particles to these media is typically many times higher.

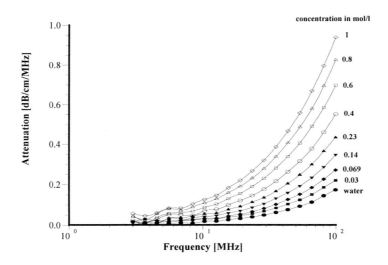

Figure 3.7 Attenuation frequency spectra for aqueous solutions of $CuSO_4$.

However, there is a large body of literature that demonstrates that both the intrinsic attenuation and sound speed of pure liquids can be dramatically changed by slight changes in chemistry, made by either adding a second liquid ("liquid mixture"), or by adding various solid materials that completely dissolve in the pure liquid. In the published literature, typically, a single paper considers only attenuation or sound speed, but not both. As examples, the handbook on molecular acoustics [32] provides a lot of data on sound speed, whereas Hunter and Darby [77] present only attenuation data for benzene, dioxane, and polysiloxane. There are very few papers showing both attenuation and sound speed for the same mixture. Figure 3.5 shows the attenuation and sound speed of a water-ethanol mixture.

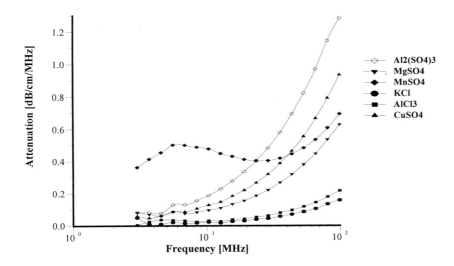

Figure 3.8 Attenuation spectra for 1M aqueous solutions of various electrolytes.

Both attenuation and sound speed vary with the water to ethanol content. However, attenuation spectra remain constant within instrument precision (0.01 dB/cm/MHz), up to 20 % of ethanol in water. Small traces of ethanol would not affect attenuation measurement. At the same time, sound speed exhibits measurable changes exceeding instrument precision (0.1 m/sec), even at 0.5% of ethanol. This clearly demonstrates that sound speed is more sensitive to any chemical variation than is attenuation.

It is interesting that the change in both attenuation and sound speed is very non-linear with respect to chemical composition. Both go through a maximum, but the position of this maximum for sound speed is much different than for attenuation. This means that initial addition of ethanol to water does not affect viscosity of water but does affect its elasticity. The viscosity of water starts to change only when the ethanol content exceeds 20%. The elastic component at this concentration already decays to the ethanol value. It is not clear why all of this happens. However, it is obvious that the acoustic properties of liquid mixtures contain much information about their structure.

Having briefly addressed liquid mixtures, we now consider the second possibility for changing the chemistry, namely adding a solid material that completely dissociates into ions on dissolving in the water, creating "ion solutions", often called "electrolytes". Researchers at the beginning of the 20th century used ultrasound to study the properties of electrolyte solutions. And in 1967 Manfred Eigen, Ronald Norrish, and George Porter received a Nobel Prize for their "relaxation spectroscopy" investigations of high-speed chemical reactions. "Relaxation spectroscopy" includes any experimental technique based on the variation of a thermodynamic parameter with time. Acoustic spectroscopy is just another version of "relaxation spectroscopy". "Dielectric spectroscopy", perhaps more familiar in Colloid Science [33], is yet another. Both Acoustic and Dielectric Spectroscopies were used in the middle of the 20th century for investigating properties of electrolyte solutions [22-24]. These studies are associated usually with the name of Manfred Eigen.

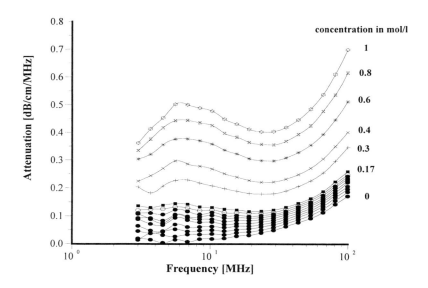

Figure 3.9 Attenuation frequency spectra for aqueous solutions of $MnSO_4$.

It is ironic that "relaxation spectroscopy" is unknown in Colloid Science, and yet it is very close in terms of technique and application. We reproduce here some results obtained originally by Eigen and others. We suggest that readers who are interested in this technique and theoretical interpretation read the original papers by Eigen and his colleagues [22-24].

Modern instruments allow investigators to collect data much more rapidly, and with less effort than was possible in the middle 20th century. All of the data presented in Figures 3.7-3.9 and Table 3.3 was collected within 1 week. This higher efficiency allowed us to study one more variable – the electrolyte concentration.

We used ten different inorganic electrolytes; all are listed in Table 3.3.

Table 3.3 Sound speed and intrinsic attenuation of various electrolytes at 1M concentration.

	sound speed m/sec	attenuation at 100 Mhz dB/cm/MHz	temperature C^0	density g/cm^3	compressibility 10^{-10} m^2/N^{-1}
water	1498	0.18	25	0.997	4.47
HCl	1509	0.18	25.3		
KCl	1547	0.18	23.7	1.04	4.02
LiCl	1552	0.18	25.4	1.02	4.07
NaCl	1559	0.18	25.6	1.04	3.96
CaCl$_2$	1572	0.18	25.8	1.06	3.82
CuSO$_4$	1555	0.94	25.8	1.13	3.66
MgSO$_4$	1623	0.63	26.6	1.1	3.45
MnSO$_4$	1593	0.69	25.7	1.12	3.52
AlCl$_3$	1640	0.22	23.9	1.08	3.44
Al$_2$(SO$_4$)$_3$	1634	1.28	26.1	1.13	3.31

Only five of the selected electrolytes exhibited any pronounced effect on the attenuation spectra (see Figure 3.8). Mono-valent electrolytes such as KCl and NaCl appear to have little affect on the attenuation spectra in the 1-100 MHz range. Manganese Sulfate (MnSO$_4$) shows the most interesting effect on the attenuation spectra especially as a function of concentration (see Figure 3.9). Interestingly, these curves reproduce results obtained 30 years ago by Jacobin and Yeager [28].

According to Eigen [22], MnSO4, CuSO$_4$ and Al$_2$ (SO$_4$)$_3$ cause the strongest known effects, and this view is continued by our measurements. For our further discussion of acoustics in colloids, it is only important to note merely that for electrolyte concentrations below 0.1 M the attenuation changes less than 0.1 dB/cm/MHz. We will see later that colloidal particles cause much greater effects on the attenuation. This means that these variations in the intrinsic background attenuation of the dispersion medium due to ion composition are usually negligible, and need only be taken into account for very high electrolytes concentrations approaching 1 Molar.

The situation with sound speed is rather different. Table 3.3 shows sound speed of 1M electrolyte solutions. It is seen that even 1:1 electrolytes affect the sound speed significantly. We will see later that the magnitude of this effect is comparable to that produced by adding colloidal particles. They are comparable even at low ion concentrations. Figure 3.6 shows the dependence of sound speed on the ion concentration for all 10 electrolytes. It is seen that it is practically linear function of the concentration. This effect can be measured even below 0.1 M, taking into account that the precision of the sound speed measurement is about 0.1 m/sec.

It is interesting that the linear dependence of the sound speed with ion concentration cannot be explained by the increase in density resulting from the added electrolyte. According to Eq. 3, the sound speed should decrease with increasing density, assuming that the compressibility remains constant. However, the experiment above indicates just the opposite trend. The data given in Table 3.3 shows that the sound speed increases with increasing density of the electrolyte solution (measured using a liquid pyncnometer).

The increase in sound speed with density suggests that the compressibility is not constant, but is also a function of the ion concentration. It follows that the compressibility of a liquid must decrease with increasing electrolyte concentration. It appears as if the ions enforce some structure in the liquid. This effect becomes stronger for the more highly charged ions (Table 3.3).

This structural effect might be important in understanding the nature of the internal parts of the double layer, where electrolyte concentrations become very high. Acoustics allows us to characterize the properties of liquids at extreme ionic strengths. It yields information about both the real and imaginary rheological components. Attenuation is related to the viscous component, whereas compressibility calculated from the sound speed is related to elasticity. The experiments described above indicate that both the real and imaginary components depend on the electrolyte concentration. This means that both the elasticity and viscosity of the liquid within the double layer are quite different than in the bulk of the solution. The viscous component of this effect has been known for a long time [33]. The concept of an elastic change within the double layer is rather new. It is not clear yet what impact this new rheological understanding will have on double layer theory.

REFERENCES.

1. Baron Rayleigh, "The Theory of Sound", Volume 1, Macmillan & Co., London, (1926)
2. Baron Rayleigh, "The Theory of Sound", Volume 2, Macmillan & Co., NY, (1896)
3. Morse, P.M. "Vibration and Sound", 1991, Acoustical Society of Amer. Publications; ISBN 0883 182874 (1991)
4. Morse, P.M. and K. Uno Ingard, Theoretical Acoustics, Princeton University Press, NJ, (1986)

5. Temkin S. Elements of Acoustics, 1st ed., John Wiley & Sons, NY (1981)
6. Horace Lamb, "Hydrodynamics", Sixth Edition, Dover Publications, NY, (1932)
7. Biot, M.A. "Theory of Propagation of Elastic Waves in a Fluid Saturated Porous Solid. I. Low Frequency range. ", J. Appl. Phys., 26, 182, pp. 168-179, (1955)
8. Biot, M.A. "Theory of Propagation of Elastic Waves in a Fluid Saturated Porous Solid. I. High Frequency range. ", J. Appl. Phys., 26, 182, pp. 179-191, (1955)
9. Biot, M.A. "Theory of propagation of elastic waves in a fluid saturated porous solid." J. Acoustic Soc. Am., 28, p.171-191, (1956)
10. Johnson, D., Koplik, J., and Dashen, R. "Theory of dynamic permeability and tortuosity in fluid saturated porous media", J. Fluid Mech, vol. 176 pp.379-402, (1987)
11. Attenborough, K. "Acoustical characterization of rigid fibrous absorbents and granular materials", J. Acoust. Soc. Amer., 73, 3, pp.785-799, (1983)
12. Shumway, "Sound Speed and Absorption Studies of Marine Sediments by a resonance Method", Geophysics, vol. 25, 3, pp. 659-682 (1960)
13. Wood, A.B. and Weston, D.E. "The Propagation of Sound in Mud", Acustica, vol.14, pp.156-162, (1964)
14. Hoverm, J. "Viscous attenuation of sound in suspensions and high porosity marine sediments", J.Acoust.Soc.Amer., vol. 67, 5, pp. 1559-1563, (1980)
15. Vidmar, P. and Foreman, T. "A Plane-Wave Reflection Loss Model Including Sediment Rigidity", J. Acoust. Soc. Amer., vol.66, 6, pp. 1830-1835, (1979)
16. Stoll, R. "Theoretical Aspects of Sound Transmission in Sediments", J. Acoust. Soc. Amer., vol.68, 5, pp. 1341-1350, (1980)
17. Hoverm, J. and Ingram, G. "Viscous Attenuation of Sound in Saturated Sand", J. Acoust. Soc. Amer, vol.66, 6, pp. 1807-1812, (1979)
18. Hampton, L. "Acoustic Proprties of Sediments", J. Acoust. Soc. Amer., vol.42, 4, pp. 882-890, (1967)
19. McLeroy, E.G. and A. DeLoach, "Sound Speed and Attenuation from 15 to 1500 kHz Measured in Natural Sea-Floor Sediments", J. Acoust. Soc. Amer., vol.44, 4, pp. 1148-1150, (1968)
20. Mackenzie, K.V. "Reflection of Sound from Coastal Bottoms", J.Acoust.Soc.Amer.,vol.32, 2, pp. 221-231 (1960)
21. Murpy, S.R., Garrison, G.R. and Potter, D.S. "Sound Absorption at 50 to 500 kHz from Transmission Measurements in the Sea", J. Acoust. Soc. Amer., vol.30, 9, pp. 871-875 (1958).
22. Eigen, M. "Determination of general and specific ionic interactions in solution", Faraday Soc. Discussions, 24, p.25 (1957)
23. Eigen, M. and deMaeyer, L. in "Techniques of Organic Chemistry", (ed. Weissberger) Vol. VIII Part 2, Wiley, p.895, (1963)

24. De Maeyer, L., Eigen, M. and Suarez, J. "Dielectric Dispersion and Chemical Relaxation", J. of Amer. Chem. Soc., 90, 12, 3157-3161, (1968)

25. Solovyev, V.A., Montrose, C.J., Watkins, M.H., and Litovitz, T.A. "Ultrasonic Relaxation in Etnanol-Ethyl Halide Mixtures", J. of Chemical Physics, vol.48, 5, pp. 2155-2162 (1968)

26. D'Arrigo,G., Mistura, L., and Tartaglia, P. " Sound Propagation in the Binary System Aniline-Cyclohaxane in the Critical Region", Physical review A, vol.1, 2, pp. 286-295, (1974)

27. Nomoto, O. and Kishimoto, T. "Velocity and Dispersion of Ultrasonic Waves in Electrolytic Solutions", Bulletin of Kobayashi Institute, vol.2, 2, pp. 58-62, (1952)

28. Jackopin, L. and Yeager, E. "Ultrasonic Relaxation in Manganese Sulfate Solutions", J. of Physical Chemistry, vol.74, 21, pp. 3766-3772 (1970)

29. Richardson, E.G. "Acoustic Experiments Relating to the Coefficients of Viscosity of Various Liquids", Faraday Soc. Discussion, 226, pp. 16-24, (1954)

30. Litovitz, T.A. and Lyon, "Ultrasonic Hysteresis in Viscous Liquids", J. Acoust. Soc. Amer., 26, 4, pp. 577- 580 (1954)

31. Chynoweth, A.G. and Scheider, W.G. "Ultrasonic Propagation in Binary Liquid Systems near Their Critical Solution Temperature", J.of Chemical Physics, vol.19, 12 , pp.1566-1569 (1951)

32. Numerical data and functional relationships in Science and Technology, Ed. K.H. Hellwege, Vol. 5, Molecular Acoustics, Ed. W. Schaaffs, Berlin, NY, (1967)

33. Lyklema, J., "Fundamentals of Interface and Colloid Science", vol. 1-3, Academic Press, London-NY, 1995-2000.

34. Bhatia, A.B., "Ultrasonic Absorption", Dover Publications Inc, NY (1985).

35. Litivitz, T.A. and Davis, C.M. "Structural and Shear Relaxation in Liquids", ed. J. Lamb, pp. 281-349 (1954)

36. Kinsler, L., Frey, A., Coppens, A., Sanders, J. "Fundamentals of Acoustics", Wiley, NY, 547 p., (2000)

37. Shortley, G. and Williams, D. "Elements of Physics", Prentice Hall Inc., NJ, (1963)

38. Costley, R. D.; Ingham, W. M.; Simpson, J. A., "Ultrasonic velocity measurement of molten liquids." Ceramic Transactions, 89, 241-251 (1998)

39. Riche, L.; Levesque, D.; Gendron, R.; Tatibouet, J., "On-line ultrasonic characterization of polymer flows."- Nondestructive Characterization of Materials VI, 37-44 (1994)

40. Jong-Rim Bae; Jeong-Koo Kim; Meyung-Ha Yi , "Ultrasonic velocity and absorption measurements for polyethylene glycol and water solutions", Japanese Journal of Applied Physics, Part 1 , VOL. 39, NO. 5B, 2946-7 (2000)

41. Swarup, S.; Chandra, S., "Ultrasonic velocity measurements in acetone solutions of some standard epoxy resins and xylene solutions of fumaric resins", Indian Journal of Physics, Part B, VOL. 61B, NO. 6, 515-21 (1987)

42. So, J. H.; Esquivel-Sirvent, R.; Yun, S. S.; Stumpf, F. B., "Ultrasonic velocity and absorption measurements for poly(acrylic acid) and water solutions", Journal of the Acoustical Society of America, VOL. 98, NO. 1, 659-60 (1995)

43. Nomura, T.; Saitoh, A.; Horikoshi, Y. "Measurement of acoustic properties of liquid using liquid flow SH-SAW sensor system", Sensors and Actuators B (Chemical), VOL. B76, NO. 1-3, 69-73 (2001)

44. Prek, M., "Experimental determination of the speed of sound in viscoelastic pipes", International Journal of Acoustics and Vibration, VOL. 5, NO. 3, 146-50 (2000)

45. Swarup, S.; Chandra, S. "Measurement of Ultrasonic Velocity in Resin Solutions." Acta Polym., 40, 8, 526-529 (1989)

46. Hoy, C. L. C.; Leung, W. P.; Yee, A. F. "Ultrasonic measurements of the Elastic Moduli of Liquid Crystalline Polymers."- Polymer, 33, 8, 1788-1791 (1992)

47. Alig, I.; Stieber, F.; Wartewig, S.; Bakhramov, A. D.; Manucarov, Yu. S. "Ultrasonic Absorption and Shear Viscosity Measurements for Solutions of Polybutadiene and Polybutadiene-Block-Polystyrene Copolymers." Polymer, 28, 9, 1543-1546 (1987)

48. Soucemarianadin, A.; Gaglione, R; and Attane, P. "High-frequency acoustic rheometer, and device for measuring the viscosity of a fluid using this rheometer" PATENT NUMBER- 00540111/EP B1, DESG. COUNTRIES- DE; ES; GB; IT; NL; SE, (1996)

49. Prugne C.; van Est J.; Cros B.; Leveque G.; Attal J. "Measurement of the viscosity of liquids by near-field acoustics", Measurement Science and Technology, 9/11, 1894-1898, (1998)

50. Hertz, T. G.; "Viscosity measurement of an enclosed liquid using ultrasound", Rev. Scientific Instruments, VOL. 62, 2 (1991)

51. Inoue, M.; Yoshino, K.; Moritake, H.; Toda, K. "Viscosity measurement of nematic liquid crystal using shear horizontal wave propagation in liquid crystal cell", Japanese Journal of Applied Physics, Part 1, VOL. 40, NO. 5B, 3528-33 (2001)

52. Prek, M., "Experimental determination of the speed of sound in viscoelastic pipes", International Journal of Acoustics and Vibration, VOL. 5, NO. 3, 146-50 (2000)

53. Belyaev, V. V., "Physical methods for measuring the viscosity coefficients of nematic liquid crystals", Physics-Uspekhi, VOL. 44, NO. 3, 255-6, 281-4 (2001)

54. Buiochi, F.; Higuti, T.; Furukawa, C. M.; Adamowski, J. C. "Ultrasonic measurement of viscosity of liquids", 2000 IEEE Ultrasonics Symposium. Proceedings. An International Symposium (Cat. No.00CH37121), Ed.: Schneider, S. C.; Levy, M.; McAvoy, B. R., vol.1, 525-8 (2000)

55. Guo Min, "A novel method to measure the viscosity factor of liquids", Wuli, VOL. 30, NO. 4, 220-2 (2001)

56. Kondoh, J.; Hayashi, S.; Shiokawa, S., "Simultaneous detection of density and viscosity using surface acoustic wave liquid-phase sensors", Japanese Journal of Applied Physics, Part 1 (Regular Papers, Short Notes & Review Papers), VOL. 40, NO. 5B, 3713-17 (2001)

57. Jain, M. K.; Schmidt, S.; Grimes, C. A. "Magneto-acoustic sensors for measurement of liquid temperature, viscosity and density", Applied Acoustics, VOL. 62, NO. 8, 1001-11 (2001)

58. All, A.; Hyder, S.; Nain, A. K. , "Intermolecular interactions in ternary liquid mixtures by ultrasonic velocity measurement", Indian Journal of Physics, Part B , VOL. 74B, NO. 1, 63-7 (2000)

59. Cohen-Bacrie, C. "Estimation of viscosity from ultrasound measurement of velocity", 1999 IEEE Ultrasonics Symposium. Proceedings. International Symposium (Cat. No.99CH37027), Ed.: Schneider, S. C.; Levy, M.; McAvoy, B. R., vol.2 , 1489-92 (1999)

60. Fritz, G.; Scherf, G.; Glatter, O. , "Applications of densiometry, ultrasonic speed measurements, and ultralow shear viscosimetry to aqueous fluids", Journal of Physical Chemistry B, VOL. 104, NO. 15, 3463-70 (2000)

61. Moritake, H.; Takahashi, K.; Yoshino, K.; Toda, K., "Viscosity measurement of ferroelectric liquid crystal", Japanese Journal of Applied Physics, Part 1, VOL. 35, NO. 9B, 5220-3 (1996)

62. Kondoh, J.; Saito, K.; Shiokawa, S.; Suzuki, H. , " Multichannel shear horizontal surface acoustic wave microsensor for liquid characterization", 1995 IEEE Ultrasonics Symposium. Proceedings. An International Symposium (Cat. No.95CH35844) , Ed.: Levy, M.; Schneider, S. C.; McAvoy, B. R. , vol.1, 445-9 (1995)

63. Kondoh, J.; Saito, K.; Shiokawa, S.; Suzuki, H. "Simultaneous measurement of liquid properties using multichannel shear horizontal surface acoustic wave micro-sensor", Japanese Journal of Applied Physics, Part 1, VOL. 35, NO. 5B, 3093-6, (1996)

64. Trinh, E. H.; Ohsaka, K , "Measurement of density, sound velocity, surface tension, and viscosity of freely suspended supercooled liquids", International Journal of Thermophysics , USA, VOL. 16, NO. 2, 545-55 (1995)

65. Sheen, S. H.; Reimann, K. J.; Lawrence, W. P.; Raptis, A. C. "Ultrasonic techniques for measurement of coal slurry viscosity", - IEEE 1988 Ultrasonics Symposium. Proceedings (IEEE Cat. No.88CH2578-3) Ed.: McAvoy, B. R., vol.1, 537-41, (1988)

66. Alig, I.; Stieber, F.; Wartewig, S.; Bakhramov, A. D.; Manucarov, Yu. S., "Ultrasonic absorption and shear viscosity measurements for solutions of polybutadiene and polybutadiene-block-polystyrene copolymers", Polymer, VOL. 28, NO. 9 , 1543-6 (1987)

67. Gaunaurd, G.; Scharnhorst, K. P.; Uberall, H. "New method to determine shear absorption using the viscoelastodynamic resonance-scattering formalism", Journal of the Acoustical Society of America, VOL. 64, NO. 4, 1211-12 (1978)

68. Hulusi, T. H. , "Effect of pulse-width on the measurement of the ultrasoinic shear absorption in viscous liquids", Acustica, VOL. 40, NO. 4, 269-71 (1978)

69. Sulun, O., "Determination of elastic constants by measuring ultra-sound velocity in Al/sub 2/(SO/sub 4/)/sub 3/ and Ce/sub 2/(SO/sub 4/)/sub 3/ solutions" Revue de la Faculte des Sciences de l'Universite d'Istanbul, Serie C (Astronomie, Physique, Chimie), VOL. 39-41, NO. 1-3, . 18-31 (1974-1976)

70. Sokolov, M., "On an acoustic method for complex viscosity measurement", Transactions of the ASME. Series E, Journal of Applied Mechanics, VOL. 41, NO. 3, 823-5 (1974)

71. Bacri, J. C., "Measurement of some viscosity coefficients in the nematic phase of a liquid crystal", Journal de Physique Letters, VOL. 35, NO. 9, . L141-2 (1974)

72. Knauss, C. J.; Leppo, D.; Myers, R. R., "Viscoelastic measurement of polybutenes and low viscosity liquids using ultrasonic strip delay lines", Journal of Polymer Science, Polymer Symposia, USA, NO. 43, 179-86 (1973)

73. Sheen, Shuh-Haw; Lawrence, William P.; Chien, Hual-Te; Raptis, Apostolos C. "Method for measuring liquid viscosity and ultrasonic viscometer", US PATENT NUMBER- 05365778, (1994)

74. Soucemarianadin, Arthur; Gaglione, Renaud; Attane, Pierre, "High-frequency acoustic rheometer and device to measure the viscosity of a fluid using this rheometer", US PATENT NUMBER- 05302878, (1994)

75. Bujard, Martial R. " Method of measuring the dynamic viscosity of a viscous fluid utilizing acoustic transducer", US PATENT NUMBER- 04862384, (1989)

76. Sachiko, S.; Toyoe, M; Yoshihiko, T, "Measuring instrument for liquid viscosity by surface acoustic wave", Patent Japan 02006728 JP, (1990)

77. Hunter, J.L. and Dardy, H.D. "Ultrahigh-frequency ultrasonic absorption cell", J. Acoust. Soc. Am., 36, 10, 1914-1917, (1964)

Chapter 4. ACOUSTIC THEORY FOR PARTICULATES

Any acoustic theory for particulates should yield a relationship between some measured macroscopic acoustic properties, such as sound speed, attenuation, acoustic impedance, angular dependence of the scattered sound, etc., and some microscopic characteristics of the heterogeneous system, such as its composition, structure, electric surface properties, particle size distribution, etc. It is also desirable that this theory satisfies a set of requirements, summarized as follows:

1. The theory must be valid for a wide particle size range, from 10 nanometer to 1 millimeter, and for ultrasonic frequencies from 1 to 100 MHz. This is roughly equivalent in magnitude to a ka range of 0.00001 to 100. (k is the compression wavenumber)
2. The theory should be valid for concentrated suspensions, and therefore must take into account particle-particle interactions.
3. The theory should reflect the multiple mechanisms of ultrasound interaction with colloids, namely: viscous, thermal, scattering, intrinsic, structural, and electrokinetic. The heuristic description of these mechanisms is given below.

Despite one hundred years of almost continuous effort by many distinguished scientists, there is still no single theory that meets all of these requirements. For example, the best known theory, abbreviated as ECAH, following the names of its creators: Epstein, Carhart, Allegra and Hawley [1, 2], meets the first requirement, completely fails the second, and takes into account only four of the six mechanisms mentioned in the third.

The ECAH theory is constructed in two stages. We call the first stage the "single particle theory", since it attempts to account for all of the ultrasound disturbances surrounding just a single particle. This stage relates the microscopic properties of both the fluid and particle to the system properties at a "single particle level". The second stage, which we refer to as the "macroscopic theory", then relates this "single particle level" to the macroscopic level at which we actually obtain our experimental raw data.

Both parts of the ECAH theory ("single particle" and "macroscopic") neglect any particle-particle interactions and are therefore valid only for dilute systems. For instance, in the ECAH "macroscopic theory" the total attenuation is regarded as simply a superposition of the contributions from each particle, and is determined only by the reflected compression wave. This part, derived by Epstein and Carhart [1], is similar to the well-known optical theorem that declares that the extinction cross section depends only on the scattering amplitude in the forward direction [3].

There have been several attempts to extend this general two-stage acoustic theory to concentrated systems by incorporating particle-particle interactions. The following are examples of three different approaches for each stage:

For the Single Particle theory:
1. Isolated particle in ECAH theory [1, 2].
2. Cell model [14].
3. Core-shell model [15, 16].

For the Macroscopic theory:
1. Scattering and multiple-scattering theories [1, 4-8].
2. Coupled phase model [9-11].
3. Phenomenological theory by Temkin [12, 13].

Obviously these extensions increase the complexity of the ECAH theory. However, the original ECAH theory, even without any modifications, is already very complex, as a result of the authors' attempt to construct a universal theory that is valid not only for any *ka* value, but also for all interaction mechanisms between the sound and the colloidal system. Modifications to implement these particle interactions, such as outlined briefly above, are therefore practically impossible.

One might ask, is it necessary to develop such a general theory? Is it possible to introduce some simplifications, while at the same time providing room for more readily implementing these particle interactions? Such simplifications are indeed possible. It turns out that there are some quite general peculiarities of ultrasound propagation through colloids that allow us to simplify the theoretical process. Historically these peculiarities prompted the introduction of six different mechanisms of sound interaction with colloids. We give here a short heuristic description of them all.

1). The "viscous" mechanism is hydrodynamic in nature. It is related to the shear waves generated by the particle oscillating in the acoustic pressure field. These shear waves appear because of the difference in the densities of the particles and the medium. The density contrast causes particle motion with respect to the medium. As a result, the liquid layers in the particle vicinity slide relative to each other. The non-stationary sliding motion of the liquid near the particle is referred to as the "shear wave". This mechanism is important for acoustics. It causes losses of acoustic energy due to shear friction. Viscous dissipative losses are dominant for small rigid particles with sizes less than 3 microns, such as oxides, pigments, paints, ceramics, cement, and graphite. The viscous mechanism is closely related to the electrokinetic mechanism which is also associated with the shear waves.

2). The "thermal" mechanism is thermodynamic in nature and is related to the temperature gradients generated near the particle surface. These temperature gradients are due to the thermodynamic coupling between pressure and temperature. Dissipation of acoustic energy caused by thermal losses is the dominant attenuation effect for soft particles, including emulsion droplets and latex droplets.

3). The "scattering" mechanism is essentially the same as in the case of light scattering. Acoustic scattering does not produce dissipation of acoustic energy. Particles simply redirect a part of the acoustic energy flow, and as a result this portion of the sound does not reach the receiving sound transducer. The scattering mechanism contributes to the overall attenuation, and this contribution is significant for larger particles with a diameter exceeding roughly 3 microns.

4). The "intrinsic" mechanism refers to losses of acoustic energy due to the interaction of the sound wave with the material of the particles and the medium, considered as homogeneous phases on a molecular level, and unrelated to the state of division of the colloidal dispersion.. It must be taken into account when the overall attenuation is low, which might happened for small particles or low volume fractions.

5). The "structural" mechanism links acoustics with rheology. It occurs when particles are joined together in some network. Oscillation of these inter-particle bonds can cause additional energy dissipation.

6). The "electrokinetic" mechanism describes the interaction of ultrasound with the double layer of the particles. Oscillation of the charged particles in the acoustic field leads to generation of an alternating electrical field, and consequently to an alternating electric current. This mechanism is the basis for electroacoustic measurements. However, it turns out its contribution to acoustic attenuation is negligible [55], which makes acoustic measurements completely independent of the electrical properties of the dispersion, including the properties of the double layer.

We can divide all of the mechanisms of ultrasound attenuation into two groups, depending on the way the acoustic energy is transformed in the colloid. The ultrasound attenuation in a colloid arises either from (1) absorption (the conversion of acoustic energy into thermal energy) or (2) scattering (the re-direction of incident acoustic energy from the incident beam). The combined effects of scattering and absorption can be described as the *extinction cross section*, Σ, which is an effective area whose product, with the incident intensity, is equal to the power lost from the sound beam [17-19]. In this respect, ultrasound is similar to light. There is a well known formula for light, "**extinction = absorption + scattering**" (Chapter 2) which is also applicable to sound. This formula is the basis for the acoustic theory and therefore we consider it in some details in the following section.

4.1. Extinction = absorption + scattering. Superposition approach.

There is a recent trend to snub this extinction equation, to consider viscous and thermal dissipations as additional forms of "scattering", and to picture all particle-particle interactions at high concentration as simply additional aspects of "multiple scattering". According to this nouveau view, the very term absorption is absorbed into a "Unified Scattering Theory "and the term extinction becomes extinct.

We think that crossing this Rubicon is misleading at best, hides a general understanding of the underlying mechanisms, and complicates rather than simplifies any extensions of the theory.

Adepts of this new philosophy offer several arguments for this new perspective.

First they claim that it is simply a semantic change. They argue that within the framework of the ECAH formulation it is impossible, in any case, to clearly separate the various mechanisms. The various components become hidden behind a veil of mathematical formalism, and this complexity defies intuitive understanding in terms of the underlying physical phenomena. This lack of clarity is then used as a reason to discard terminology and philosophical underpinnings that have a long and honorable history.

In rebuttal to this first claim, we take strong exception to this semantic argument. It not only completely alters the historical perspective, but the very language is a contradiction in terms. According to Webster, we find the following definitions:

Scatter: *to separate and go in several direction; to reflect and refract in an irregular, diffuse matter.*

Absorption: *a taking in and not reflecting; partial loss of power of light or radio waves passing through a medium.*

It is obvious that Webster's definition stresses "absorption" as being quite different from "scattering", emphasizing that it is "not reflecting".

This difference was obvious, and seemed important to the founder of the scattering theory, Lord Rayleigh. He wrote more than one hundred years ago in his book "The Theory of Sound. Vol.2" the following:

"...When we inquire into the mechanical question, it is evident that sound is not destroyed by obstacles as such. In the absence of dissipative forces, what is not transmitted must be reflected. Destruction depends upon viscosity and upon conduction of heat; but the influence of these agencies is enormously augmented by the contact of solid matter exposing a large surfaces. At such a surface the tangential as well as the normal motion is hindered, and a passage

of heat to and fro takes place, as the neighboring air is heated and cooled during its condensations and rarefactions. With such rapidity of alternation as we are concerned with in the case of audible sounds, these influences extend to only a very thin layer of the air and of the solid, and are thus greatly favored by a fine state of division... "

It is clear that Lord Rayleigh considered sound absorption as especially peculiar for in colloids, systems with the finite state of division and extended surfaces.

Secondly, the adepts appeal to the light propagation through the colloids. The general formula **extinction=scattering + absorption,** since it expresses only a basic conservation of energy, must be valid for light and sound. However, in the case of light the "scattering" term certainly dominates discussion in scientific talks and publications. The terms "extinction" and "absorption" are rarely mentioned, and only then in special handbooks [3].

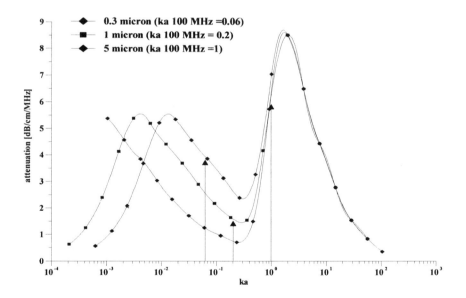

Figure 4.1 Attenuation spectra calculated for monodisperse colloids with three different sizes showing superposition of viscous absorption and rigid particle scattering. The arrows show the point corresponding to a frequency of 100 MHz.

This neglect of absorption phenomena in optics might be often justified, but it hardly justifies carrying over this bias in case of sound. The role of absorption is very different in acoustics as compared to optics. In optics, absorption is an electrodynamic effect, whereas in acoustics it is either hydrodynamic or thermodynamic. For optics, absorption presents a difficult problem because it requires information about the imaginary component of the refractive index and this information is generally not available. For optics, this normally leaves only one choice - to ignore the absorption of light. In this case, the description of light propagation thorough a colloid is reduced to a consideration of scattering alone. In many real colloids, this simplification works well enough, because the absorption is often negligible.

In acoustics the situation is dramatically different. The absorption of ultrasound is easy to calculate. No special properties of the particles are required. As we will see later, the absorption of ultrasound by solid rigid particles depends only on their density, which is readily available or can be easily measured. In the case of ultrasound, absorption is not a complicating factor. Rather it is very important source of information about the particles, especially sub-micron particles. Ignoring this term means ignoring the major advantage of ultrasound over light as the characterization technique.

Finally, we offer several arguments in support of our view of retaining the historical viewpoint of Rayleigh and others.

1. For acoustics, sub-micron particles do not scatter ultrasound at all in the frequency range under 100 MHz. They only absorb ultrasound. There is no need to develop a general complex scattering-absorption theory for such sub-micron particles.
2. For acoustics there is a very simple way to eliminate the nonlinear effects of multiple scattering. (We define scattering here in the classical sense, as the interaction of the compression waves scattered by particles with other particles). The effects of multiple scattering are completely eliminated by following Morse's suggestion to measure the intensity of the incident ultrasound after transmission. The intensity of this ultrasound is not affected by multiple scattering, but it may be affected by particle-particle interactions through viscous or thermal absorption mechanisms. We prove this statement experimentally later.
3. The ultrasound attenuation in pure liquids and gels is via absorption only. There is no scattering there. This allows us to interpret acoustic spectra in rheological terms and to use n acoustic spectrometer as a high frequency rheometer.
4. If we combine scattering and absorption in a single mathematical model, such as the ECAH or modification to it, we are forced to use at all times an extended set of input parameters, many of which may be unknown in a given case. Even in the case where a given mechanism may be

unimportant, the relevant physical properties may still be required because of the complicated perhaps very nonlinear characteristics of a "unified" approach. Separating scattering and absorption opens the way to minimize the number of required input parameters. This issue is discussed in details at the end of this Chapter.
5. Separation of scattering and absorption phenomena provides more insight as to the nature of the attenuation phenomena. The unified approach is like a "black box" . We input information and get an answer without any understanding of the processes going on.
6. Calculation of the PSD from attenuation spectra is a classical ill-defined problem. The likelihood of multiple solutions can be minimized by carefully using all *a'priori* and independent information. Such information can be more readily employed in helping to solve the inverse problem if the mechanisms concerning the sound attenuation can be linked to all available *a'prior* independent information. The unified or "black box" approach does not easily provide format for this purpose

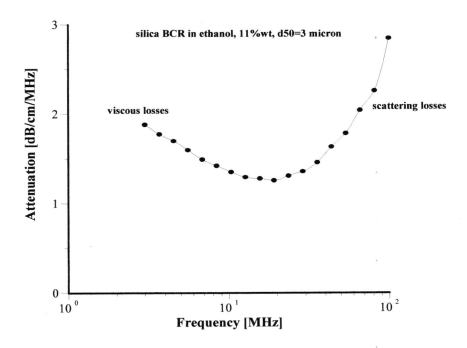

Figure 4.2 Measured attenuation spectra for particle size standard silica BCR-70.

Taking into account all these argument, we have decided to proceed in this book with the classical meaning of the terms "scattering" and "absorption", following Rayleigh, Morse, Sewell, and other distinguished scientists.

There is one more important argument supporting our decision. This last argument requires more detail consideration and justification.

7. In the case of light it is practically impossible to separate absorption and scattering [3] in measurement, whereas in the case of ultrasound it easily achievable.

For ultrasound, absorption and scattering are distinctly separated in the frequency domain. As a general rule, for a given particle size, the absorption of ultrasound is dominant at low frequencies, whereas scattering is dominant at high frequencies, with an overlap region existing within a narrow intermediate frequency range.

This is illustrated in Figure 4.1 (where the sum of the viscous absorption and scattering is plotted as a function of ka) for particles of three different sizes. The low frequency peaks correspond to the viscous absorption for different size particles, the high frequency peak corresponds to the scattering losses.

Figure 4.2 shows the measured attenuation spectrum for silica BCR-70. This material is a certified standard with a median size of 2.9 microns [52]. This attenuation spectrum demonstrates very clearly the transition from viscous absorption losses at low frequency, to scattering losses at high frequency. In this case, the transition range occurs in the vicinity of 20 MHz.

As with light, ultrasound scattering decreases rapidly with decreasing particle size. For ultrasound frequencies below 100 MHz, scattering becomes unimportant for particles below 1 micron. Thus for samples containing such small particles it makes no sense to employ a complicated theory that combines both absorption and scattering; an absorption theory would be sufficient. In the opposite case of large particles with $ka > 1$, ultrasound absorption is negligible and any theory need take into account only scattering.

Although Figure 4.1 considered only viscous type absorption, this separation between absorption and scattering is also observed for the thermal loss mechanism. Anson and Chivers [20] studied seventy two liquids and showed that thermal losses are important up to $ka=0.5$. Other mechanisms dominate for higher ka, except in those emulsions where only the thermal properties differ substantially between the two phases.

This separation of the loss mechanisms in the frequency domain allows us to express the total attenuation, α_T, as the sum of the attenuation due to the absorption, α_{ab}, and the attenuation due to the scattering, α_{sc}. In addition, we should add a background intrinsic attenuation α_{in} because the dispersion medium attenuates ultrasound as does any other liquid (see Chapter 3). Thus:

$$\alpha_T = \alpha_{ab} + \alpha_{sc} + \alpha_{int} \tag{4.1}$$

A further simplification can be considered from the fact that ultrasound absorption is important for small ka, that is for wavelengths larger than the particle size. It is the so-called Rayleigh region [21, 22], or long wavelength limit. At this limit the two general components of ultrasound absorption (viscous and thermal) are additive. This then allows us to consider these two effects separately, and to neglect their possible coupling. The total attenuation now becomes:

$$\alpha_T = \alpha_{vis} + \alpha_{th} + \alpha_{sc} + \alpha_{int} \tag{4.2}$$

The potential contribution of any structural losses in a structured colloid would lead to the modification of the viscous losses, and to additional energy dissipation as a result of oscillation of the inter-particle bonds. This last contribution can be accounted for by adding a further term to the total attenuation:

$$\alpha_T = \alpha_{vis} + \alpha_{th} + \alpha_{st} + \alpha_{sc} + \alpha_{int} \tag{4.3}$$

The simple superposition expression for the measured attenuation is not universally applicable. There are some exceptions [23]. However, equation 4.3 is valid for a very large number of practical concentrated colloids.

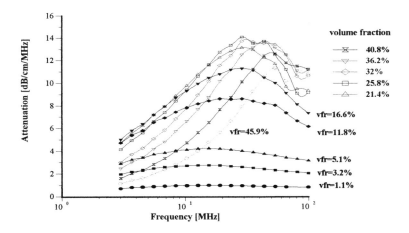

Figure 4.3. Measured attenuation spectra for 0.3 micron diameter rutile in water.

The most important advantage of this superposition approach is that, compared to the ECAH theory, it allows us to take particle–particle interactions into account more easily. Several independent groups have determined the range of volume fractions for which such particle-particle interactions become important. The conclusion from these studies is that the viscous loss mechanism is the most sensitive to such interactions.

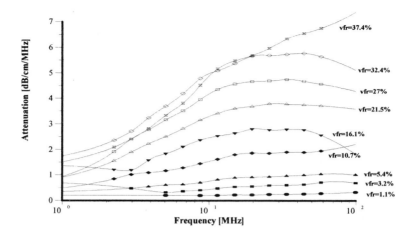

Figure 4.4 Measured attenuation spectra for 0.1 micron neoprene latex in water.

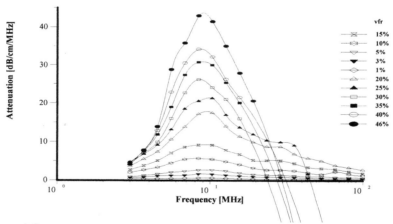

Figure 4.5 Measured attenuation spectra for 120 micron quartz particles in water.

For example, Uric [24, 25] showed that the attenuation of rigid heavy particles becomes a nonlinear function of volume fraction above 10%vl, reaching a maximum attenuation at 15%vl. This was confirmed later by Hampton [26], Blue and McLeroy [27], Marlow *et al* [53], and Dukhin and Goetz [14]. It is interesting that we find this non-linear behavior to be independent of particle shape, and that it occurs at the same critical concentration, even for profoundly non-spherical particles [27]. Perhaps this happens because, for long wavelengths, the particles behave essentially as point sources; shape effects are therefore not pronounced. This same phenomenon is found in light scattering [3].

The thermal loss mechanism, which is thermodynamic in nature, is less dependent on particle-particle interactions. This was shown to be true for several different polymer latices by Dukhin and Goetz [28, 29], and also for several emulsions by McClements, Hemar *et al* [15, 16, 30].

The importance of particle-particle interactions, as it relates to the scattering mechanism, depends very much on the method of measurement. In principle, we can measure the sound scattered at some angle to the incident beam, or, instead, consider only the decrease in the intensity of the incident wave as it travels through the colloid. It is not widely understood that this choice of measurement technique plays a very important role in defining the necessary theory. In fact, the effect of "multiple scattering" can be minimized by measuring the attenuation of the incident wave as was clearly pointed out by Morse [17]:

"...whether multiple scattering is important or not, the attenuation of the incident wave in traversing a distance z of region R is given by factor $Exp(-N_p \Pi z)$, where Π is the total scattering cross-section, N_p is the number concentration of particles. When multiple scattering is important, some energy is absorbed by the scattered wave before it emerges from R, and some of the scattered wave is rescattered, but the formulas for the Σ and attenuation still tells us what happens to the incident wave, the part of the wave which has not been affected by scattering..."

Hence, by choosing to measure the attenuation of the incident beam, the scattering mechanism becomes much less dependent on particle-particle interactions. The result is that the variation of the transmitted wave intensity remains a linear function, of the volume fraction, up to much higher volume fractions than would be possible if directly monitoring the off-axis scattered sound. For example, Busby and Richardson [31] showed that the attenuation for large 95 micron glass spheres remains a linear function of the ultrasound path length up to 18%vl. Atkinson and Kytomaa [32, 33] concluded that the attenuation of 1 mm particles remains linear to 30%vl.

We have also collected data that confirms these observations by using a variety of disperse systems, each of which exhibits only one of the several possible mechanisms of ultrasound attenuation. Accordingly, Figure 4.3 shows the attenuation spectra for a dispersion of 0.3 micron rutile particles, for which we would expect mainly viscous losses. For comparison, Figure 4.4 presents the attenuation spectra for a dispersion of 0.16 micron latex particles, for which the thermal losses are dominant. Finally, Figure 4.5 illustrates the attenuation for a dispersion of 120 micron quartz particles, for which scattering would be dominant.

The viscous attenuation in Figure 4.3 is strongly volume fraction dependent, leading also to a shift in the critical frequency from about 25 MHz to 60·MHz. This frequency shift is much less pronounced for the thermal attenuation shown in Figure 4.4. Furthermore, we see that the critical frequency for the scattering attenuation in Figure 4.5 is virtually unchanged with volume fraction.

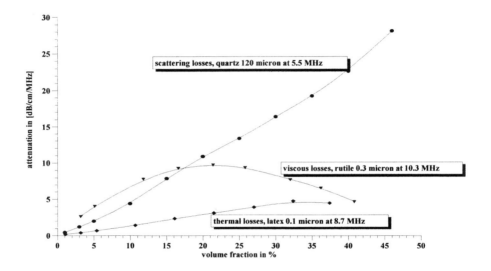

Figure 4.6 Measured attenuation at one particular frequency as a function of volume fraction for viscous, thermal and scattering losses.

To better illustrate the difference in volume fraction dependence, we can extract the viscous, thermal and scattering attenuation at a single frequency and plot this versus volume fraction (Figure 4.6). The viscous attenuation is a linear function of volume fraction up to only about 10%vl. In contrast, the thermal attenuation remains linear up to about 30%vl. Finally, the scattering attenuation remains linear up to 46%vl. Linear dependence with volume fraction is a strong indication of the absence of particle-particle interactions. Conversely, deviation from linearity implies the presence of such interactions. The boundary, between the linear and nonlinear regions, determines the volume fraction limit for any dilute case theory.

In summary, particle-particle interactions become an important consideration for viscous losses at fairly low volume fraction, whereas such interactions are relatively unimportant for scattering losses, even at very high volume fraction. Thus we can conclude that ultrasound absorption is the most important mechanism to address when developing a more general theory that takes into account hydrodynamic, thermodynamic and specific particle interactions. Development of an extended scattering theory to account for such particle interactions, the so-called "multiple scattering" problem, is of much less importance for acoustics. However, multiple scattering is indeed a major concern when attempting to analyze similar concentrated systems using optical systems

The general structure of this expanded ultrasound absorption theory remains the same as in the ECAH theory. Again there are two stages: "single particle" and "macroscopic". There are currently several versions of the "single particle" theory that addresses particle-particle interactions. They are based on the "cell model", or the "effective media model", or their combination in the "core-shell model". For both absorption mechanisms (viscous and thermal) the "coupled phase model" can be used as the "macroscopic" theory in place of the Epstein version of the optical theorem.

There is a substantial difference between the ways absorption and scattering are measured. Absorption irreversibly converts acoustic energy to heat. The measured intensity of the ultrasound after propagation through colloid, I_t, is what remains of the incident power, I_i, after subtraction of the energy absorbed by the particles and liquid. For scattering, the situation is more complicated because the energy disturbed by the particles is simply redirected, although remaining as acoustic energy. This difference provides many options when performing a scattering type measurement.

For instance, it is possible to measure the angular dependence of scattered sound, a possibility that simply does not exist for absorbed sound. Such angular dependence measurements were made by Kol'tsova and Mikhailov [34] for very large graphite particles in gelatin media, and by Faran [35] for metal cylinders. However, the measurement of such angular dependence of the scattered sound is not common.

The more typical way to measure the scattered sound is to determine the attenuation and/or sound speed. Interpretation of a scattering measurement made in this mode is somewhat more complicated than characterizing simple absorption, because part of the scattered power may still reach the detector together with the remainder of the incident beam. For this reason we speak of three components of scattered sound: coherent, incoherent and multiple scattered. There are different ways to separate these components, either in theory or in the experimental design of an instrument. What we want to stress here, however, is that absorption and scattering are very different, not only in their nature and frequency dependencies, but also in the ways the measurement can be performed.

In summary, we conclude that there is a strategic approach for deriving acoustic theory which is an alternative to the ECAH theory. We give up the idea of considering simultaneously all the mechanisms of ultrasound attenuation for all ka, but are rewarded with the ability to incorporate particle-particle interactions into the theory of absorption. As a result, acoustics can be more easily applied to the characterization of real concentrated colloidal systems. Consequently, in the following section we will describe this new theory, which we call the "superposition theory". The term "superposition" reflects superposition of the various mechanisms of the ultrasound extinction on an additive basis.

As mentioned previously, the "single particle theory" might be derived on the basis of either a "cell model" or a "core-shell model". We prefer to use a "cell model" for the viscous losses, since this mechanism is hydrodynamic in nature and the use of a cell model is well established for such effects. At present there is no corresponding "cell model" theory for thermal losses, but in principle the "core-shell model" theory can be used. It is not yet clear which approach is the most advantageous for characterizing thermal losses.

The macroscopic part of the "superposition" acoustic theory is a combination of the "scattering theory" and the "coupled phase model". We suggest that the "scattering theory" be used only for characterizing ultrasound scattering, whereas the "coupled phase model" provides a better framework to describe ultrasound absorption.

4.2. Acoustic theory for dilute systems

The fundamental basis of the ECAH theory was originally developed by Epstein and Carhart [1]. Allegra and Hawley [2] later generalized it for a more representative particle rheology. They also performed the first vigorous experimental tests of the theory, which were more adequate than some earlier studies made by Stakutis, Morse, Dill, Beyer, Hartmann and Focke [36, 37]. Following Allegra and Hawley, we now present a short overview of the ECAH theory.

The description of the propagation of a sound wave through a heterogeneous media requires three potentials: compression ϕ_c; thermal ϕ_T; and viscous ϕ_η. A system of wave equations determines the space-time dependencies of these potentials:

$$(\nabla^2 + k^2)\phi_c = 0 \qquad (4.4)$$
$$(\nabla^2 + k_T^2)\phi_T = 0 \qquad (4.5)$$
$$(\nabla^2 + k_s^2)\phi_\eta = 0 \qquad (4.6)$$

where:

$$k = \frac{\omega}{c} + j\alpha \tag{4.7}$$

$$k_T = (1+j)\sqrt{\frac{\omega \rho C_p}{2\tau}} \tag{4.8}$$

$$k_s = \sqrt{\frac{\omega^2 \rho}{\mu_L}} \tag{4.9}$$

These equations are applied to both the dispersed and continuous phase. The wave equation for a viscous fluid can be derived by replacing the Lame constant, μ_L, with $-j\omega\eta$.

The general solution of this system of equations can be presented, in the case of spherical symmetry, as an expansion of Legendre polynomials. For the reflected compression wave, for example, this expansion is:

$$\phi_c^r = \sum_{n=0}^{\infty} j^n (2n+1) A_n h_n(k_c r) P_n(\cos\theta) \tag{4.10}$$

Altogether, there are six sets of constants similar to A_n, that describe the general solution for the three potentials within the medium plus three potentials inside the spherical particle. There are also a set of six boundary conditions necessary to calculate these constants:

$$v_m^r(r=a) = v_p^r(r=a) \tag{4.11}$$

$$v_m^\theta(r=a) = v_p^\theta(r=a) \tag{4.12}$$

$$T_m(r=a) = T_p(r=a) \tag{4.13}$$

$$\tau_m \frac{\partial T_m}{\partial r}\bigg|_{r=a} = \tau_p \frac{\partial T_p}{\partial r}\bigg|_{r=a} \tag{4.14}$$

$$P_m^{rr}(r=a) = P_p^{rr}(r=a) \tag{4.15}$$

$$P_m^{r\theta}(r=a) = P_p^{r\theta}(r=a) \tag{4.16}$$

The final system of algebraic equations used to determine these constants is very complex. We refer those who are interested to the original Allegra-Hawley paper [2].

Equations 4.4 through 4.16 represent the "single particle" portion of the ECAH theory. The "macroscopic" portion of this theory was derived by Epstein [1]. He showed that the total attenuation is determined only by the constants, A_n, of the reflected compression wave. Unfortunately, the equations for the system of boundary conditions are interlinked, and it is impossible to calculate each of the individual values of A_n. Nevertheless, the following expression allows one to calculate the total attenuation, if the constants A_n are known:

$$\alpha = -\frac{3\varphi}{2k_c^2 a^3} \sum_{n=0}^{\infty} (2n+1) \operatorname{Re} A_n \tag{4.17}$$

Allegra and Hawley stressed that this expression is only valid for dilute suspensions, where the effect of the individual particles is additive. Also, particle-

particle interactions were also neglected when the general solution was constructed from Legendre polynomials, automatically eliminating terms that would become zero at an infinite distance from the particle. However, these are the very terms that become important in concentrated colloidal systems where long distance fields are indeed affected by the presence of other particles.

Although the ECAH acoustic theory is the most widely known, it is not the only theory. Temkin [12, 13] created an alternative theory, in which he suggested that the attenuation and sound speed be defined in terms of the particle-to-fluid velocity and temperature ratios. Consequently, these two parameters can be considered as local driving forces for viscous and thermal losses. Accordingly, he derived the following equations for sound speed and attenuation:

$$2\alpha \frac{c_s(\omega=0)}{c_s(\omega)} = \frac{L_w |\mathrm{Im}\, u_p^*|}{1+L_w} + \frac{(\gamma_m^h - 1)L_w \frac{C_p^p}{C_p^m}}{1+L_w \frac{C_p^p}{C_p^m}} |\mathrm{Im}\, \frac{T_p^* - 1}{1+L_w T_p^* \frac{C_p^p}{C_p^m}}| \quad (4.18)$$

$$\frac{c_s(\omega=0)}{c_s(\omega)} - \alpha^2 = 1 + \frac{L_w [\mathrm{Re}\, u_p^* - 1]}{1+L_w} + \frac{(\gamma_m^h - 1)L_w \frac{C_p^p}{C_p^m}}{1+L_w \frac{C_p^p}{C_p^m}} \mathrm{Re}\, \frac{T_p^* - 1}{1+L_w T_p^* \frac{C_p^p}{C_p^m}} \quad (4.19)$$

where u_p^* is the particle-to-fluid complex velocity normalized by the colloid velocity, γ^h is a specific heat ratio, T_p^* is the particle temperature normalized by the liquid heat transfer rate, and L_w is the mass load related to the weight fraction of the solid φ_w simply by:

$$L_w = \frac{\varphi_w}{1-\varphi_w} \quad (4.20)$$

Temkin's theory has one significant advantage over the ECAH theory, in that it yields an expression for the sound speed. In particular, the low frequency sound speed limit, $c_s(\omega=0)$, is important. There are several other derivations for this low frequency sound speed. The earliest is Wood's equation [38]:

$$\frac{c_m^2}{c_s^2(\omega=0)} = \frac{\rho_s}{\rho_m}(1-\varphi+\varphi\frac{\rho_m c_m^2}{\rho_p c_p^2}) \quad (4.21)$$

where the density of the colloid $\rho_s = \varphi\rho_p + (1-\varphi)\rho_m$.

Temkin generalized this expression by including thermodynamic effects for the elastic particles, assuming no mass transfer, electric or chemical affects. Temkin's final expression is:

$$\frac{c_m^2}{c_s^2(\omega=0)} = \gamma_m \frac{\rho_s}{\rho_m}(1-\varphi+\varphi\frac{\rho_m c_m^2}{\rho_p c_p^2}\frac{\gamma_p^h}{\gamma_m^h}) - ((\gamma_m^h-1)\frac{(1-\varphi+\varphi\frac{\beta_p}{\beta_m})^2}{1+\varphi_w(\frac{C_p^p}{C_p^m}-1)} \quad (4.22)$$

The thermodynamic corrections to sound speed are usually negligible because the specific heat ratio, γ_m, is very close to unity for almost all fluids. As a result, the simple Wood equation is sufficient for many practical colloidal systems.

Temkin, however, stressed that his theory, like the ECAH theory, is valid only for dilute systems.

4.3. Ultrasound absorption in concentrated dispersions

Consideration of ultrasound absorption is simplified by the fact that, for most colloidal systems, it occurs at ultrasound frequencies having a wavelength that is much longer than the particle size. This was justified earlier with some theoretical and experimental data (Figures 4.1-4.3). The long wavelength restriction has been used in many studies, as early as the Rayleigh scattering theory [21, 22]. This very convenient restriction allows us to consider the hydrodynamic and thermodynamic effects as both independent and additive. In turn, this allows us to now introduce both hydrodynamic and thermodynamic particle interactions into the theory. Inclusion of hydrodynamic effects is essential above 10%vl, as was shown first by Urick [24, 25], and later confirmed by Hampton [26], Blue and McLeroy [27], and Dukhin and Goetz [14]. Thermodynamic effects are less sensitive to these particle-particle interactions, but also need to be included above 30%vl, as demonstrated originally by Dukhin and Goetz [28], and more recently by McClements et al [15, 16, 30].

The "superposition" theory is built in two stages, similar to the ECAH theory: a "single particle" theory and "macroscopic" theory. The long wave restriction allows us to include particle-particle interactions on both levels.

The "cell model" is used as a "single particle" theory for the viscous losses. There is a version of the cell model [14] which is valid for the non-stationary mode given a polydisperse system. The "single particle" theory for thermal losses was created by Hemar et al [15, 16] using the "core-shell model" which is a combination of the "cell model" and "effective medium model" (Chapter 2).

The "coupled phase model" is the logical candidate for the "macroscopic" theory. We present here the version valid for a colloidal system that is polydisperse and structured.

For structured colloidal systems there is one other dissipative mechanism for ultrasound absorption. This has only recently been incorporated into the absorption theory. It is based on the transient network model that is widely used in rheology.

The theory of ultrasound absorption is derived only for spherical particles. There are some experimental indications, however, (Blue and McLeroy [27]) that the equivalent sphere model is reasonably adequate even for particles with pronounced shape.

4.3.1. Coupled phase model

Let us consider an infinitesimal volume element in the dispersed system. There is a differential force acting on this element that is proportional to the pressure gradient of the sound wave, ∇P. This external force is applied to both the particles and the liquid, and is distributed between the particles and liquid according to the volume fraction, φ.

Both particles and liquid move with an acceleration created by the sound wave pressure gradient. In addition, because of inertia effects, the particles move relative to the liquid, and this causes viscous friction between the particles and liquid.

The balance of these forces can be represented by the following two equations, written for particles and liquid respectively:

$$-\varphi \nabla P = \varphi \rho_p \frac{\partial u_p}{\partial t} + \gamma (u_p - u_m) \qquad (4.23)$$

$$-(1-\varphi)\nabla P' = (1-\varphi)\rho_0 \frac{\partial u_0}{\partial t} - \gamma(u_p - u_0) \qquad (4.24)$$

where u_m and u_p are the velocities of the medium and particles in a laboratory frame of reference, t is time, γ is a friction coefficient that is proportional to the volume fraction, φ, and the particle hydrodynamic drag coefficient, Ω

$$\gamma = \frac{9\eta\varphi \, \Omega}{2a^2} \qquad (4.25)$$

$$F^h = 6\pi \eta a \Omega (u_p - u_m) \qquad (4.26)$$

where η is the dynamic viscosity, and a is the particle radius.

Equations 4.23 and 4.24 are well known in the field of acoustics. They have been used in several papers to calculate sound speed and acoustic attenuation [9, 10, 14]. They are valid without any restriction as to the volume fraction. Importantly, it is known that they yield a correct solution as the concentration is reduced to the dilute case.

The two equations together are normally referred to as the "coupled phase model". The word "model" usually suggests the existence of some alternative formulation, but it is hard to imagine what one can change since these equations essentially express Newton's second law. Perhaps, the word "model" is not strong enough in this case.

The equations can be solved for the speed of the particle relative to the liquid. The time and space dependence of the unknowns, u_m and u_p, are presented as a monochromatic wave $A e^{j(\omega t - kx)}$, where j is a complex unit, and k is a compression complex wavenumber. As a result, the relationship between the gradient of pressure and the speed of the particle relative to the fluid can be expressed as:

$$\gamma(u_p - u_m) = \frac{\varphi(\rho_p - \rho_s)}{\rho_s + i\omega\varphi(1-\varphi)\frac{\rho_p \rho_m}{\gamma}} \nabla P \qquad (4.27)$$

where $\rho_s = \varphi\rho_p + (1-\varphi)\rho_m$, and φ is the volume fraction of the solid particles

Dukhin and Goetz [14] have suggested a generalization of the coupled phase model for polydisperse colloids. Assume that we have polydisperse system having N size fractions. Each fraction can be described by a certain particle diameter, d_i, volume fraction, φ_i, drag coefficient, γ_i, and particle velocity, u_i (in a laboratory frame of reference). Assume also that the density of the particles, ρ_p, are the same for all fractions and that the total volume fraction of the dispersed phase is φ. The liquid is characterized by a dynamic viscosity, η, a density, ρ_m, and a velocity, u_m (again in a laboratory frame of reference).

The coupled phase model suggests that we apply the force balance equation to each fraction of the dispersed system, including the dispersion medium. We did this earlier for just one fraction, but now we apply the same principle to the N fractions and obtain the following system of $N+1$ equations:

$$-\varphi_1 \nabla P = \varphi_1 \rho_p j\omega u_1 + \gamma_1(u_1 - u_m)$$
$$\cdots\cdots\cdots\cdots\cdots\cdots\cdots\cdots\cdots\cdots\cdots\cdots\cdots \quad \text{N equations for particles} \qquad (4.28)$$
$$-\varphi_i \nabla P = \varphi_i \rho_p j\omega u_i + \gamma_i(u_i - u_m)$$

$$-(1-\varphi) \nabla P = (1-\varphi)\rho_m j\omega u_m - \sum_i \gamma_i(u_i - u_m) \quad \text{an equation for liquid} \quad (4.29)$$

where j is a unity imaginary number, P and ω are the pressure and frequency of the ultrasound, and:

$$\gamma_i = \frac{18\eta\varphi_i \Omega_i}{d_i^2} \qquad (4.30)$$

$$F_i^h = 3\pi\eta d_i \Omega_i (u_i - u_m) \qquad (4.31)$$

We can solve this system of $N+1$ equations [14]. But to do this, we must first reformulate all equations to introduce the desirable quantities, $x_i = u_i - u_m$, and eliminate the parameter, u_m, using Equation 4.31 that specifies the liquid velocity in a form:

$$u_m = -\frac{\nabla P}{j\omega\rho_m} + \frac{\sum_i \gamma_i x_i}{(1-\varphi)j\omega\rho_m} \qquad (4.32)$$

This reformulated system of N equations is now:

$$(\frac{\rho_p}{\rho_m} - 1)\nabla P = (j\omega\rho_p + \frac{\gamma_i}{\varphi_i})x_i + \frac{\rho_p}{(1-\varphi)\rho_m}\sum_i \gamma_i x_i \qquad (4.33)$$

This system can be solved using the principle of mathematical induction. We guess the solution for N fractions, and then prove that the same solution works for $N+1$ fractions. As a result we obtain the following expression for velocity of the i-th fraction particle relative to the liquid:

$$u_i - u_m = \frac{(\frac{\rho_p}{\rho_m} - 1)\nabla P}{(j\omega\rho_p + \frac{\gamma_i}{\varphi_i})(1 + \frac{\rho_p}{(1-\varphi)\rho_m}\sum_{i=1}^{N} \frac{\gamma_i}{j\omega\rho_p + \frac{\gamma_i}{\varphi_i}})} \qquad (4.35)$$

This expression indicates that the coupled phase model will allow us to calculate the relative particle speed, $u_i - u_m$, for each fraction, without using the superposition assumption. This is important for the electroacoustic theory (Chapter 5).

For the acoustic theory, it is enough that we now have the ability to calculate the complex compression wavenumber, k, of the polydisperse colloidal system, without using any superposition assumption. But to derive k we must first add the following equation to the coupled phase model [9,10,14]:

$$-\frac{\partial P^*}{\partial t} = M_s(1-\varphi_i)\nabla u_m + \sum_{i=1}^{N} M_s \varphi \nabla u_i \qquad (4.36)$$

The two equations 4.28 and 4.29 yield then a solution for the complex wavenumber of the polydisperse colloid:

$$\frac{k^2 M_s^*}{\omega^2 \rho_s} = \frac{1 - \frac{\rho_p}{\rho_s}\sum_{i=1}^{N}\frac{\varphi_i}{1-\frac{9j\rho_m\Omega_i}{4s_i^2\rho_p}}}{1 + (\frac{\rho_s}{\rho_p} - 2)\sum_{i=1}^{N}\frac{\varphi_i}{1-\frac{9j\rho_m\Omega_i}{4s_i^2\rho_p}}} \qquad (4.37)$$

where

$$s_i^2 = \frac{a_i^2 \omega \rho_m}{2\eta_m} \qquad (4.38)$$

and $\rho_s = \rho_p \varphi + \rho_m (1-\varphi)$ \qquad (4.39)

At this point we believe it would be helpful to create a heuristic picture of the physical phenomena that occur when an ultrasound pulses passes through a system of dispersed particles. Such a description, given below, will help to provide answers to some general questions.

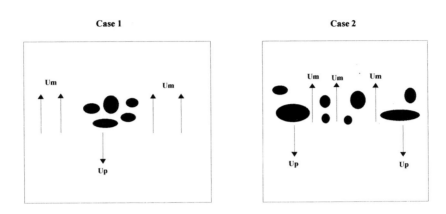

Figure 4.7 Two possible scenarios for sedimentation of particles in a liquid.

Let us consider an element of the concentrated dispersed system within the sound wave. The size of this element is selected such that it is much larger than the particle size, and also larger than the average distance between the particles. As a result, this element contains many particles. At the same time this element is much smaller than the wavelength of sound.

This dispersion element moves with a certain velocity and acceleration in response to the gradient of acoustic pressure. As a result, an inertial force is applied to this element. At this point we can introduce the principle of equivalency between inertia and gravity. The effect of the inertial force, created by the sound wave, is equivalent to the effect of the gravitational force.

This gravitational force is exerted on both the particles and the liquid in the dispersion element. The densities of the particles and the liquid are different, as are the forces. The force acting on the particles depends on the ratio of the densities.

The question arises as to what density we should consider. To answer this, let us consider the forces acting on a given particle in the gravitational field. The first force is the weight of the particle, and is proportional to its density, ρ_p. This force will be partially balanced by the pressure of the other particles and the surrounding liquid. This pressure is equivalent to the pressure generated by an effective media having a density equal to the density of the dispersed system. It is clear that this force is proportional to the difference in density between that of the particles, ρ_p, and that of the dispersed system, ρ_s. This statement can be illustrated by the sedimentation in a peat bog, consisting of mainly cellulose based material dispersed in water. The effective density of this effective media is much less than that of water, hence a person would drown in such a material if not pulled out.

This happens because there are two different scenarios for sedimentation, as shown in Figure 4.7. Case 1 corresponds to the situation where the group settles as a single entity, and the liquid envelopes this settling group from the outside. Case 2 corresponds to a different scenario, in which the liquid is forced to move through the group of particles. A difference exists for the two cases between the forces exerted on the particles within the group. In the first case the friction only occurs between the settling group and the liquid. In the second case, however, there is an additional frictional force resulting from the liquid pushing through the group of particles.

Ultrasound propagation through colloidal suspensions corresponds to the second case, because the motion of the particles and the liquid on the edges of the sound pulse is negligible when the width, w, of the sound pulse is much larger than wavelength, λ, at high ultrasound frequencies.

$$w >> \lambda \qquad (4.40)$$

It is in the center section of the sound pulse that the liquid is forced to move through the group of particles.

Thus, the balance of forces exerted on the particles in a given element of the dispersion consists of the effective gravitational force, the buoyancy force, and the frictional force related to the movement of the liquid through the array of particles.

4.3.2. Viscous loss theory

The coupled phase model provides a link between macroscopically measured parameters such as attenuation, α_s, and sound speed, c_s, and the properties of a single particle. The particles are considered to be rigid, and consequently characterized only by a density, ρ_p, and a drag coefficient, Ω_i, for each particle size fraction. This drag coefficient can be calculated using the Happel cell model (Section 2.5.2.). The cell model theory relates the particle size to the drag coefficient, that in turn is linked using the coupled phase model to the experimentally measured attenuation and sound speed. All required mathematical expressions are given in the Sections 2.5.2 and 4.2.2.

It is instructive to see how the theoretical attenuation changes with respect to various aspects of the colloidal suspension. For example, Figure 4.8 illustrates a decrease in the critical frequency (the frequency that corresponds to the maximum attenuation) as the particle size is increased. According to theory, this critical frequency is inversely proportional to the square of the particle size. In this example, the two monodisperse systems have a size ratio of 2:1 and the corresponding ratio of critical frequencies is seen to be 1:4 *(i.e.* 4 – 16 MHz).

Figure 4.9 illustrates the effects of polydispersity. Broadening of the particle size distribution leads to a flattening of the attenuation curve.

Together, Figures 4.10 and 4.11 illustrate the influence of modality on the attenuation spectra. The two particle size distributions shown in Figure 4.11 have the same median size and the same polydispersity. However, their third and fourth moments are obviously quite different, as is reflected in the shape of the particle size distribution. The corresponding theoretical attenuation curves for these two size distributions are shown in Figure 4.10. Clearly, there is a pronounced difference between the two attenuation spectra. This strongly suggests that acoustic spectroscopy can be used to study the differences in the shape of the size distribution of practical dispersions.

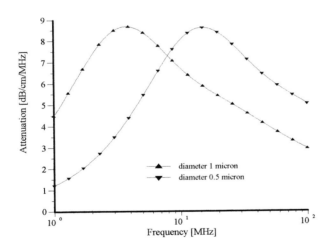

Figure 4.8 Effect of particle size on the theoretical viscous attenuation for a 20%vl slurry of monodisperse particles having a density contrast of 2.

Figure 4.12 illustrates the effect of particle-particle interactions on the attenuation spectra. It is seen that an increase in the volume fraction increases the critical frequency. In the least concentrated dispersion, there are two geometric parameters that determine the critical frequency value: the particle radius and the viscous layer thickness, δ_η. The viscous layer thickness is defined as the distance from the particle surface over which the shear wave, generated by the oscillating particle, decays by a factor $1/e$ as it passes into the bulk liquid. It is thus useful to think of the particle as being surrounded by a viscous layer with a thickness, δ_η, given by [1,2,14]:

$$\delta_\eta = \sqrt{\frac{2\eta}{\rho_m \omega}} \qquad (4.41)$$

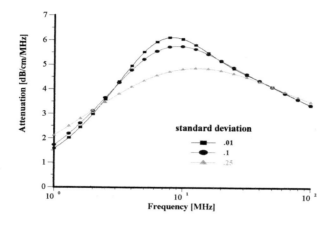

Figure 4.9 Effect of the width of the PSD on the theoretical viscous attenuation for a 15%vl slurry of particles having a mean size of 0.5 microns and a density contrast of 2.

The viscous layer thickness decays with frequency. There is a certain frequency at which the viscous layer thickness becomes equal to the particle radius. It turns out that this frequency is approximately equivalent to the critical acoustic frequency for a dilute suspension. This gives the following approximate expression for the critical frequency:

$$\omega_{cr} \approx \frac{2\eta}{a^2} \qquad (4.42)$$

For the case of a concentrated suspension with hydrodynamically interacting particles, there is one additional important geometrical parameter: the distance between particles. The distance between particles becomes smaller with increasing volume fraction. At some point it becomes less than the particle radius. For the volume fractions exceeding this critical value, the inter-particle distance overtakes the particle radius as the parameter which determines the critical frequency. This new condition for calculating the critical frequency can be formulated as an equality between the viscous layer thickness and the inter-particle distance for the given volume fraction. The inter-particle distance is smaller than the particle radius at this high volume fraction, and correspondingly the critical frequency is higher, as is illustrated in Figure 4.12.

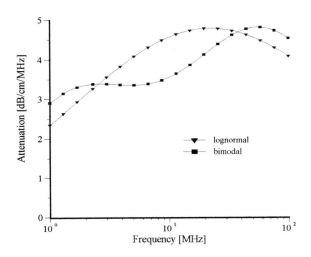

Figure 4.10 Attenuation spectra corresponding to two particle size distributions from Figure 4.11.

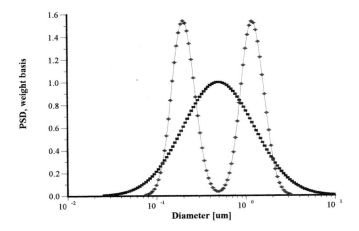

Figure 4.11 Particle size distributions with the same median size and width, but different higher order moments.

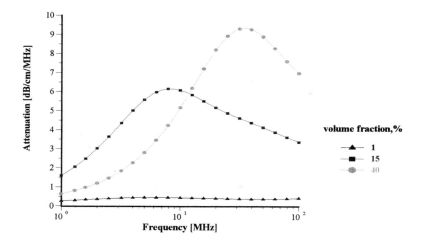

Figure 4.12 Effect of particle volume fraction on the attenuation spectra for a colloidal suspension of monodisperse 0.5 micron particles.

Without an adequate theory, this volume fraction effect might be mistakenly attributed to a variation in the particle size, since a decrease in particle size would cause a similar shift in the critical frequency. Ignoring particle-particle interactions would therefore lead to an underestimate of the particle size in concentrated colloidal dispersions. In the example presented in Figure 4.12, the critical frequency shifts almost ten-fold because of volume fraction induced particle-particle interactions. Simply applying the ECAH theory to this attenuation spectrum in order to calculate the particle size of the 40%vl dispersion would result in a particle size of only one third the actual size. Hence, to obtain correct particle sizing results using acoustics, it is of critical importance to take hydrodynamic particle-particle interactions into account through a suitable theory.

4.3.3. Thermal loss theory.

Isakovich [39] initially developed a theory for thermal losses, again only for dilute colloidal suspensions. Thermal losses dominate any ultrasound propagation in emulsions, polymer latices, or other systems with small (<1 micron) soft particles having low density contrast, such as wax suspensions. Isakovich's theory yields the following expression for thermal loss attenuation:

$$\alpha_{th} = \frac{3\varphi T c_m \rho_m \tau_m T}{2a^2} [\frac{\beta_m}{C_p^m \rho_m} - \frac{\beta_p}{C_p^p \rho_p}]^2 \operatorname{Re} H \qquad (4.43)$$

where the function, H, equals:

$$H^{-1} = \frac{1}{1-jz_m} - \frac{\tau_m \tanh z_p}{\tau_p \tanh z_p - z_p} \qquad (4.44)$$

$$z_{m,p} = a(1+j)\sqrt{\frac{\omega \rho_{p,m} C_p^{p,m}}{2\tau_{p,m}}} \qquad (4.45)$$

This original Isakovich theory is valid only for dilute dispersions, because it does not take into account particle-particle interactions. The term "dilute" is somewhat misleading in this instance, since experimental studies have shown that there are often no thermodynamic particle-particle interactions up to 30%vl. This means that Isakovich's theory can be applied, without any additional complexity, to a wide variety of practical systems like emulsions, polymer latices, and soft particles. Additional corrections would only be required at exceptionally high volume loading.

This surprising difference between "viscous" and "thermal" losses is related to the difference between the "viscous depth" and the "thermal depth". These two parameters are a measure of the decay of the shear and thermal waves, respectively, in the liquid. Particles oscillating in the sound wave generate these shear and thermal waves that are then damped as they move away from the particle. The characteristic distance for the shear wave amplitude to decay is the "viscous depth", δ_η (see section 4.2.2). The corresponding distance for the thermal wave to decay is the "thermal depth", δ_T, given by:

$$\delta_T = \sqrt{\frac{2\tau_m}{\rho_m \omega C_p^m}} \qquad (4.46)$$

The relationship between δ_η and δ_T has been considered by McClements, who calculated the "thermal depth" and "viscous depth" versus frequency [40]. For aqueous suspensions, it is relatively easy to show that the "viscous depth" is 2.6 times larger than the "thermal depth" [29]. As a result, the viscous layers surrounding the particles overlap each other at a much lower volume fraction than the corresponding thermal layers. The overlap of these boundary layers is a measure of the corresponding particle-particle interactions. Obviously, there are no particle-particle interactions when the corresponding boundary layers are sufficiently separated.

Thus, for a given frequency an increase in the dispersed volume fraction leads first to an overlap of the viscous layers because they extend further into the liquid. Only then, at higher volume fractions, will the thermal layers overlap. In general, we can qualitatively conclude that at lower volume fractions, the hydrodynamic interactions between particles are more important than any thermodynamic interactions.

Quantitatively, this factor of 2.6 for the ration of δ_η and δ_T implies that there is a large difference between these two mechanisms in so far as the critical volume fraction at which the boundary layers begin to overlap (with resultant particle-particle interactions). The dilute case theory is valid for any volume fraction smaller than this critical volume fraction. The ratio of these critical volume fractions, for the viscous and thermal losses, can be estimated from:

$$\varphi_{vis-th} = (\frac{a\sqrt{\pi f} + 1/2.6}{a\sqrt{\pi f} + 1})^3 \qquad (4.47)$$

where a is particle radius in microns, and f is the frequency in MHz.

It is interesting that this important feature of the thermal losses works for virtually all liquids. In order to illustrate this, it is convenient to introduce a parameter referred to as the "depth ratio", which is completely independent of the properties of the particle and is uniquely defined by the media alone:

$$depth\ ratio = \frac{\delta_\eta}{\delta_T} \qquad (4.48)$$

Figure 4.13 shows the values of this depth ratio parameter for many liquids taken from Anson and Chivers [20] and the "Materials Database" of Dispersion Technology, Inc. Without exception, this depth ratio is larger than that of water for all fluids studied. Therefore thermal losses are much less sensitive to particle-particle interactions than viscous losses for virtually all known liquids. For emulsions, etc. we conclude that Isakovich's theory is valid for a much wider volume fraction range than one might otherwise expect.

However, there is still a definite need for a thermal loss theory that would be valid for very high volume concentrations, up to 50%vl. To extend acoustic theory to these highly concentrated colloidal suspensions, Hemar et al [15, 16] modified Isakovich's theory to take into account the overlap of thermodynamic layers. They used the "core-shell model" for the "single particle theory", and a "scattering model" for the "macroscopic theory". A more detailed description of the core-shell model can be found in Section 2.5.2., but here we repeat some essential features.

The core-shell model replaces a real suspension with a model in which a single particle is placed inside of a spherical shell of the dispersion medium. The radius of this shell, b_{sh}, differs from the radius of the dispersion medium cell, b, in the cell model:

$$b_{sh} = \frac{a}{\sqrt{3\varphi}}; \quad b = \frac{a}{\sqrt[3]{\varphi}} \qquad (4.49)$$

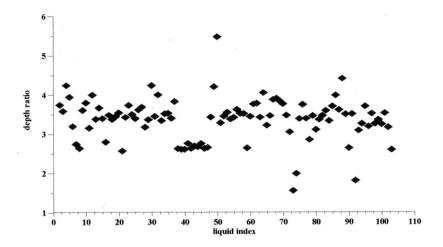

Figure 4.13 Depth ratio parameter for various liquids. The liquid index determines the liquid in the DT Material Database.

There is another difference between this "core-shell model" and the straight forward "cell model"; one applies an effective media approach to describe the dispersion beyond the liquid core. This effective media is characterized with "effective" values for the density, ρ_{ef}, the compressibility, β_{ef}, the heat conductance, τ_{ef}, and the heat capacity, $C_p{}^{ef}$. In turn these effective values are related to the properties of the dispersion medium and the dispersed phase using the following equations:

$$\rho_{ef} = (1-\varphi)\rho_m + \varphi\rho_p \tag{4.50}$$

$$\beta_{ef} = \varphi\beta_p + (1-\varphi)\beta_m \tag{4.51}$$

$$C_p^{ef} = (1-\varphi)\frac{\rho_m}{\rho_{ef}}C_p^m + \varphi\frac{\rho_p}{\rho_m}C_p^p \tag{4.52}$$

$$\tau_{ef} = \tau_m \frac{1 + 2\varphi\dfrac{\tau_p - \tau_m}{\tau_p + 2\tau_m} - 2(1-\varphi)(0.21\varphi - 0.047\varphi^2)(\dfrac{\tau_p - \tau_m}{\tau_p + 2\tau_m})^2}{1 - \varphi\dfrac{\tau_p - \tau_m}{\tau_p + 2\tau_m} - 2(1-\varphi)(0.21\varphi - 0.047\varphi^2)(\dfrac{\tau_p - \tau_m}{\tau_p + 2\tau_m})^2} \tag{4.53}$$

This core-shell model yields a new expression for the function, H: (equation 4.44)

$$H = \frac{\tau_m \beta_m}{\beta_s} \frac{z_p - \tanh(n_p a)}{EC + FD} \frac{g_m - g_{ef}}{g_p - g_m} [2\beta_{ef} z_m \frac{b_{sh}}{a} (z_{ef} \frac{b_{sh}}{a} + 1) + C(1 + z_m) + D(1 - z_m)]$$

(4.54)

where:

$$g_m = \frac{\beta_m}{\rho_m C_p^m}; \quad g_p = \frac{\beta_p}{\rho_p C_p^p}; \quad g_{ef} = \frac{\beta_{ef}}{\rho_{ef} C_p^{ef}}$$

(4.55)

$$z_{ef} = a(1+j)\sqrt{\frac{\omega \rho_{ef} C_p^{ef}}{2\tau_{ef}}}$$

(4.56)

$$C = EXP[z_m(\frac{b_{sh}}{a} - 1)][\tau_m(z_m \frac{b_{sh}}{a} - 1) + \tau_{ef}(z_{ef} \frac{b_{sh}}{a} + 1)]$$

$$D = EXP[-z_m(\frac{b_{sh}}{a} - 1)][\tau_m(z_m \frac{b_{sh}}{a} + 1) - \tau_{ef}(z_{ef} \frac{b_{sh}}{a} + 1)]$$

$$E = \tau_p z_p + [\tau_m(z_m + 1) - \tau_p]\tanh z_p$$

$$F = \tau_p z_p - [\tau_m(z_m - 1) + \tau_p]\tanh z_p$$

The introduction of the thermal layer overlap through this "core-shell model" substantially improves the ability of acoustics to adequately characterize the properties of soft particles such as latices and emulsions.

4.3.4. Structural loss theory

In addition to hydrodynamic and thermodynamic interaction, particles might experience specific colloid-chemical interactions, especially in concentrated dispersions. These specific interactions can affect the propagation of ultrasound through the colloidal suspension, and provide an additional contribution to the ultrasound attenuation. We refer to this contribution as "structural loss". The name reflects the fact that this specific interaction usually implies some underlying macroscopic structure in the concentrated suspension.

Structural losses are closely related to the rheology of concentrated colloids. According to Voigt's rheological model [41], a structured suspension exhibits both dissipative and elastic behavior. The main difference between the structural losses in classical rheology and those in acoustics is the frequency range over which the loss is measured. Although acoustics utilizes much higher frequencies than classical rheology, the main principles and models are the same.

On a macroscopic level, Voigt's model suggests that we regard stress as a function of the strain and the strain rate. The Rouse-Bueche-Zimm [41] model translates this idea to a microscopic level, by representing the colloidal suspension as N+1 beads joined together, by N springs. The viscous resistance of the beads moving through the liquid is inversely proportional to the Stokes' drag coefficient, and causes viscous dissipation. The springs between the beads represent the stress-strain relationship. This model is successfully used to describe classical rheological behavior. We apply it to acoustics here, following

McCann [42]. He added a Hookean factor to the force balance, but only used the first term, in which force is proportional to displacement. We take the next logical step, following Voigt's model, which states that the stress is proportional not only to displacement, but also to the rate of displacement.

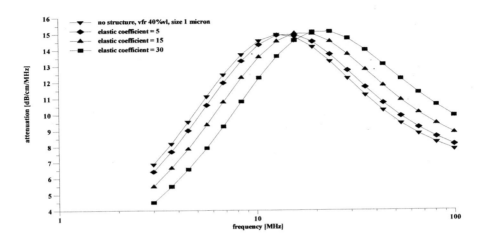

Figure 4.14 Influence of the elastic Hookean coefficient, H_1, on the attenuation of sound.

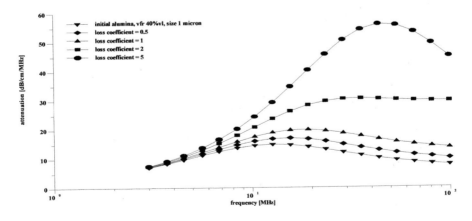

Figure 4.15 Influence of the loss Hookean coefficient, H_2, on the attenuation of sound.

We can reformulate the coupled phase model, including additional forces acting on the particles due to the specific interactions. The resulting coupled phase theory reflects both viscous and structural losses. It is based on balancing the forces on both the particles and the liquid, as well as applying the law of conservation of mass. According to this generalized coupled phase model, expressions for the force balance of the ith fraction of the particles, moving with a speed, u_i^p are:

$$-\varphi_i \nabla P = \varphi_i \rho_p \frac{\partial^2 x_i^p}{\partial t^2} + F_i^h + F_i^{hook} \tag{4.57}$$

and, for the liquid, moving with a speed, u_0:

$$-(1-\varphi)\nabla P = (1-\varphi)\rho_0 \frac{\partial^2 x_0}{\partial t^2} - \sum_{i=1}^{N} F_i^h \tag{.4.58}$$

These force balance equations assume that the total volume force caused by a sound pressure wave, P, is redistributed between the particles and the liquid according to their volume fractions (φ_i, for the ith fraction and φ for the total solid content). This volume force is compensated by an inertia force, a hydrodynamic friction, F^h and a "specific" Hookean force, F^{hook}.

$$F_i^h = 6\pi\eta a_i \Omega_i \frac{\partial x_i^p}{\partial t} \tag{4.59}$$

$$F_i^{hook} = H_1 \Delta x_i^p + H_2 \frac{\partial x_i^p}{\partial t} \tag{4.60}$$

Two terms are then necessary to adequately describe the contribution of the non-Hookean springs to the force balance. The first term is an elastic force proportional to the displacement of the particle with a coefficient H_1. The second term is a dissipative force proportional to the particle velocity with a coefficient H_2. The coefficients, H_1 and H_2, are assumed to be the same for all particles.

The resulting expression for the complex compression wavenumber of the dispersion, k is [29]:

$$\frac{k^2 M_s^*}{\omega^2} = \frac{(1-\varphi)(\rho_m + \sum_{i=1}^{N}\frac{\gamma_i(D_i - j\omega\gamma_i)}{j\omega D_i})}{[1-\varphi+\sum_{i=1}^{N}\frac{j\omega\varphi_i\gamma_i}{D_i}]^2 - \sum_{i=1}^{N}\frac{\omega^2\varphi_i^2}{D_i}[(1-\varphi)\rho_m + \sum_{i=1}^{N}\frac{\gamma_i(D_i - j\omega\gamma_i)}{j\omega D_i}]} \tag{4.61}$$

where

$$D_i = -\omega^2 \rho_p^i \varphi_i + j\omega\gamma_i + j\omega H_1 + H_2 \tag{4.62}$$

$$\gamma_i = \frac{9\eta\varphi_i\Omega_i}{2a_i^2} \tag{4.63}$$

The attenuation of the ultrasound, α, is related to the complex wavenumber through:

$$\alpha_s = -\operatorname{Im}(k) \tag{4.64}$$

Expression 4.61 is more general than the original "coupled phase theory" (Equation 4.37). First, it is valid for polydisperse systems. This is important because practical systems might be polydisperse not only in size, but also with respect to density, or perhaps other physical properties. This more general theory allows us to study systems containing mixed dispersed phases, i.e. containing particles of different chemical nature and different internal structure.

Second, this theory yields an expression for the complex wavenumber in pastes and gels. Attenuation in gels is caused just by the oscillation of the polymer network, or in our terminology it is purely "structural loss". By assuming that the drag coefficient, γ, in Equation 4.61 is equal to zero, we obtain the following expression for gels:

$$\frac{k^2 M_s^*}{\omega^2} = \frac{\rho_m(H_1 - \omega^2 \rho_p \varphi) + j\omega \rho_m H_2}{(1-\varphi)(H_1 - \omega^2 \rho_p \varphi) - \omega^2 \varphi^2 \rho_m + j(1-\varphi)\omega H_2} \quad (4.66)$$

This equation can be used to calculate the attenuation and sound speed in pastes and gels, but as of yet has not been tested experimentally. It is seen that attenuation occurs only when the second coefficient, H_2, is non- zero. The situation here is analogous to a damped spring; the dissipation of energy depends on the rate at which the spring is compressed, but not on the displacement *per se*.

Figures 4.14 and 4.15 illustrate the effect of structure on the theoretical attenuation spectra of a 40%vl alumina dispersion having a median size of 1 micron. It is seen that the first Hookean coefficient simply shifts the critical frequency from about 10 MHz to 20 MHz, keeping the shape of the curve more or less intact, and the peak attenuation constant. Since the particle size is reciprocally proportional to the square root of this critical frequency, the influence of the structure would have to be very substantial in order to create large errors in the calculated particle size.

The influence of the elastic coefficient cannot affect the quality of the fit between theory and experiment. For instance, it cannot explain any excess in experimental attenuation over that predicted by the theory. The elastic structure does not change the average amplitude of the attenuation spectra. The excess attenuation sometimes observed in experiments can be explained only with the second coefficient, as follows from Figure 4.15.

In principle, this second coefficient can be extracted from the attenuation spectra as an adjustable parameter, as will be shown in Chapter 8.

4.3.5. Intrinsic loss theory

"Intrinsic loss" reflects an acoustic energy dissipation process that occurs at a molecular level. We can easily speak of the intrinsic loss for a homogeneous material, but it is meaningful in a heterogeneous system as well. It is important, however, to understand that the intrinsic loss in a heterogeneous system is not related to how finely divided one phase is within the other. Roughly speaking, the intrinsic attenuation of a heterogeneous system is a volume weighted average of the intrinsic attenuation of each component. More precisely, we can compute the intrinsic loss in a heterogeneous colloid system [29] as follows:

$$\alpha_{int} = \frac{(1-\varphi)\frac{\alpha_m}{c_m} + \varphi\frac{\rho_m \alpha_p}{\rho_p \alpha_m}}{\sqrt{\frac{1-\varphi}{c_m^2} + \frac{\varphi \rho_m}{\rho_p c_p^2}}} \sqrt{\frac{\rho_s}{\rho_m}} \quad (4.67)$$

where α_m and α_p are intrinsic attenuations of the liquid medium and the particle material respectively.

The attenuation of any liquid (medium) is easily measured with an acoustic spectrometer. Usually, the attenuation of the internal phase material is negligible for solid particles. For soft particles, such as emulsions, it is easy to measure, if the internal phase is available as a pure liquid. If not, we need to use this as an additional adjustable parameter and determine the value from the acoustic data.

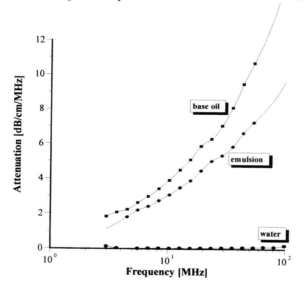

Figure 4.16 Attenuation spectra measured for water, oil and a water-in-oil emulsion.

The intrinsic attenuation of the colloid, α_{int}, can be considered as a background, independent of the particle size distribution, against which any additional attenuation that results from the state of division of the two phases, can be judged. Typically, the medium makes the largest contribution to the overall intrinsic attenuation of the system and we can sometimes neglect entirely the contribution of the particulates to the total intrinsic loss in the colloidal suspension. However, there are some interesting exceptions.

Figure 4.16 shows the attenuation spectra of a 30% water-in-oil emulsion together with the attenuation for each component of this emulsion, namely the oil and the water. At first glance it seems strange that the attenuation of the emulsion is actually less than that of the base oil, which comprises the continuous phase of the emulsion. However, this makes perfect sense when we realize the intrinsic attenuation of the colloid as a whole is much less than that of the oil media because it is reduced significantly by the relatively low intrinsic attenuation of the water particulates, in accordance with equation 4.67. Although not immediately obvious, the division of the water phase into small droplets did cause the attenuation of emulsion to exceed the intrinsic attenuation of this water in oil system.

4.4. Ultrasound scattering

Rayleigh initially developed the theory of ultrasound scattering for both the long wavelength condition and for solid particles. Anderson generalized this for spherical fluid droplets, also within the long wavelength limit. Atkinson and Kytomaa [32, 33], Hay and Mercer [43], and Faran [35] made attempts to extend the scattering theory to larger ka (shorter wavelength), and to combine it with the asymptotic value of the viscous loss at high frequency. McClements provides a comparison of several attempts to create a multiple scattering theory as developed by Waterman and Truel [4], Lloyd and Berry [5], Ma *et al* [6], and Twersky and Tsang [7].

Morse [17] gives the most comprehensive review of single scattering, and we chose his review as the basis for this section because it provides a single scattering theory applicable for any ka. Importantly, Morse points out that there is a simple way to eliminate multiple scattering effects. Multiple scattering is not important if the experiment measures just the changes in the intensity of the incident beam as it passes through the colloidal suspension. We can demonstrate that this is true, by measuring the attenuation of 120 micron quartz particles. As seen in Figure 4.6 the attenuation remains a linear function of volume fraction up to 46%vl, a good indication of the absence of multiple scattering.

When sound traverses a region filled with small obstacles, each obstacle produces a scattered wave, and these wavelets reinforce in some directions and interfere in others. The wave incident on each scatterer is affected by the presence of the others, which in turn affects the shape of the scattered wave. This gives rise to **coherent**, **incoherent**, and **multiple** scattering and, in the case where the scatterers are regularly spaced, to **diffraction**.

The scattered wave is the sum of all the wavelets arising from the various small obstacles, each having a unique phase and amplitude, determined by the particular obstacle's size, shape and position. If the obstacles that are scattering sound are randomly distributed, no mutual reinforcement occurs, except in the direction of the incident wave. In this forward direction the reinforced scattering modifies the incident wave. It then behaves as though it were traveling through a region having a different index of refraction.

Thus, as far as average terms are concerned, the scattering region behaves as though it has a uniform density and compressibility, which differs from that of the external medium. This difference will produce scattering, or refraction, of the sound incident on the region. In the long wavelength limit the scattered intensity depends on the squares of the density and compressibility- in other words on the square of the number of scattering particles, N, per unit volume in R_{obs}. This cooperative portion of the scattered wave, in which each scatterer adds its contribution to the wave amplitude, is called the **coherent** scattered wave. Thus the intensity is proportional to N^2. The rest of the scattering, the intensity of which is proportional to N rather than to N^2, is called the **incoherent** scattered wave.

One can calculate various scattering terms if appropriate expressions for *scattering cross sections* of the obstacles in the scattering region are available. The most extensive study of this subject appears to have been performed by Morse *et al* [17]. Here we provide an overview of this work, with some additions from more recent developments.

Consideration of the scattering problem usually starts with a theoretical modeling of the sound radiation by an oscillating object. The simplest model is the radiation from a circular piston which is mounted in a rigid baffle. The general model for the pressure field produced by the vibration of such a piston is a solution called the *Rayleigh integral*. For the full circular piston it is:

$$P(r,\theta,t) = \frac{jkP_0}{\pi} e^{j\omega t} \int_0^\pi d\psi \int_0^a \sigma \frac{e^{-jkR}}{R_{obs}} d\sigma \qquad (4.68)$$

where:

$$R_{obs} = \sqrt{r^2 + \sigma^2 - 2r\sigma \sin\theta \cos\psi} \qquad (4.69)$$

and furthermore, where u_0 is the amplitude of the normal velocity of the piston oscillation, a is the piston radius, r, θ and ψ are the polar coordinates associated with the center of the piston, R_{obs} is the distance from the center of the piston to the observation point, and P_0 is pressure amplitude that would exist at the face of the piston, if the piston diameter were infinite.

When the observation point is far away from the piston face ($r >> a$), the expression for R_{obs} can be simplified to:

$$R_{obs} = r - a\sin\theta \cos\psi + O(a/r)^2 \qquad (4.70)$$

Including only the first two terms for R_{obs}, we obtain the following expression for the pressure far away from the piston face:

$$P(r,\theta,t) = \frac{jaP_0}{r} \frac{J_1(ka\sin\theta)}{\sin\theta} e^{j(\omega t - kr)} = \frac{jR_0 P_0}{r} D(\theta) e^{j(\omega t - kr)} \quad (4.71)$$

where R_0 is the *Rayleigh distance*. It is defined as a ratio of the piston surface area, S, to the wavelength, λ, defined by:

$$R_0 = \frac{S}{\lambda} \quad (4.72)$$

Thus, for a spherical piston:

$$R_0 = \frac{\pi a^2}{\lambda} = \frac{ka^2}{2} \quad (4.73)$$

There is an interesting interpretation of the Rayleigh distance that comes from a comparison of the current reciprocating piston, to the radiation from a pulsating sphere of radius R_0, and pressure amplitude P_0, at its surface:

$$P(r,\theta,t) = \frac{R_0 P_0}{r} e^{j(\omega t - kr)} \quad (4.74)$$

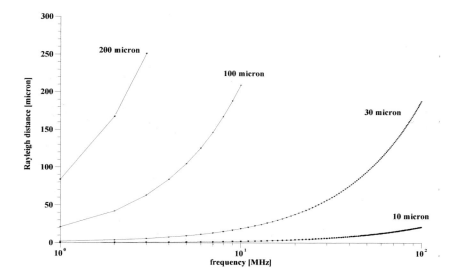

Figure 4.17 Rayleigh distance as a function of ultrasound frequency for particles of different sizes.

A comparison of the two radiation pressure fields (Equation 4.71 and 4.74) shows that, on axis, the piston signal appears to come from a spherical source of radius R_0. This interpretation mildly suggests that the piston radiation starts out as a collimated plane-wave beam, where the amplitude is P_0 from the face out to $r=R_0$, beyond which point the beam spreads spherically.

Thus, R_0 roughly marks the end of the near-field and the beginning of the far-field. This model is a gross oversimplification and must not be taken too literally, but sometimes it is a useful interpretation of the radiation field. A more detailed analysis of the near-field [54] concluded that, in a sense, the far-field begins much closer to the surface, at about $R_0/4$.

At the same time, the far-field formula 1/r for pressure dependence with distance from the piston face is well established at about $R=2\ R_0$. The total on-axis pressure in the far-field region seems to be the sum of two plane-wave signals of the same amplitude, one coming directly from the center of the piston, the other from the edge of the piston. The latter is often called the "edge wave".

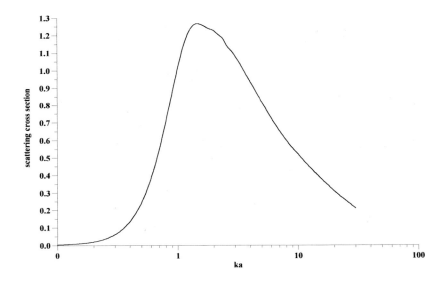

Figure 4.18 Scattering cross section for a rigid sphere.

Interpreting the axial fields as being made of two plane waves makes some physical sense; one can eliminate the edge wave by making the piston radius large enough. In the limit of an infinitely large value for *a,* the edge wave would never arrive at the axis. The solution is then just an ordinary plane wave, as one would expect, from the vibration of the infinite plane.

Far-field and near-field radiation regions determine the long-wavelength and short-wavelength frequency ranges. At low frequencies (long wavelength) the near field region extends only a very short distance from the sound source. Observation is usually performed in the far-field, where sound radiates out with equal intensity in all directions, and the amount radiated is small. As the wavelength decreases, more power is radiated, and the intensity has more directionality; the piston begins to cast a "shadow" and a smaller proportion of the energy is sent out on the side of the piston opposite the radiating element. For very short waves the intensity is large and uniform in the forward direction, and is zero in the shadow behind the cylinder.

Figure 4.17 illustrates the dependence of the Rayleigh distance on frequency within the typical ultrasound range of 1-100 MHz. It is seen that even for large particles of several hundred microns, the Rayleigh distance does not exceed one millimeter. Normally the ultrasound propagation path through the colloidal suspension is much longer and, consequently, any observed scattering will correspond to the far-field.

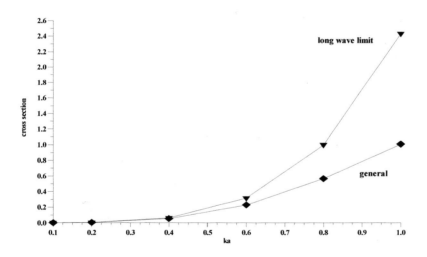

Figure 4.19 Scattering cross section calculated using the general theory and the long wave limit.

This general description is characteristic of all wave motion when it strikes an obstacle. When the wavelength is large compared to the size of the obstacle, the presence of the obstacle is of no consequence and can, effectively, be neglected. On the other hand, when the wavelength is very short compared to the size of the obstacle, the motion resembles the motion of particles, with the waves traveling in straight lines and the obstacle casting sharp-edged shadows. When the wavelength is about the same size as the obstacle, complicated interference effects can sometimes occur, and the analysis of the behavior of the waves becomes quite involved.

When a sound wave encounters an obstacle, some of the wave is deflected from its original course. It is usual to define the difference between the actual wave, and the undisturbed wave that would exist if no obstacle were present, as the scattered wave.

If the obstacle is very large compared to the wavelength, half of this scattered wave spreads out more or less uniformly in all directions from the particle, and the other half is concentrated behind the obstacle in such a manner as to interfere, destructively, with the unchanged wave behind the obstacle, thus creating a sharp-edged shadow. This is the situation in geometrical optics (as well as acoustics); the half of the scattered wave spreading out uniformly is called the *reflected* wave, and the half responsible for the shadow is called the *interfering* wave.

If the obstacle is very small compared to the wavelength, the scattered wave propagates out in all directions, and no sharp-edged shadow exists.

Notice that, in spite of the various peculiarities of the angular distribution of the intensity, the total scattered intensity turns out to be a fairly smooth function of ka.

4.4.1. Rigid sphere.

The intensity of the scattered wave, I_s, and the total power scattered, Π_s, are:

$$I_s = \frac{a^2 I}{r^2} \frac{1}{k^2 a^2} \sum_{m,n=0}^{\infty} (2m+1)(2n+1) \sin\delta_m \sin\delta_n \cos(\delta_m - \delta_n) P_m(\cos\theta) P_n(\cos\theta) \quad (4.75)$$

$$\Pi_s = 2\pi a^2 I \frac{2}{k^2 a^2} \sum_{m=0}^{\infty} (2m+1) \sin^2\delta_m \quad (4.76)$$

where the phase angles, δ_m, and amplitudes, B_m, are the same as for a radiating sphere and are determined from:

$$m n_{m-1}(ka) - (m+1) n_{m+1}(ka) = (2m+1) B_m \cos\delta_m \quad (4.77)$$

$$(m+1) j_{m+1}(ka) - m j_{m-1}(ka) = (2m+1) B_m \sin\delta_m \quad (4.78)$$

where $j_m(ka)$ is the spherical Bessel function, and $n_m(ka)$ is the spherical Neumann function. Phase angles δ_m are tabulated in Morse [17].

In the extreme cases of either long wavelength or short wavelength, equations 4.75 and 4.76 can be simplified to:

$$I_s(ka \ll 1) = \frac{16\pi^4 a^6 I}{9\lambda^4 r^2}(1 - 3\cos\theta)^2 \quad (4.79)$$

$$I_s(ka \gg 1) = I[\frac{a^2}{4r^2} + \frac{a^2}{4r^2}\cot^2(\frac{\theta}{2}) J_1^2(ka\sin\theta)] \quad (4.80)$$

$$\Pi_s(ka \ll 1) = \frac{112\pi^5 a^6}{9\lambda^4} I \quad (4.81)$$

$$\Pi_s(ka \gg 1) = 2\pi a^2 I \quad (4.82)$$

where a is a sphere radius, r and θ are polar coordinates, and $k=\omega/c=2\pi/\lambda$.

In the short wave limit, the total power scattered is that contained in an area of primary beam, equal to twice the cross section of the sphere. This result is the same as that for the light scattering "extinction paradox" [3]. Half of this sound is reflected equally in all directions from the sphere, and the other half is concentrated into a narrow beam, which tends to interfere with the primary beam and cause a shadow.

The total scattering cross section versus frequency for the full ka range is shown in Figure 4.18. It looks like a bell-shaped curve, similar to the absorption losses. Figure 4.19 compares the general theory with the long wavelength asymptote that is reciprocally proportional to the fourth power of the frequency (Rayleigh solution). It is clear that the long wavelength solution is valid only up to $ka=0.5$.

4.4.2. Rigid Cylinder.

Consider a plane wave traveling in a direction perpendicular to the axis of an infinitely long cylinder having a radius, a.

The first order approximations for the scattered intensity at long and short wavelengths are, respectively:

$$I_s(ka \ll 1) = \frac{\pi^4 a^4 I}{\lambda^3 r}(1-2\cos\psi)^2 \tag{4.83}$$

$$I_s(ka \gg 1) = I[\frac{a}{r}\sin\frac{\psi}{2} + \frac{1}{\pi k r}\cot^2(\frac{\psi}{2})\sin^2(ka\sin\psi)] \tag{4.84}$$

The total power scattered by the cylinder per unit length is obtained by multiplying I_s by r, and integrating over ψ, from 0 to 2π.

$$\Pi_s = 4aI\frac{1}{ka}\sum_{m=0}^{\infty}\varepsilon_m \sin^2\delta_m \tag{4.85}$$

where the phase angles δ_m have been defined in connection to the radiation of the cylinder [17].

Thus, the total scattered power for these two extreme cases is given by the following simple expressions:

$$\Pi_s(ka \ll 1) = \frac{6\pi^5 a^4}{\lambda^3}I \tag{4.86}$$

$$\Pi_s(ka \gg 1) = 4aI \tag{4.87}$$

These approximate formulas can be used for illustrating a particle shape effect.

4.4.3. Non-rigid sphere.

Consider now the case of a scattering sphere composed of a material that allows the sound wave to propagate through it. To describe this motion one needs the sphere's density, ρ_p, and the compressibility, β_p. In this non-rigid case the sphere may now dissipate the sound by absorption, in addition to scattering it. The total extinction is now not just equal to the scattering alone; there will be a contribution from absorption. The scattering and the absorption can be described separately by using two different cross sections; Π_s for the scattering and Π_a for the absorption.

For scattering the cross section is given by:

$$\Pi_s(ka \ll 1) = \frac{64\pi^5 a^6}{9\lambda^4} I[\frac{(\beta_p^p - \beta_m^p)^2}{\beta_m^{p\,2}} + 3\frac{(\rho_p - \rho_m)^2}{(2\rho_p + \rho_m)^2}] \qquad (4.88)$$

and for absorption the cross section is:

$$\Pi_a(ka \ll 1) = \frac{8\pi^2 a^3}{3\lambda} I \operatorname{Im}[\frac{\beta_p^p - \beta_m^p}{\beta_m^p} + 3\frac{\rho_p - \rho_m}{2\rho_p + \rho_m}] \qquad (4.89)$$

It follows, from the Equation 4.89, that the absorption exists only if the compressibility and/or the density of the sphere material are complex. This is another way to characterize intrinsic attenuation by the material of the particle, which had been adopted earlier by Morse [17].

For larger spheres or shorter wavelengths, these approximate formulas are not valid, and the exact formulas from [17] must be used.

4.4.4. Porous sphere.

Finally, consider that the sphere material, itself, is not homogeneous but porous, and can be characterized with a porosity, Ω_e, and flow resistance, Φ_e. The material of the pore walls is assumed to be stationary. The effective compressibility and density of the fluid in the pores are given by the approximate expressions:

$$\beta_{ef}^p = \beta_p^p \Omega_e$$

$$\rho_{ef} = \rho_p + j\frac{\Phi_e}{\omega} \qquad (4.90)$$

In this approximation, the compressibility has no imaginary part, although the inclusion of thermal transfer would indeed produce a small imaginary term. Substitution of the effective compressibility and the effective density in the expressions for scattering and absorption cross section [17], would then yield corresponding values for this system of porous spheres.

When the framework of the pore walls is not held at rest, another correction must be added, because of the motion of the solid material. If the pore wall material is effectively incompressible, then the radially symmetric term, ($m=0$), in the series for the pressure inside the sphere is unaffected by this motion; the term (β_e - β_m)/β_m remains unchanged. However, the ($m=1$) term, representing linear motion along the z axis, is changed because the solid part of the sphere as a whole will move back and forth. The mass of a unit volume of the sphere, exclusive of the fluid contained in pores, is $\rho_p (1-\Omega_e)$. In this case, for the $(m=1)$ term, the impedance corresponding to the fluid flow can be calculated as for a parallel circuit with the resistance Φ_e of the porous material being shunted by the reactance, $-j\omega\rho_m (1-\Omega_e)$, of the solid material that is forced into motion by the drag of the fluid moving through the pores. In other words, the Φ_e term must be replaced by

$$[\frac{1}{\Phi_e} + \frac{j}{\omega\rho_m(1-\Omega_e)}]^{-1} \qquad (4.91)$$

The spherically symmetric term, $(m=0)$, is unaffected by this approximation, because the velocity is out of phase with the pressure at $r=a$. The axial component, $(m=1)$, on the other hand, is changed by the motion of the solid material, both in regard to the phase relationship and to the magnitude ratios.

As for the rigid and non-rigid sphere cases, these approximate formulas are again not valid for larger spheres or shorter wavelengths.

4.4.5. Scattering by a group of particles

The solution to the problem of sound scattered by a group of particles becomes much simpler if we address just the attenuation of the incident sound pulse as it propagates a distance through the colloidal suspension. Whether multiple scattering is important or not, the intensity of the ultrasound pulse that has traversed a distance x, decays as an exponential function of x; the attenuation coefficient becomes a scale factor. The total amount of the scattered power is, simply, the product of Π_s and the number of particles that the pulse encounters while propagating through this distance x. The resulting attenuation coefficient is:

$$\alpha_{sc}[\frac{neper}{cm}] = \frac{3\varphi \Pi_s}{4\pi Ia^3} \qquad (4.92)$$

This simple expression allows us to use expressions for Π_s derived by Morse [17] for different particles. The attenuation, for non-rigid particles, at the long wavelength limit, for instance is given by:

$$\alpha_{sc} = \frac{\varphi\omega^4 a^3}{2c_m^4}[\frac{1}{3}(1-\frac{\rho_m c_m^2}{\rho_p c_p^2})^2 + (\frac{\rho_p - \rho_m}{2\rho_p + \rho_m})^2] \qquad (4.93)$$

This is a well-known expression for single scattering, which is identical to the explicit result of the ECAH theory. This expression is valid up to approximately $ka\approx 1$, or, in terms of the wavelength, for diameters which are less than 1/6 the wavelength (i.e. $1/2\pi$). For large particles the more general expression for Π_s [17] must be used.

There is yet one more advantage to the elimination of multiple scattering by measuring the attenuation of the incident beam intensity: the ability to treat polydisperse colloidal suspensions using a superposition assumption. The total attenuation can then be presented as a simple sum of the contributions from each of the different size fractions.

4.4.6. Ultrasound resonance by air bubbles

Acoustic waves propagating through a fluid that contains air bubbles cause the bubbles to resonate. A resonating air bubble vibrates and then re-radiates acoustic energy out of the sound beam. In addition it creates temperature oscillations that then result in thermal absorption. Viscous absorption by air bubbles is relatively much less important.

Scattering by these pulsating bubbles occurs within a frequency range that is defined by a resonance frequency, ω_{cr}, defined by:

$$\omega_{cr}^2 = \frac{3\gamma P_{st}}{a^2 \rho_m} \tag{4.94}$$

where P_{st} is the total hydrostatic pressure inside the bubble.

For instance, the resonance frequency for bubbles with a diameter of 100 microns is about 60 KHz.

In addition to the scattering loss by radiation, there is also absorption of energy because of the high thermal conductivity of the fluid.

The total mechanical impedance Z_{bub} of such a bubble is:

$$Z_{bub} = (R_{sc} + R_{ab}) + j(\omega m_r - stif/\omega) \tag{4.95}$$

where $stif$ is effective stiffness of the bubble given by:

$$stif = 12\pi a \gamma P_{st} \tag{4.96}$$

The scattering resistance, R_{sc}, and absorption resistance, R_{ab}, are given by:

$$R_{sc} = 4\pi a^2 \rho_m c_m (ka)^2 \tag{4.97}$$

$$R_{ab} = 6.4 * 10^{-4} \pi a^3 \rho_m \omega^{3/2} \tag{4.98}$$

The corresponding scattering cross section, Σ_{sc}, and extinction cross section, Σ, are given by:

$$\Sigma_{sc} = \frac{64\pi^3 a^6 \omega^2 \rho_m^2}{Z^2} \tag{4.99}$$

$$\Sigma = \frac{16\pi^2 a^4 \rho_m c_m (R_{sc} + R_{ab})}{Z^2} \tag{4.100}$$

For very large bubbles, with a radius of say 650 microns, the extinction cross section is about 7000 times greater than the geometrical cross section at the 5 KHz frequency. Despite this enormous extinction, such a large bubble would not affect the attenuation in the ultrasound range because the extinction cross section decreases rapidly above the resonance frequency, and very quickly approaches the geometrical cross section.

4.5. Input parameters

To calculate the particle size from an acoustic spectrum requires information about the materials of which the particle and the dispersion medium are composed. For the most general case the list of the required parameters is quite extensive (Table 4.1). Fortunately, it is not necessary in each case to know all of these parameters. The peculiarity of ultrasound interaction with colloidal particles allows one to quite dramatically reduce this list for a specific colloidal dispersion.

For example, submicron rigid particles cause only viscous absorption of ultrasound. Thus, there is no need for any thermodynamic parameters of either the particles or the media; neither is sound speed data required. However, sound speed information does become important for larger, rigid, particles that generate ultrasound scattering. The sound speed of most solid materials is usually in the range of 5000 to 6000 m/s and can be found in the literature, or can be obtained from the Material Database of Dispersion Technology.

Table 4.1 The complete set of properties for both particle and medium that affect the ultrasound propagation through a colloidal suspension

Dispersion Medium	Dispersed Particle	Units
Density	Density	Kg m^{-3}
Shear viscosity (microscopic)		Pa s
Sound speed (or compressibility)	Sound speed (or compressibility)	M s^{-1}
Intrinsic attenuation	Intrinsic attenuation	Db/cm/MHz
Heat capacity at constant pressure	Heat capacity at constant pressure	J kg^{-1} K^{-1}
Thermal conductivity	Thermal conductivity	W m^{-1} K^{-1}
Thermal Expansion	Thermal Expansion	K^{-1}

Submicron sized soft particles, such as latices and microemulsions, require information about the thermodynamic properties: the thermal conductivity, τ, the heat capacity, C_p and the thermal expansion, β. Fortunately, τ and C_p are almost identical for all liquids except water. The variation of these parameters for some 100 liquids from Anson and Chivers [20] and the Dispersion Technology Material Database has been given in Chapter 2 (Figures 2.1 and 2.2). Thus, in practical terms, the number of required parameters is reduced to just one - thermal expansion. This parameter plays the same role with regard to "thermal losses" as density plays for "viscous losses."

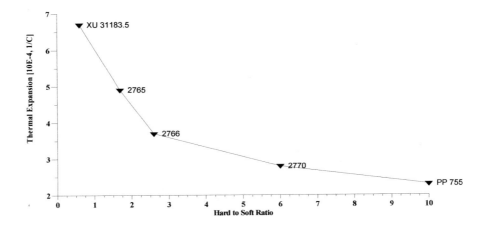

Figure 4.20 Thermal expansion coefficient calculated from the attenuation spectra for a series of latices.

The thermal expansion is known for many liquids, but it presents a problem for polymer latices because the chemical composition of latex particles, and the exact process of their preparation, can dramatically affect the thermal expansion coefficient. Initially, this was a serious obstacle to the use of acoustics to characterize such systems. This is no longer a problem because the thermal expansion coefficient can now be determined from the attenuation spectrum. Hence, instead of using it as a given input parameter, it can be treated as another adjustable output parameter. To demonstrate this we measured a set of five polymer latices that had been intentionally synthesized to have different rheological properties, as characterized by a "hard to soft ratio" parameter. The thermal expansion coefficient for these latices, calculated from the measured attenuation spectra, correlates with the measured particle rheology (Figure 4.20). Thus the attenuation spectra of polymer latices contain sufficient information to characterize not only the particle size, but also the thermal expansion coefficient.

In some cases, particularly for dilute measurements or in viscous media, the intrinsic attenuation of the media and/or the particles may also be required to obtain precise results. This can easily be determined for the medium by measuring it directly using the acoustic spectrometer. For liquid droplets (emulsions), the intrinsic attenuation is also simple to determine as long as a homogeneous sample of the droplet material is available, as is most often the case. For solid particles, the intrinsic attenuation can almost always be ignored.

Table 4.2 Input parameters required to interpret acoustic data for particular colloid classes.

Colloid	Properties of the Particle	Properties of the Liquid
Rigid submicron particles	Density	Density, Shear viscosity
Soft submicron particles	Thermal expansion (might be calculated from attenuation)	Thermal expansion
Large soft particles	Density, sound speed	Density, sound speed
Large rigid particles	Nothing	Nothing
		Intrinsic attenuation for all colloids (might be easily measured)

The sound speed in soft particles becomes important for large sizes when scattering is dominant.

The properties necessary to interpret acoustic data are summarized in Table 4.2. They are necessarily simplifications that are not completely general or universal; they are the result of our experience in dealing with a large number of practical colloidal suspensions over many years. They are empirical, not theoretical, and some exceptions are to be expected. In some cases more input parameters may be required than indicated here, see for instance Babick *et al* [23].

In addition to knowing a specific set of material properties for the particles and the suspending medium, it is also necessary to know the relative amount of both materials. This is usually characterized by a weight fraction of the disperse phase and is normally considered as an input parameter. It is usually quite simple to measure, even if unknown, by using a pyncnometer, or by weighing a sample of the suspension before and after drying in oven. Determination of the weight fraction is even possible when the sample has two separate disperse phases (Chapter 8).

There is, however, another approach to determining the weight fraction of a suspension. It is very tempting to try to extract weight fraction from the acoustic data. For example, Alba has suggested using the weight fraction as an additional adjustable parameter when fitting the attenuation spectra [51]. However, based on our experience such an approach can be misleading, because there is usually insufficient information in the attenuation spectra to extract data for both the particle size distribution and weight fraction. There are certainly exceptions, but they are very rare.

The attenuation is not the only source of acoustic data that can be collected with ultrasound based instruments. Sound speed and acoustic impedance are two other, independent, experimental parameters that can also be used as a source of weight fraction information. There are four factors, however, that complicate the calculation of weight fraction, from either the sound speed or the acoustic impedance.

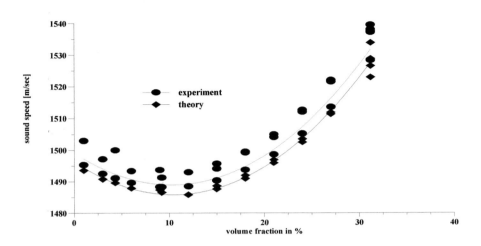

Figure 4.21 Measured sound speed for silica (Ludox) at different volume fractions.

First, these two parameters are much more sensitive to variations in temperature than attenuation [50]. For instance, a 1^0 C temperature change causes a 2.4 m/sec variation in the sound speed of water. This can introduce a large uncertainty into the calculated weight fraction. It is possible, of course, to eliminate this factor by either or measuring or stabilizing the temperature during the measurement.

The second factor is more difficult to tackle. Sound speed might be very sensitive to small variations of the chemical composition of the liquid, as is shown in Chapter 3. It would be a mistake to assign the variation in sound speed caused by the chemical composition of the sample to the weight fraction of the dispersed phase.

The third problem is related to the low sensitivity of sound speed to the solids content at certain volume fraction ranges. For instance, silica (Ludox) exhibits very little variation of the sound speed over the concentration range of 7% to 15%vl, as is shown in Figure 4.21. In this case acoustic impedance is a much better candidate for determining the weight fraction. Acoustic impedance is the product of the sound speed and density. So, even when the sound speed remains practically constant, the acoustic impedance will still change as a function of weight fraction, due to variation of the density.

The fourth problem is related to the particle size. Sound speed and acoustic impedance are independent of particle size only for very small or very large particles. In general, both of them exhibit a dependence on particle size at the same frequencies as attenuation. This dependence is much weaker than in the case of attenuation, but still might be sufficient to produce a substantial error in the weight fraction.

Overall, the most reliable way to measure particle size using acoustics is to assume that the weight fraction will be used as an input parameter that has been determined independently. Acoustics does provide some means to determine the weight fraction, and eliminate it from the list of the required input parameters, but this approach must be used with caution.

REFERENCES.

1. Epstein, P.S. and Carhart R.R., "The Absorption of Sound in Suspensions and Emulsions", J. Acoust. Soc. Amer., 25, 3, 553-565 (1953)
2. Allegra, J.R. and Hawley, S.A. "Attenuation of Sound in Suspensions and Emulsions: Theory and Experiments", J. Acoust. Soc. Amer., 51, 1545-1564 (1972)
3. Bohren, C. and Huffman, D. "Absorption and Scattering of Light by Small Particles", J. Wiley & Sons, 530 p. (1983)
4. Waterman, P.S. and Truell, R., "Multiple Scattering of Waves", J. Math. Phys., 2, 512-537 (1961)
5. Lloyd, P. and Berry, M.V. "Wave propagation through an assembly of spheres", Proc. Phys. Soc., London 91, 678-688 (1967)
6. Ma, Y., Varadan, V.K., and Varadan, V.V. "Comments on ultrasonic propagation in suspensions", J. Acoust. Soc. Amer., 87, 2779-2782 (1990)
7. Twersky, V., "Acoustic bulk parameters in distribution of pair-correlated scatterers", J. Acoust. Soc. Amer., 64, 1710-1719 (1978)
8. Tsang, L., Kong, J.A. and Habashy, T. "Multiple scattering of acoustic waves by random distribution of discrete spherical scatterers with the quasicrystalline and Percus-Yevick approximation", J. Acoust. Soc. Amer., 71, 552-558 (1982)
9. Harker, A.H. and Temple, J.A.G., "Velocity and Attenuation of Ultrasound in Suspensions of Particles in Fluids", J. Phys .D.: Appl. Phys., 21, 1576-1588 (1988)
10. Gibson, R.L. and Toksoz, M.N., "Viscous Attenuation of Acoustic Waves in Suspensions", J. Acoust. Soc. Amer., 85, 1925-1934 (1989)
11. Evans, J.M. and Attenborough, K. "Coupled Phase Theory for Sound Propagation in Emulsions", J. Acoust. Soc. Amer., 102, 1, 278-282 (1997)
12. Temkin, S. "Viscous attenuation of sound in dilute suspensions of rigid particles", The Journal of the Acoustical Society of America, vol. 100, 2, pp.825-831, (1996)
13. Temkin, S. "Sound Propagation in Dilute Suspensions of Rigid Particles" The Journal of the Acoustical Society of America, vol. 103, 2, pp.838-849, (1998)

14. Dukhin, A.S. and Goetz, P.J. "Acoustic Spectroscopy for Concentrated Polydisperse Colloids with High Density Contrast", Langmuir, 12, 21, 4987-4997 (1996)

15. Hemar, Y, Herrmann, N., Lemarechal, P., Hocquart, R. and Lequeux, F. "Effect medium model for ultrasonic attenuation due to the thermo-elastic effect in concentrated emulsions", J. Phys. II 7, 637-647 (1997)

16. McClements, J.D., Hemar, Y. and Herrmann, N. "Incorporation of thermal overlap effects into multiple scattering theory", J. Acous. Soc. Am., 105, 2, 915-918 (1999)

17. Morse, P.M. and Uno Ingard, K., "Theoretical Acoustics", 1968 McGraw-Hill, NY, Princeton University Press, NJ, 925 p, (1986)

18. Blackstock, D. "Fundamentals of Physical Acoustics", J. Wiley & Sons, NY, 541p (2000)

19. Kinsler, L., Frey, A., Coppens, A., and Sanders, J. "Fundamentals of Acoustics", J. Wiley & Sons, NY, 547 p, (2000)

20. Anson, L.W. and Chivers, R.C. "Thermal effects in the attenuation of ultrasound in dilute suspensions for low values of acoustic radius", Ultrasonic, 28, 16-25 (1990)

21. Rayleigh, J.W. "The Theory of Sound", Volume 1, Macmillan & Co., London, (1926)

22. Rayleigh, J.W. "The Theory of Sound", Vol.2, Macmillan and Co., NY, second edition 1896, first edition 501 p. (1878)

23. Babick, F., Hinze, F., and Ripperger, S. "Dependence of Ultrasonic Attenuation on the Material Properties", Colloids and Surfaces, 172, 33-46 (2000)

24. Urick, R.J. "A Sound Velocity Method for determining the Compressibility of Finely Divided Substances", J. Appl. Phys., 18, 983 (1947)

25. Urick R.J. "Absorption of Sound in Suspensions of Irregular Particles", J. Acoust. Soc. Amer., 20, 283 (1948)

26. Hampton, L. "Acoustic Properties of Sediments", J. Acoust. Soc. Amer., vol.42, 4, pp. 882-890 (1967)

27. Blue, J.E. and McLeroy, E.G. "Attenuation of Sound in Suspensions and Gels", J. Acoust. Soc. Amer., 44, 4, 1145-1149 (1968)

28. Dukhin, A.S, Goetz P.J. and Hamlet, C.W. "Acoustic Spectroscopy for Concentrated Polydisperse Colloids with Low Density Contrast", Langmuir, 12, 21, 4998-5004, (1996)

29. Dukhin, A.S., and Goetz, P.J. "Acoustic and Electroacoustic Spectroscopy for Characterizing Concentrated Dispersions and Emulsions", Adv. in Colloid and Interface Sci., 92, 73-132 (2001)

30. Chanamai, R., Hermann, N. and McClements, D.J. "Influence of thermal overlap effects on the ultrasonic attenuation spectra of polydisperse oil-in-water emulsions", Langmuir, 15, 3418-3423 (1999)

31. Busby, J. and Richrdson, E.G. "The Propagation of Utrasonics in Suspensions of Particles in Liquid", Phys. Soc. Proc., v.67B, 193-202 (1956).

32. Atkinson, C.M and Kytomaa, H.K., "Acoustic Wave Speed and Attenuation in Suspensions", Int. J. Multiphase Flow, 18, 4, 577-592 (1992)

33. Atkinson, C.M and Kytomaa, H.K., "Acoustic Properties of Solid-Liquid Mixtures and the Limits of Ultrasound Diagnostics-1.Experiments", J. Fluids Engineering, 115, 665-675 (1993)

34. Kol'tsova, I.S. and Mikhailov, I.G. "Scattering of Ultrasound Waves in Heterogeneous Systems", Soviet Physics-Acoustics, 15, 3, 390-393 (1970)

35. Faran, J.J. "Sound Scattering by Solid Cylinders and Spheres", J. Acoust. Soc. Amer., 23, 4, 405-418 (1951)

36. Stakutis, V.J., Morse, R.W., Dill, M. and Beyer, R.T. "Attenuation of Ultrasound in Aqueous Suspensions", J. Acous. Soc. Amer., 27, 3 (1955)

37. Hartmann G.K and Focke, A.B. "Absorption of Supersonic Waves in Water and in Aqueous Suspensions", Physical Review, 57, 1, 221-225 (1940)

38. Wood, A.B. "A Textbook of Sound", Bell, London, (1940)

39. Isakovich, M.A. Zh. Experimental and Theoretical Physics, 18, 907 (1948)

40. McClements, D.J. "Ultrasonic Characterization of Emulsions and Suspensions", Adv. Colloid Int. .Sci., 37, 33-72 (1991)

41. Mason, W.P. "Dispersion and Absorption of Sound in High Polymers", in Handbuch der Physik., vol.2, Acoustica part1, (1961)

42. McCann, C. "Compressional Wave Attenuation in Concentrated Clay Suspensions", Acustica, 22, 352-356 (1970)

43. Hay, A.E. and Mercer, D.G. "On the Theory of Sound Scattering and Viscous Absorption in Aqueous Suspensions at Medium and Short wavelength", J. Acoust. Soc. Amer., 78, 5 1761-1771 (1985)

44. Chow, J.C.F., "Attenuation of Acoustic Waves in Dilute Emulsions and Suspensions", J. Acoust. Soc. Amer., 36, 12, 2395-2401 (1964)

45. Allison, P.A. and Richardson, E.G. "The Propagation of Ultrasonics in Suspensions of Liquid Globules in Another Liquid", Proc. Phys. Soc., 72, 833-840 (1958)

46. Hayashi, T., Ohya, H., Suzuki, S. and Endoh, S. "errors in Size Distribution Measurement of Concentrated Alumina Slurry by Ultrasonic Attenuation Spectroscopy", J. Soc. Powder Technology Japan, 37, 498-504 (2000)

47. Anderson, V.C. "Sound Attenuation from a Fluid Sphere", J. Acoust. Soc. Amer., 22, 4, 426-431 (1950)

48. Riebel, U. et al. "The Fundamentals of Particle Size Analysis by Means of Ultrasonic Spectrometry" Part. Part. Syst. Charact., vol.6, pp.135-143, (1989)

49. Hipp, A.K., Storti, G. and Morbidelli, M. "On multiple-particle effects in the acoustic characterization of colloidal dispersions", J. Phys., D: Appl. Phys. 32, 568-576 (1999)

50. Chanamai, R., Coupland, J.N. and McClements, D.J. "Effect of Temperature on the Ultrasonic Properties of Oil-in-Water Emulsions", Colloids and Surfaces, 139, 241-250 (1998)

51. Alba, F. "Method and Apparatus for Determining Particle Size Distribution and Concentration in a Suspension Using Ultrasonics", US Patent No. 5121629 (1992)

52. Wilson, R., Leschonski, K., Alex, W., Allen, T., Koglin, B., Scarlett, B. "BCR Information", Commission of the European Communities, EUR 6825, (1980)

53. Marlow, B.J., Fairhurst, D. and Pendse, H.P., "Colloid Vibration Potential and the Electrokinetic Characterization of Concentrated Colloids", Langmuir, 4,3, 611-626 (1983)

54. Rogers, P.H. and Williams, A.O. "Acoustic field of circular plane piston in limits of short wavelength or large radius", J. Acoust. Soc. Am., 52, 3, 865-870 (1972)

55. Strout, T.A., "Attenuation of Sound in High-Concentration Suspensions: Development and Application of an Oscillatory Cell Model", A Thesis, The University of Maine, (1991)

Chapter 5. ELECTROACOUSTIC THEORY

Electroacoustic phenomena, first predicted by Debye in 1933 [1] for electrolytes, arise from coupling between acoustic and electric fields. Debye realized that, in the presence of a longitudinal sound wave, any differences in the effective mass or friction coefficient between anions and cations would result in different displacement amplitudes. In turn, this difference in displacement would create an alternating electric potential between points within the solution. Indeed, this phenomenon is measurable and can yield useful information about the properties of ions. It is usually referred to as an "Ion Vibration Potential" (IVP). The first experimental observations of this IVP were reported by Yeager [2] in 1949, and by Derouet and Denizot [12] in 1951. A thorough theoretical treatment of the IVP phenomenon was given in 1947 by Bugosh et al [11]. There was a lot of interest in this effect in the 1950's and 1960's, because it was considered to be a very promising tool for characterizing ion solvation [2-12]. Zana and Yeager [10] summarized the results of this two decade effort in a review, which we follow here, in our presentation of IVP in this book. Sadly, this phenomenon has been largely forgotten, and virtually all of the interest in electroacoustic phenomena has shifted from pure electrolyte solutions to colloids.

Hermans [14] and Rutgers [15] in 1938 were the first to report a Colloid Vibration Potential (CVP). Since that time there have been several hundred experimental and theoretical works published. For brevity, we mention here only a few key papers. Enderby and Booth [16, 17] developed the first theory for CVP in the early 1950's. The first quantitative experiments were made in 1960 by Yeager's group [10]. In the early 1980's Cannon with co-authors [18] discovered an inverse electroacoustic effect, that they termed ElectroSonic Amplitude (ESA). The first commercially available electroacoustic instrument was developed by Pen Kem, Inc [19]. There are now several commercially available instruments based on both electroacoustic effects. These are manufactured by Colloidal Dynamics, Dispersion Technology, and Matec.

After the basic works by Enderby and Booth [16, 17], the electroacoustic theory for colloids has been developing in two quite different directions. The original Enderby/Booth theory was very complex; a result of considering both surface conductivity and low frequency effects. In addition, it did not take particle-particle interactions into account, and consequently was valid only for dilute colloids. Hence, this Enderby/Booth theory required modifications and simplifications. The first extension of the Enderby-Booth approach was performed by Marlow, Fairhurst and Pendse [20]. They attempted to generalize it for concentrated systems using a Levine cell model [21]. This approach leads to somewhat complicated mathematical formulas, and perhaps this was the reason that it was abandoned. An alternative approach to electroacoustic theory was later suggested by O'Brien [22]. He introduced the concept of a dynamic electrophoretic mobility, μ_d, and suggested a simple relationship between this parameter and measured electroacoustic parameters, such as Colloid Vibration Current (CVI) or Electric Sonic Amplitude (ESA). Later, O'Brien stated that his relationship is valid for concentrated system as well as dilute systems [34].

According to the O'Brien relationship, the average dynamic electrophoretic mobility, μ_d, is defined as:

$$\mu_d = \frac{ESA(\omega)\rho_m}{A(\omega)F(Z_T,Z_s)\varphi(\rho_p - \rho_m)} \qquad (5.1)$$

where the ESA is normalized by the applied external electric field, $A(\omega)$ is an instrument constant found by calibration, and $F(Z_T, Z_s)$ is a function of the acoustic impedances of the transducer and the dispersion under investigation.

A similar expression can be used for the CVI mode:

$$\mu_d = \frac{CVI(\omega)\rho_m}{A(\omega)F(Z_T,Z_s)\varphi(\rho_p - \rho_m)} \qquad (5.2)$$

Here the CVI is assumed to be normalized by the gradient of pressure (grad P), which, in the case of CVI, is the external driving force.

According to O'Brien, a complete functional dependence of ESA (CVI) on key parameters, such as ζ-potential, particle size and frequency, is incorporated into the dynamic electrophoretic mobility. O'Brien stated that for all considered cases the coefficient of proportionality between ESA (CVI) and μ_d is frequency independent, and, in addition, is independent of particle size and ζ-potential. This feature made the dynamic electrophoretic mobility an important and central parameter of the electroacoustic theory.

The first theory that relates this dynamic electrophoretic mobility to the properties of the dispersion medium and dispersed particles was initially created by O'Brien, but it neglected particle-particle interactions and, was therefore limited to the dilute case. We shall call this version the "dilute O'Brien theory". Later, Rider and O'Brien [23], Ohshima [24], and Ennis, Shugai and Carnie [25, 26] suggested modifications to extend this approach to concentrated systems.

Figure 5.1 is a block diagram intended to classify the complicated relationships between the various electroacoustic theories.

Because it appeared to yield a desirable electroacoustic theory for the concentrated case, the O'Brien approach appeared superior to the Enderby-Booth approach. However, one important question remained unanswered. In principle, these two approaches must give the same result, but it had not been clear if this was indeed the case. It is obvious that such a comparison needed to be done. It would provide strong support for both theories if the two approaches merged.

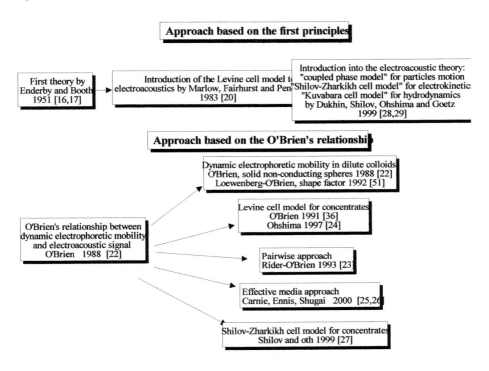

Figure 5.1 Various versions of the electroacoustic theory for colloids.

The situation is even more complicated because O'Brien's approach has been developed mostly for ESA, whereas the Enderby-Booth approach was primarily used to explain the (inverse) effect of CVP or CVI. For dilute systems this difference does not present any problem. However, it does for concentrates, because the inertial frame of references can be different for the ESA and CVI effects.

All of these issues have been addressed recently by an international group of scientists: Shilov from Ukraine, Ohshima from Japan, and A.Dukhin and Goetz from the USA [27-29]. This group has generalized the Enderby-Booth approach to the point that it should now be equivalent to the latest version of O'Brien's approach. What is of utmost importance, is that the group has also developed an electroacoustic analog of Smoluchowski's equation. As described earlier (Chapter 2), Smoluchowski's equation is known to be valid for any concentration and any particle shape, provided that the DL is thin ($\kappa a >> 1$) and the surface conductivity is negligible ($Du << 1$). By making these same two assumptions, this group derived a general theory for electroacoustics, that is valid for any concentration or particle shape. The theory requires no other assumptions. We shall call this theory the "Smoluchowski dynamic electroacoustic limit" (SDEL). The SDEL is a low frequency asymptotic solution and is a natural test for all possible electroacoustic theories. Hence, any other proposed theory must reduce to the SDEL under the conditions of $\kappa a >> 1$ and $Du << 1$, and at sufficiently low frequencies. The SDEL serves as a verification criterion for any proposed electroacoustic theory in the same manner that Smoluchowski's equation serves for electrophoretic theories.

Later Shilov and others further generalized the CVI theory to include particle inertia, surface conductivity, Maxwell-Wagner DL relaxation, thermodiffusion, and barodiffusion [27-30]. This modern CVI theory satisfies the Smoluchowski dynamic electroacoustic limit.

We are not aware of any direct comparison between this SDEL and any theory based on the O'Brien approach. Experimental tests by Ennis, Shugai and Carnie [25, 26] of the latest theory for the dynamic mobility indicate a good correlation even at very high volume fractions for the thin DL. This strongly suggests that the theory may be in compliance with the SDEL; otherwise one would not expect such a good correlation with experiment.

In summary, the connection, between the modern version of the O'Brien approach to ESA theory and modern versions of the Enderby-Booth approach to CVI theory remain uncertain. Although both are able to independently fit corresponding experimental data, the path of theoretical affinity between the two approaches has yet to be resolved.

We emphasize here the CVI theory for the following reasons:
- It is known that in the radio frequency domain electric fields might affect the surface properties and cause strange kinetic and thermodynamic effects [31, 32], including oscillation of the ζ-potential. The use of ultrasound as the driving force is therefore a better choice than an electric field as required in the ESA method.
- When using ultrasound there is an opportunity to calibrate an absolute power using the internal reflection inside of the transducer. This calibration procedure will be discussed in detail in Chapter 7. The ESA method does not offer this option. As a result the ESA calibration is more difficult and less reliable compared to CVI.
- The frame of references is well defined for CVI, but problematic for ESA.

We suggest those readers who are interested in the details of the electroacoustic theory of ESA read the original papers [33-40] or a review written by Hunter [41].

5.1 The Theory of Ion Vibration Potential (IVP)

Bugosh at al [11] developed the original Ion Vibration Current (IVI) theory in 1947. The basis of this theory is a system of $(2N+1)$ equations used to characterize the electro-diffusion effects that occur when a plane longitudinal ultrasound wave propagates through a solution of N electrolyte species. For a wave traveling in the x direction, this can be described as follows:

$$ez_i E - \gamma_i (u_i - u_m) - \frac{ez_i q\kappa E}{3\varepsilon_m kT[1+\sqrt{q(1+j\omega/\omega_{MW})}]} - \frac{ez_i \kappa E}{6\pi\eta} - \frac{kT\partial n_i}{n_i \partial x} - v_i \rho_m \frac{\partial \rho_m}{\partial t} = m_i \frac{du_i}{dt}$$

electric   friction   relaxation   electrophoretic   diffusion   pressure   reaction

(5.3)

$$\frac{\partial n_i}{\partial t} + \frac{\partial (n_i u_i)}{\partial x} = 0 \tag{5.4}$$

$$\frac{\partial E}{\partial x} - \frac{4\pi e}{\varepsilon_m} \sum n_i z_i = 0 \tag{5.5}$$

where e is the value of the charge of an electron, k is the Boltzman constant, n is the ion concentration, the index, i, specifies the ion species, z is the ion valency, u is the ion velocity, E is the electric field strength, v and m are the volume and mass, respectively, of the solvated ion, and q is a parameter of the relaxation force (calculated by Bugosh for an electrolyte of two ion species).

Hermans [14] analysis of Eq.5.3 showed that the contributions of the electrophoretic, relaxation and diffusion terms are negligible. The final expression for a simple binary electrolyte normalizes the IVP by dividing it with the oscillation velocity amplitude U_m:

$$\frac{IVP}{U_m} = \frac{c_m}{N_A e}[\frac{t^+ W^+}{z^+} - \frac{t^- W^-}{z^-}]\frac{1}{[\frac{\omega^2}{\omega_{MW}^2}+1]^{1/2}} \tag{5.6}$$

where, N_A is Avogadro's number, $z^\pm$ are the valencies of the anion and cation, c_m is the sound speed of the media, $t^\pm$ are the transport numbers of the cation and anion, $W^\pm$ are the apparent molar mass of the anion and cation, ω is the radian frequency of acoustic field, and ω_{MW} is the Maxwell-Wagner frequency as given by:

$$\omega_{MW} = \kappa^2 D_{eff} = \frac{K_m}{\varepsilon_0 \varepsilon_m} \tag{5.7}$$

where K_m is the conductivity of the media.

The apparent molar mass of an ion is given by:

$$W^\pm = N_A(m^\pm - \rho_m v^\pm) \tag{5.8}$$

and $t^{\pm} = \dfrac{K_{\pm}^{0}}{K_{+}^{0} + K_{-}^{0}}$ \hfill (5.9)

where $K_{\pm}^{0}$ are the limiting conductances of the cations or anions, and $v^{\pm}$ are the effective volumes, occupied by cation and anion, correspondingly.

The Ion Vibration Current (IVI) depends on the concentration of electrolyte because of changes in the Maxwell-Wagner frequency, and this is proportional to the medium conductivity K_m. The IVI is a linear function of concentration at frequencies that sufficiently exceed the Maxwell-Wagner frequency. At low frequency, the IVI becomes independent on the electrolyte concentration.

Zana and Yeager [10] provide an interesting discussion of the physical meaning of the apparent molar masses $W^{\pm}$. Only those solvent molecules whose volume, due to the electrostriction effect, differ from that in the bulk, contribute to the apparent molar mass. Solvent molecules that are part of the solvation shell, without appreciable electrostriction, or difference in packing, contribute equally to the terms m_i and v_i in Equation 5.6, and their contributions cancels each other. This makes the IVP/IVI method a unique tool for collecting information about the solvation shell of ions. For a more detailed theoretical analysis, and a large volume of experimental data, we suggest that the interested reader go to the original review.

5.2 The Low frequency electroacoustic limit - Smoluchowski limit, (SDEL)

There is a way to derive an expression for CVI using a well-known Onsager reciprocal relationship [42, 43, 44]. This relationship is certainly valid in the stationary case, but much less is known about its validity for alternating fields. This uncertainty cautions us to use the relationship only in the limiting case of very low frequencies, or at least frequencies much lower than both the characteristic hydrodynamic frequency, ω_{hd}, and the electrodynamic Maxwell-Wagner frequency, ω_{MW}. It thus follows that:

$$\omega << \omega_{hd} = \dfrac{\eta}{\rho_m a^2} \hfill (5.10)$$

$$\omega << \omega_{MW} = \dfrac{K_m}{\varepsilon_0 \varepsilon_m} \hfill (5.11)$$

where K_m is the electric conductivity of the medium, and ε_0 and ε_m are dielectric permittivities of the vacuum and medium, respectively.

The Onsager relationship provides the following link between the quasi-stationary streaming potential CVP, the effective pressure gradient, ∇P_{rel}, which moves liquid relative to the particles, the electroosmotic current, $<I>$, and the electroosmotic flow, $<V>$:

$$\frac{<V>}{<I>_{<\nabla P>=0}} = \frac{<CVP_{\omega \to 0}>}{<\nabla P_{rel}>_{<I>=0}} \quad (5.12)$$

Let us use the Onsager relationship in consideration of the CVP effect for a macroscopically small element of the suspension's volume (i.e. the element, which is small with respect to the length of ultrasonic wave, but which contains a majority of particles). To use Equation 5.12, we need to know the effective gradient of pressure $<\nabla P_{rel}>$, which provides the velocity of liquid passing around the particles in a vibrated element of the suspension's volume. This value can be easily obtained following the "coupled phase model" [see Section 4.2.1] that characterizes the particles motion in a sound field in concentrated colloids. In the quasi-stationary (low-frequency) case, the effective gradient of pressure is equal to the specific (per unit volume of suspension) friction force exerted on the particles, which is $\gamma(u_p - u_m)$. This force is a part of the pressure gradient that moves the particles relative to the liquid. In the extreme case of low frequency, Eq.4.27 leads to the following expression for this effective pressure gradient:

$$\nabla P^{\omega \to 0}_{rel} = \frac{\varphi(\rho_p - \rho_s)}{\rho_s} \nabla P \quad (5.13)$$

In addition, we can use the fact that the expression on the left hand side of Equation 5.12 is the electrophoretic mobility, μ, divided by the conductivity of the system, K_s. As a result, we obtain the following expression for CVI:

$$CVI_{\omega \to 0} = CVP * K_s = \mu \frac{\varphi(\rho_p - \rho_s)}{\rho_s} \nabla P \quad (5.14)$$

This expression specifies the Colloid Vibration Current at the low frequency limit. This means that μ is the usual stationary case electrophoretic mobility. We can thus use the Smoluchowski law for electrophoresis [45] in the form that is valid for concentrated systems:

$$\mu = \frac{\varepsilon_m \varepsilon_0 \zeta}{\eta} \frac{K_s}{K_m} \quad (5.15)$$

Here we used two conditions, 2.54 and 2.55, that restrict the applicability of Smoluchowski law. As a result the asymptotic value of the CVI at low frequency is:

$$\frac{CVI_{\omega \to 0}}{\nabla P} = \frac{\varepsilon_m \varepsilon_0 \zeta \varphi K_s}{\eta K_m} \frac{(\rho_p - \rho_s)}{\rho_s} \quad (5.16)$$

This yields the following asymptotic value for the dynamic electrophoretic mobility that is defined according to the O'Brien relationship (5.2):

$$\mu_d = \frac{\varepsilon_m \varepsilon_0 \zeta}{\eta} \frac{K_s}{K_m} \frac{(\rho_p - \rho_s)\rho_m}{(\rho_p - \rho_m)\rho_s} \quad (5.17)$$

The density dependent multiplier appears because, according to O'Brien's relationship, the dynamic electrophoretic mobility is defined in relationship to the density contrast between the particle and the media. This multiplier disappears in dilute systems, but is very important in concentrates, since it conveys additional volume fraction dependence.

Equation 5.17 is very important because it provides a test for the electroacoustic theory. The theory is supposed to be valid for small particles with a thin DL ($\kappa a >> 1$), and negligible surface conductivity ($Du << 1$). If these three conditions are met, Equation 5.17 is then valid for any volume fraction, and any particle shape and particle size. This makes it analogous to the Smoluchowski theory of microelectrophoresis.

There should be a similar equation for ESA, perhaps with a somewhat different density and volume fraction dependence, reflecting the difference in the inertial frame of references between CVI and ESA, but as yet there is none.

5.3 The O'Brien theory

O'Brien's theory was a substantial step forward in our understanding of the electroacoustic phenomena. He provided a link between electroacoustics and classical electrokinetics by introducing the concept of the dynamic electrophoretic mobility, μ_d, through a special relationship between this parameter and an experimentally measured electroacoustic signal:

$$ESA(CVI) = A(\omega)F(Z_T, Z_s)\varphi \frac{\rho_p - \rho_m}{\rho_m} \mu_d \qquad (5.18)$$

The Electroacoustic signals (ESA, or CVI) in this equation are normalized by the corresponding driving force (E or grad$P$).

This first theory for the dynamic electrophoretic mobility was derived assuming a thin DL and employing the longwave restriction for frequency. The theory does not take into account particle-particle interactions, and is thus valid only in dilute dispersions.

$$\mu_d = \frac{2\varepsilon_0 \varepsilon_m \varsigma}{3\eta} G(s)(1 + F(Du, \omega')) \qquad (5.19)$$

where

$$G(s) = \frac{1 + (1+j)s}{1 + (1+j)s + j\frac{2s^2}{9}(3 + 2\frac{\rho_p - \rho_m}{\rho_m})} \qquad \text{for thin DL, } \kappa a >> 1 \qquad (5.20)$$

$$G(s) = \frac{1}{1 + (1-j)s + j\frac{2s^2}{9}(1 + 2\frac{\rho_p}{\rho_m})} \qquad \text{for thick DL, } \kappa a << 1 \text{ and } \frac{e\varsigma}{kT} < 1 \qquad (5.21)$$

$$F(Du, \omega') = \frac{(1 - 2Du) + j\omega'(1 - \frac{\varepsilon_p}{\varepsilon_m})}{2(1 + Du) + j\omega'(2 + \frac{\varepsilon_p}{\varepsilon_m})} \qquad \text{for thin DL } \kappa a >> 1 \qquad (5.22)$$

$$s^2 = \frac{a^2 \omega \rho_m}{2\eta_m}; \quad \omega' = \frac{\omega}{\omega_{MW}}$$

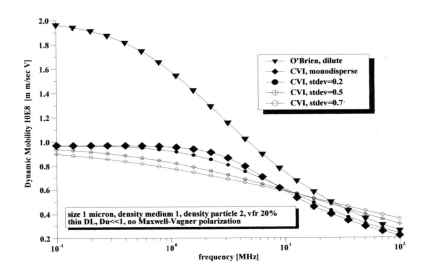

Figure 5.2 Frequency dependence of the dynamic electrophoretic mobility of colloid.

One can see that there are two factors that determine the frequency dependence of the dynamic electrophoretic mobility. The factor G, reflects the frequency dependence related with the inertia effects, whereas factor F, represents the influence of the Maxwell-Wagner polarization of the DL [46]. Neither of these two kinds of frequency dependencies is important at low frequency; in the limit when the G and F factors equate to 1 and $\frac{1-2Du}{2+2Du}$, respectively. The value of these factors deviates from the low frequency limit when the frequency approaches some critical value. For the factor G, this critical frequency is ω_{hd}, and for the factor F, it is ω_{MW}.

For aqueous colloids, the inertial factor, G, plays a more important role than the DL polarization factor, F. The inertial factor dramatically reduces the magnitude of the dynamic mobility of larger particles and high frequencies (Figure 5.2). In addition to reducing the amplitude, the inertial factor also causes a lag in the particle motion relative to the external driving force, and this interposes a phase shift on the dynamic mobility. This phase shift reaches a maximum value of 45^0 at the high frequency limit. This phase dependence on particle size can be used to determine particle size and is the basis of the method used in the AcoustoSizer by Colloidal Dynamics, Inc.

The initial dilute case version of the O'Brien theory has been expanded later by Loewenberg and O'Brien [51], by considering the effect of particle shape on the dynamic mobility.

After O'Brien's dilute case theory, the main effort in the field of theoretical electroacoustics has been concentrated on incorporating particle-particle interactions into the theory, and expanding it to high volume fractions. There are four different approaches within the framework of the O'Brien theory that have been tried: the cell model [36], particles pair interaction [39], the Percus-Yevick approximation [47], and effective media [25, 26]. All

The first approach using the cell model is considered unsuccessful [41].

The second approach taken by Rider and O'Brien [39] allowed them to expand the volume concentration limit up to 10%.

Application of the Percus-Yevick approximation allowed O'Brien to develop a theory for emulsions [47] for volume concentrations up to 60%. In this case, the dynamic electrophoretic mobility is now given by:

$$\mu_d = \frac{\varepsilon_0 \varepsilon_m \varsigma}{\eta} \frac{H}{1+[2Du+j\omega'\frac{\varepsilon_p}{\varepsilon_m}]\frac{1-\varphi}{2+\varphi}} \frac{1 - \frac{2s_j^2}{(2+\varphi)(3+3s_j+s_j^2)} - \frac{3\varphi}{2+\varphi}F}{1+\varphi\frac{\rho_p - \rho_m}{\rho_m}HF} \quad (5.23)$$

where

$$s_j = (1+j)s$$

$$H = \frac{3+3s_j+s_j^2}{3+3s_j+\frac{s_j^2}{3}(1+2\frac{\rho_s}{\rho_m})} \quad (5.24)$$

$$F = \frac{1}{3} + \frac{2+4s_j+8\ell^{2s_j}s_j^2 J}{3+3s_j+s_j^2} \quad (5.25)$$

$$J = \int_1^\infty [g(r)-1]r\ell^{-2s_j r}dr \quad (5.26)$$

where r is the distance between near-neighbors, and the function, g, is the pair distribution function.

Finally, the effective media approach was taken by Ennis, Shugai and Carnie [25, 26]. They managed to fit electroacoustic spectra even at very high solids loading up to 40%vl for dense particles.

However, the present authors are encouraged to resuscitate the Enderby-Booth theoretical approach based on six unresolved issues, namely:
- all these recent developments of O'Brien's theory are for the ESA mode, and it is not yet clear if they are valid for CVI;

- O'Brien's approach applies the superposition principle for calculating the electroacoustic signal of polydisperse colloids. This creates an internal contradiction in the theory since superposition does not work when particles interact by definition;
- There is no definition of the frame of reference. This becomes important since ESA and CVI might require a different inertial frame of reference;
- O'Brien's theory completely neglects thermodynamic effects, even for emulsions, and it is not yet clear why thermodynamics is the dominant loss mechanism in the attenuation spectra for soft particles, yet negligible in computing the electroacoustic effects, for both ESA and CVI;
- it is not clear if these theories, as they are, solely based on O'Brien's theoretical base, satisfy the low frequency Smoluchowski electroacoustic limit (Equation.5.17); and
- It is not clear why the "cell model" appears to fail in electroacoustics, while being very successful when applied in traditional stationary hydrodynamics and electrokinetics.

The following section presents a new electroacoustic theory for CVI that does not use O'Brien's relationship. Instead it derives expressions for the electroacoustic signal from first principles.

5.4 The Colloid Vibration Current in concentrated systems

A new theory for Colloid Vibration Current has been created in close collaboration with Prof. Vladimir Shilov, and would have been impossible without his major contribution.

We retain the O'Brien expression for introducing dynamic mobility as follows:

$$CVI = A(\omega)F(Z_T, Z_s)\varphi \frac{\rho_p - \rho_m}{\rho_m} \mu_d \nabla P \qquad (5.27)$$

We also retain the same structure for the dynamic electrophoretic mobility expression, presenting as separate multipliers both the inertial effects (function G) and the electrodynamic effects (function $1+F$). However, in contrast to the dilute case expressed by Equations 5.20 and 5.21, functions G and F for concentrated systems depend on the particle concentration. There is also an additional density dependent multiplier, $\frac{(\rho_p - \rho_s)\rho_m}{(\rho_p - \rho_m)\rho_s}$, which is equal to the ratio of the particle velocity relative to the liquid, and to the particle velocity relative to the center of mass of the dispersion. The convenience of the introduction of such a multiplier, which differs from unity only for concentrated suspensions, follows from the exact structure of Smoluchowski's asymptotic solution for μ_d, given by Equation 5.17. The corresponding equation, which in a convenient way reflects simultaneously limiting transformations both to Smoluchowski's asymptotic solution (5.17), and to O'Brien's asymptotic solution (5.19) is given as follows:

$$\mu_d = \frac{2\varepsilon_0 \varepsilon_m \varsigma (\rho_p - \rho_s) \rho_m}{3\eta(\rho_p - \rho_m)\rho_s} G(s,\varphi)(1 + F(Du, \omega', \varphi)) \qquad (5.28)$$

The generalization for the case of polydisperse systems is given by:

$$\mu_d = \frac{2\varepsilon_0 \varepsilon_m \varsigma (\rho_p - \rho_s) \rho_m}{3\eta(\rho_p - \rho_m)\rho_s} \sum_{i=1}^{N} G_i(s_i,\varphi)(1 + F_i(Du_i, \omega', \varphi)) \qquad (5.29)$$

The new values of the functions G and F are given with the following equations:

$$G_i(s,\varphi) = \frac{\overline{4j\varphi(1-\varphi) s_i I(s_i)(\rho_p - \rho_m(\frac{3H_i}{2I_i} + 1))}^{9\varphi_i h(s_i)\rho_s}}{1 - \frac{\rho_p}{1-\varphi} \sum_{i=1}^{N} \frac{\varphi_i(\frac{3H_i}{2I_i} + 1)}{\rho_p - \rho_m(\frac{3H_i}{2I_i} + 1)}} \qquad (5.30)$$

$$F_i(Du_i, \omega', \varphi) = \frac{(1 - 2Du_i)(1-\varphi) + j\omega'(1 - \frac{\varepsilon_p}{\varepsilon_m})(1-\varphi)}{2(1 + Du_i + \varphi(0.5 - Du_i)) + j\omega'(2 + \frac{\varepsilon_p}{\varepsilon_m} + \varphi(1 - \frac{\varepsilon_p}{\varepsilon_m}))} \qquad (5.31)$$

where $s_i^2 = \frac{a_i^2 \omega \rho_m}{2\eta_m}$; $\omega' = \frac{\omega}{\omega_{MW}}$, $\omega_{MW} = \frac{K_m}{\varepsilon_0 \varepsilon_m}$, φ_i and $Du_i = \kappa^\sigma / K_m a_i$ is the volume fraction and Dukhin number for the ith fraction of the polydisperse colloid, correspondingly, and φ is the total volume fraction of disperse particles. Special functions H and I are given below.

These expressions are restricted to the case of a thin DL and are valid for a broad frequency range, including the Maxwell-Wagner relaxation range. They take into account both hydrodynamic and electrodynamic particle interaction, and are valid for polydisperse systems without making any superposition assumption.

The remaining text in this section provides the more interested reader with a step-by-step derivation and justification of the above, perhaps otherwise daunting, expressions.

5.4.1 CVI and Sedimentation Current

There is a simple analogy that helps to create a clearer picture for illustrating the physical nature of the Colloid Vibration Potential or Colloid

Vibration Current. This analogy stresses the similarity between Electroacoustics and Sedimentation. Sedimentation current is well known from classical colloid chemistry textbooks [48]. Simply put, charged particles sediment because of gravity, and will develop a sedimentation potential between two vertically spaced electrodes. If we externally short-circuit these electrodes, the current that flows is referred to as the sedimentation current. We can extend this simple concept to include Colloid Vibration Current by simply replacing the acceleration of gravity with an analogous acceleration caused by the applied acoustic field.

Figure 5.3 illustrates a particle having a double layer and moving relative to the liquid. This motion includes the counter ions of the double layer. In this example of a negative particle, we need consider only the positive counter ions opposing the negative charged particle surface. The hydrodynamic surface current, I_s, reduces the number of positive ions near the right-side particle pole, and enriches the double layer with extra ions near the left-side pole. As a result, the double layer shifts from the original equilibrium. A negative surface charge dominates at the right pole, whereas an extra positive diffuse charge dominates at the left pole. The net result is that the particle motion has an induced a dipole moment.

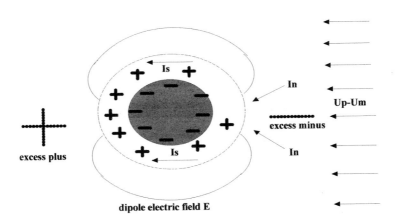

Figure 5.3 Polarization of the particle double layer with the relative liquid flow.

This induced dipole moment generates an electric field usually referred to as the Colloid Vibration potential (CVP). This CVP is external to the particle double layer. It affects ions in the bulk of the electro-neutral solution beyond the double layer, and generates an electric current, I_n. This electric current serves a very important purpose. It compensates for the surface current, I_s and makes the whole system self-consistent.

The next step is to add a quantitative description to this simple qualitative picture. To do this, we need a relationship between the CVI and $(u_p - u_m)$ as defined in Section 4.2.1. This problem is solved using the Shilov-Zharkikh cell model [49]. The advantages of this cell model over the Levine cell model [21] have been given in papers [27, 29].

According to both the Shilov-Zharkikh and Levine cell models, the macroscopic electric current density $\langle I \rangle$, local electric current density I, and the local gradient of electric potential $-\nabla \phi$ are related with the following expression:

$$CVI = \langle I \rangle = \frac{I_n (r=b)}{\cos\theta} = -K_m \frac{1}{\cos\theta} \frac{\partial \phi}{\partial r}\bigg|_{r=b} \tag{5.32}$$

Under these conditions the CVI is identical to the macroscopic electric current density $\langle I \rangle$.

We start our derivation of the CVI using two simplifications. First, we assume a thin DL, thus:

$$\kappa a \gg 1 \tag{5.33}$$

Second, we assume a negligible surface conductivity, κ^σ. This requires the Dukhin number to be sufficiently small:

$$Du = \frac{\kappa^\sigma}{K_m a} \ll 1 \tag{5.34}$$

This initial simplification will be removed later (Section 5.4.3), but is convenient for the present description.

The thin double layer simplification allows us to describe the distribution of the electric potential ϕ using the Laplace equation:

$$\Delta \phi = 0 \tag{5.35}$$

The general spherically symmetric solution of this equation yields:

$$\varphi = -Er \cos\theta + \frac{d}{r^2}\cos\theta \tag{5.36}$$

Equation 5.36 contains only two unknown constants E and d. Two boundary conditions are required for their calculation, that of the particle surface and that of the cell.

The particle surface boundary condition reflects continuity of the bulk current, $I = -K_m \nabla \phi$ and surface current, I_θ and is defined by:

$$K_m \nabla_n \phi = div_s I_\theta \tag{5.37}$$

There is only one essential component of the surface current when the double layer is thin and the surface conductivity is low. It is caused by hydrodynamic involvement of the electric charge ρ_{dl} in the thin diffuse layer, as given by:

$$I_\theta = \int_0^\infty \rho_{dl}(x) u_\theta(x) dx \tag{5.38}$$

For a thin double layer it can be considered to be essentially flat.

We can then apply a Taylor series expansion to the tangential speed, u_θ, near the particle surface. This is described by the following equation, in which x is the distance from the particle surface.

$$u_\theta(x) = u_\theta(x=0) + x \frac{\partial u_\theta}{\partial x}\bigg|_{x=0} \tag{5.39}$$

The first term is equal to zero because, at the surface, the liquid does not move relative to the particle. As a result the surface current within the thin double layer can be expressed as:

$$I_s = \frac{\partial u_\theta}{\partial x}\bigg|_{x=0} \int_0^\infty \rho_{dl}(x) x\, dx \tag{5.40}$$

At this point we again employ a peculiarity of the thin double layer. There is a known relationship between the electric charge density in the double layer and the ζ-potential:

$$\int_0^\infty \rho_{dl}(x) x\, dx = -\varepsilon_m \varepsilon_0 \zeta \tag{5.41}$$

Since the surface current is given by:

$$I_s = -\varepsilon_m \varepsilon_0 \zeta \frac{\partial u_\theta}{\partial r}\bigg|_{r=a} \tag{5.42}$$

it then follows that substitution of Equation 5.42 into 5.37 yields the first boundary condition.

The cell boundary condition, correspondent to the definition of the CVI effect, specifies a zero value for the macroscopic electric current in the suspension, expressed by a zero value for the macroscopic electric field, $\overline{E}$, and, hence, in the frame of the Shilov-Zharkich cell model, by a zero value of the variation of the electric potential on the cell surface:

$$\overline{E} = \frac{\varphi(r=b)}{b\cos\theta} = 0 \tag{5.43}$$

In the framework of the Levine cell model, the relationship between the macroscopic and local electric field is given, contrary to Equation 5.43, by:

$$\overline{E}_{Levine} = -\frac{1}{\cos\theta}\frac{\partial \varphi}{\partial r}\bigg|_{r=b}$$. A comparison of this expression with Equation 5.32 (which connects the macroscopic electric current with the local electric field), reveals an onerous contradiction in attempting to employ the Levine cell model. It follows that $\overline{K}_{Levine} = \overline{I}/\overline{E}_{Levine} = K_m$. From this we must conclude that the conductivity of the suspension, K_s, must equal the conductivity of the media K_m,

which is obviously false. Clearly the conductivity of the suspension must be a function of the properties of particles and their double layers. Therefore the Levine cell model is inappropriate for this task.

To compute the CVI we calculate the unknown constants E and d using Equations 5.37 and 5.43, substitute these constants into the general solution of Equation 5.36 for ϕ, then calculate a value of ϕ at $r = b$, and finally substitute this result into Equation 5.32. Following these steps we obtain the following expression for CVI:

$$CVI = \frac{3\varepsilon_m \varepsilon_o \zeta}{a} \frac{a^3}{b^3 + 0.5a^3} \frac{1}{\sin\theta} \frac{\partial u_\theta}{\partial r}\bigg|_{r=a} \quad (5.44)$$

This expression relates CVI with an unknown, albeit derivative of the tangential component of the liquid motion relative to the particle surface.

$$CVI = \frac{3\varepsilon_m \varepsilon_o \zeta}{a} \frac{\varphi}{1 + 0.5\varphi} \frac{1}{\sin\theta} \frac{\partial u_\theta}{\partial r}\bigg|_{r=a} \quad (5.45)$$

where ε_m and ε_0 are the dielectric permittivities of the medium and vacuum, ζ is the electrokinetic potential, a is the particle radius, φ is the volume fraction, r and θ are the spherical coordinates associated with the particle center, and u_r and u_θ are the radial and tangential velocities of the liquid motion relative to the particle.

The next step in the development of this CVI theory is to calculate the hydrodynamic field, assuming that the speed of particle with respect to the liquid is given by Equation 4.27 for a monodisperse colloid, or Equation 4.35 for a polydisperse colloid. This has been solved by A.Dukhin et al [28] using a Happel cell model, and later using the Kuwabara cell model [50], under the assumption of an incompressible liquid. This condition is valid only when the wavelength, λ, is much larger than the particle size (the so-called long wavelength requirement):

$$\lambda \gg a \quad (5.46)$$

The final expressions for the drag coefficient and tangential velocity are:

$$\gamma = \omega \rho_m \varphi \left[\frac{3}{2I}(-\frac{dh}{dx} + \frac{h}{x})_{x=a} - j\right] \quad (5.47)$$

$$\frac{du_\theta}{dr}\bigg|_{r=a} = \frac{3(u_p - u_m)h(s)}{2I} \quad (5.48)$$

where: $s = a\sqrt{\omega/2\nu}$, ν is kinematic viscosity, and η is dynamic viscosity.

The values of the special functions $h(x)$ and $I(x)$ are presented in Chapter 2, Section 2.5.2.

Substituting the drag coefficient into Equation 4.27 and applying the result to 5.45 leads to the following expressions for CVI and dynamic mobility:

$$CVI = \frac{9\varepsilon_m \varepsilon_0 \zeta (\rho_p - \rho_s)}{4\eta} \frac{\varphi}{\rho_s} \frac{1}{1+0.5\varphi} \frac{jh(\alpha)}{1.5H - \frac{(1-\varphi)\rho_p - \rho_s}{\rho_s} I} \nabla P \qquad (5.49)$$

$$\mu_d = \frac{9\varepsilon_m \varepsilon_0 \zeta (\rho_p - \rho_s)\rho_m}{4\eta} \frac{1}{(\rho_p - \rho_m)\rho_s} \frac{1}{1+0.5\varphi} \frac{jh(s)}{H - \frac{2(1-\varphi)\rho_p - \rho_s}{3\rho_s} I} \nabla P \qquad (5.50)$$

$$G = \frac{9j}{2s} \frac{h(s)}{3H + 2I\left[1 - \frac{\rho_p}{\rho_s}(1-\varphi)\right]} \qquad F = \frac{1-\varphi}{2+\varphi}$$

These expressions are valid for non-conducting particles, with a thin double layer and a low surface conductivity ($Du<<1$). The frequency dependence of CVI and dynamic mobility given by Equations 5.49 and 5.50, correspondingly, is caused by the factor G only, which reflects the formation of hydrodynamic fields. Factor F, that reflects the formation of electric fields, in the case of Equation 5.50 is frequency independent. This happens because we did not yet take into account the non- stationary process for the storage of sound-induced ionic charge in the DL, when deriving Equation 5.35 and 5.37, which reflect the balance of the electric fields and currents.

5.4.2 CVI for polydisperse systems

Let us assume now that we have polydisperse system with N conventional fractions. Each fraction of particles has a certain particle diameter, d_i, volume fraction, φ_i, drag coefficient, γ_i, and particle velocity, u_i, in the typical laboratory frame of reference. We also assume the density of the particles to be the same for all fractions, ρ_p. The total volume fraction of the dispersed phase is φ. The liquid is characterized by a dynamic viscosity, η, a density ρ_m and a velocity, u_m.

The coupled phase model allows us to calculate the difference $u_i - u_m$ for each fraction, without the need to invoke the superposition assumption (Equation 4.35), viz:

$$u_i - u_m = \frac{(\frac{\rho_p}{\rho_m} - 1)\nabla P}{(j\omega\rho_p + \frac{\gamma_i}{\varphi_i})(1 + \frac{\rho_p}{(1-\varphi)\rho_m} \sum_{i=1}^{N} \frac{\gamma_i}{j\omega\rho_p + \frac{\gamma_i}{\varphi_i}})} \qquad (5.51)$$

The next step is a calculation of the corresponding electric field. We can apply Equation 5.44 to calculate the electric current generated by the i-th fraction of the polydisperse system, CVI_i, keeping in mind that the number of dipoles in this fraction is φ_i/φ times smaller:

$$CVI_i = \frac{\varphi_i}{\varphi} \frac{3\varepsilon_m\varepsilon_o\zeta}{a_i} \frac{a_i^3}{b_i^3 + 0.5a_i^3} \frac{1}{\sin\theta} \frac{\partial u_{i\theta}}{\partial r}\bigg|_{r=a} \quad (5.52)$$

This equation has been derived using the cell model. Usually, a cell model is formulated only for a monodisperse system. We have proposed [28, 50], and described briefly (Section 2.5.2) a way to generalize the cell model to account for polydispersity. This yields the following simplification:

$$CVI_i = \frac{3\varepsilon\varepsilon_o\zeta\varphi_i}{a_i} \frac{\varphi}{1+0.5\varphi} \frac{1}{\sin\theta} \frac{\partial u_{i\theta}}{\partial r}\bigg|_{r=a} \quad (5.53)$$

The derivative from the tangential velocity contains a dependence on the speed of the particle motion in the sound wave given by Equation 5.48:

$$\frac{1}{\sin\theta} \frac{\partial u_{i\theta}}{\partial r}\bigg|_{r=a} = \frac{3(u_i - u_m)h(s_i)}{a_i I(s_i)} \quad (5.54)$$

where h and I are special functions (see Section 2.5.2).

The total CVP generated by all particles is then calculated as the superposition of all the fractional CVP_i. The total electroacoustic current equates to CVP multiplied by the complex conductivity of the dispersion K_s. The final expression for CVI and the dynamic mobility of the polydisperse system is then given by:

$$CVI = \frac{9\varepsilon_m\varepsilon_o\zeta(\rho_s - \rho_m)}{4\eta\rho_s} \frac{\varphi}{1+0.5\varphi} \frac{\sum_{i=1}^{N} \frac{\rho_s\varphi_i h(s_i)}{j(1-\varphi)s_i I(s_i)(\rho_p - \rho_m(\frac{3H_i}{2I(s_i)}+1))}}{1 - \frac{\rho_p}{1-\varphi}\sum_{i=1}^{N} \frac{(\frac{3H_i}{2I(s_i)}+1)\varphi_i}{\rho_p - \rho_m(\frac{3H_i}{2I(s_i)}+1)}} \quad (5.55)$$

$$\mu_d = \frac{2\varepsilon_m\varepsilon_o\zeta(\rho_s - \rho_m)\rho_s}{3\eta(\rho_p - \rho_m)\rho_m} \frac{\sum_{i=1}^{N} \frac{9\rho_s\varphi_i h(s_i)}{4j\varphi(1-\varphi)s_i I(s_i)(\rho_p - \rho_m(\frac{3H_i}{2I(s_i)}+1))}}{1 - \frac{\rho_p}{1-\varphi}\sum_{i=1}^{N} \frac{(\frac{3H_i}{2I(s_i)}+1)\varphi_i}{\rho_p - \rho_m(\frac{3H_i}{2I(s_i)}+1)}} \frac{3}{2+\varphi} \equiv$$

$$\quad (5.56)$$

$$\equiv \frac{2\varepsilon_m\varepsilon_o\zeta(\rho_s - \rho_m)\rho_s}{3\eta(\rho_p - \rho_m)\rho_m} \sum_{i=1}^{N} G(s_i.\varphi)(1+F_i)$$

where the special functions h, H and I are given in Section 2.5.2, $H_i = H(s_i)$, and $I_i = I(s_i)$. The comparison of (5.56) with the form (5.29) gives rise to the conclusion that for the considered case of negligible value for the surface conductivity ($Du \ll 1$) and a negligible contribution of the Maxwell-Wagner dispersion ($\omega' \ll 1$), we have $G(s,\varphi) = \sum_{i=1}^{n} G_i(s_i,\varphi)$ and $F_i = \frac{3}{2+\varphi} - 1 = \frac{1-\varphi}{2+\varphi}$.

The procedure used for the derivation of Eqs. 5.55 and 5.56 might be also applied for taking into account the surface conductivity of the particles and (or) the Maxwell-Wagner dispersion. Such a generalization would require modification of the function F_i only, without any revision of $G_i(s_i,\varphi)$.

5.4.3 Surface conductivity.

So far the surface current that generates CVP has been attributed only to the hydrodynamic component.

$$I_{si} = -\varepsilon_m \varepsilon_o \zeta \frac{\partial u_\theta}{\partial r}\bigg|_{r=a_i} \tag{5.57}$$

This restricts the theory to the situation of negligible surface conductivity and small Dukhin number $Du \ll 1$. Now, as stated earlier, we are going to expand the theory to both include surface conductivity effects, and remove the restriction on the Dukhin number. In the presence of substantial surface conductivity, the expression for the surface current must take into account not only the hydrodynamic component, but also the electrodynamic components of the surface current:

$$I_{si} = -\varepsilon_m \varepsilon_o \zeta \frac{\partial u_\theta}{\partial r}\bigg|_{r=a_i} + \frac{\kappa^\sigma}{a_i} \frac{\partial \phi}{\partial r}\bigg|_{r=a} \tag{5.58}$$

The addition of this second component to the surface current modifies the expression for the dynamic mobility:

$$\mu_d = \frac{2\varepsilon_o \varepsilon_m \zeta (\rho_p - \rho_s)\rho_m}{3\eta(\rho_p - \rho_m)\rho_s} \sum_{i=1}^{N} G(s_i,\varphi_i)(1 + F_i(Du_i,\varphi)) \tag{5.59}$$

The surface conductivity might be assumed to be the same for all fractions, especially in the case of a thin DL. Under such an assumption, the substitution of Eq. 5.58 into the surface condition Eq. 5.37, and following the application of Eqs. 5.37 and 5.43 for the determining the constants in the general solution, Eq. 5.35, and substitution of the resulting potential distribution into Eq. 5.32, leads to:

$$\mu_d = \frac{2\varepsilon_m \varepsilon_o \zeta (\rho_s - \rho_m)\rho_s}{3\eta(\rho_p - \rho_m)\rho_m} \frac{\sum_{i=1}^{N} \frac{9\rho_s \varphi_i h(s_i)}{4j\varphi(1-\varphi)s_i I(s_i)(\rho_p - \rho_m(\frac{3H_i}{2I(s_i)} + 1))} \frac{3}{2+\varphi+2Du_i(1-\varphi)}}{1 - \frac{\rho_p}{1-\varphi} \sum_{i=1}^{N} \frac{(\frac{3H_i}{2I(s_i)} + 1)\varphi_i}{\rho_p - \rho_m(\frac{3H_i}{2I(s_i)} + 1)}}$$

$$\tag{5.60}$$

where $Du_i = \kappa^\sigma/K_m\, a_i$ is the Dukhin number for the ith fraction of the polydisperse colloid.

5.4.4 Maxwell-Wagner relaxation. Extended frequency range.

So far we have neglected variation of the surface charge density within the DL. This was justified for low frequencies much smaller than the Maxwell-Wagner relaxation frequency defined by:

$$\omega \ll \omega_{MW} = \frac{K_m}{\varepsilon_0 \varepsilon_m}$$

Consequently, the boundary condition of the continuity bulk and surface currents was defined earlier by Equation 5.37 as:

$$K_m \nabla_n \phi = div_s I_\theta$$

To remove the restriction on frequency and to incorporate the Maxwell-Wagner relaxation into the CVI theory, we need to add a field-induced charge density variation to the electrodynamic boundary condition:

$$K_m \nabla_n \phi = div_s I_\theta + \frac{\partial \sigma^d}{\partial t} \qquad (5.61)$$

where σ^d is the surface charge density induced and stored in the DL.

The appearance of this additional surface function, namely $\partial \sigma^d/\partial t$, itself requires an additional surface boundary condition to complete the mathematical formulation of the problem. The electrostatic condition of the continuation of the normal components of electrostatic displacement and the tangential components of electric field strength serve this purpose:

$$\varepsilon_p E_r(r = a - 0) - \varepsilon_m E_r(r = a + 0) = \sigma^d \qquad (5.62)$$

$$E_\theta(r = a - 0) - E_\theta(r = a + 0) = 0 \qquad (5.63)$$

Substituting an induced surface charge density into the charge conservation condition Eq.5.61 yields the following surface condition that express the continuity of the complex currents near the particle's surface:

$$(K_m - j\omega\varepsilon_m)E_r(r = a + 0) - (K_p^{ef} - j\omega\varepsilon_p)E_r(r = a - 0) = \varepsilon_m \varepsilon_0 \zeta\, div_s \frac{\partial u_\theta}{\partial r}\bigg|_{r=a} \qquad (5.64)$$

where

$$K_p^{ef} = Du\, K_m = \frac{2\kappa^\sigma}{a} \qquad (5.65)$$

The tangential components of the electric fields have the same value on both sides of the particle's surface:

$$E_\theta(r = a + 0) = E_\theta(r = a - 0) \qquad (5.66)$$

Solution of the Laplace equations for the potential distributions outside and inside the particles surface together with the surface conditions Eqs. 5.64, 5.65 and 5.43, and following substitution of the resulting potential distribution into Eq. 5.32, leads to the final expression for the frequency dependent electrodynamic contribution to CVI:

$$F(Du,\omega',\varphi) = \frac{(1-2Du)(1-\varphi) + j\omega'(1-\frac{\varepsilon_p}{\varepsilon_m})(1-\varphi)}{2(1+Du+\varphi(0.5-Du)) + j\omega'(2+\frac{\varepsilon_p}{\varepsilon_m}+\varphi(1-\frac{\varepsilon_p}{\varepsilon_m}))} \qquad (5.67)$$

This expression for the function, F is only valid for a thin DL, but for any surface conductivity, frequency, and volume fraction.

5.5 Qualitative analysis.

It is helpful to create some heuristic understanding of the physical phenomena that take place when an ultrasound pulse passes through a dispersed system. This also provides answers to some general questions. For instance, why do we need a density contrast in the case of ESA, when the particles already move relative to the liquid under the influence of an electric field? Why do we need a density contrast to generate a CVI at low frequency when the particles already move in phase with the liquid?

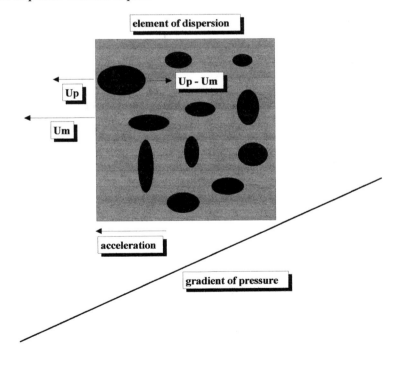

Figure 5.4 An element of the colloid in the ultrasound pressure field.

As far as we can find there are no simple published answers to these questions. To find them we utilize again the analogy between the sedimentation potential and electroacoustic phenomena. Marlow has already used this analogy [19, 20], and we will give further justification for his approach.

Let us consider an element of the concentrated dispersed system in the sound wave (Figure 5.4). The size of this element is selected such that it is much larger than the particle, and also larger than the average distance between particles. As a result, this element contains many particles. At the same time this element is much smaller than wavelength.

This dispersion element moves with a certain velocity and acceleration in response to the gradient of acoustic pressure. As a result, an inertia force is applied to this element. At this point we can use the principle of equivalency between inertia and gravity. The effect of the inertia force created by the sound wave is equivalent to the effect of the gravity force.

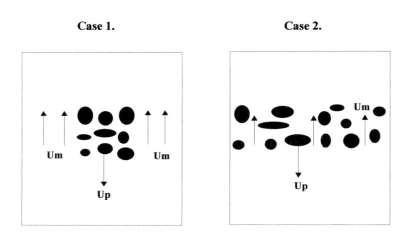

Figure 5.5 Two possible ways in which particles can sediment.

This gravity force is exerted on both the particle and liquid inside of the dispersion element. The densities of the particles and liquid are different, as are the forces. The force acting on the particles depends on the ratio of the densities.

The question then arises as to what density we should take into account. To answer this, let us consider the forces acting on a given particle in the gravity field. The first force is the weight of the particle, which is proportional to its density, ρ_p. This force will be partially balanced by the pressure of the surrounding liquid and other particles. This pressure is equivalent to the pressure generated by an effective media with a density equal to the density of the dispersed system. This may be clearer when one imagines a large particle surrounded by smaller ones.

We are coming to the well known conclusion that the resulting force, which moves the particles with respect to the liquid, is proportional to the density difference between particle, ρ_p, and dispersed system, ρ_s.

It is important to mention here that this force is not necessarily a buoyant force. A buoyant force would be proportional to the difference between particle density, ρ_p, and the dispersion medium, ρ_m, as given by Archimedes law. Figure 5.5 illustrates this difference for the example of the sedimentation of a small spherical group of spherical particles in the liquid. Case 1 corresponds to the situation when the array settles as one entity, and the liquid envelopes this settling entity from the outside. Case 2 corresponds to the different situation in which the liquid is forced to move through the array. There will be a difference between the forces exerted on the particles within the array. There is an additional force in Case 2 caused by the liquid pushing through the array of particles.

Electroacoustic phenomena correspond to Case 2. It happens because the width of the sound pulse, w, is much larger than wavelength, λ, at high ultrasound frequencies.

$$w >> \lambda \tag{5.68}$$

The balance of forces exerted on the particles in a given element of the dispersion consists of the effective gravity force, the buoyancy force, and the friction force, related to the filtration of the liquid through the array of particles. As a result the particles move relative to the liquid with a speed, $(u_p - u_m)$, that is proportional to the density difference between the particle and the system, $(\rho_p - \rho_s)$.

The motion of the particles relative to the liquid disturbs their double layers, and as a result generates an electroacoustic signal. This electroacoustic signal is zero when the speed of the particle is equal to the speed of the medium. This occurs when the density of the particle is equal to the density of the dispersed system. This implies that the electroacoustic signal must be proportional to $(\rho_p - \rho_s)$. This conclusion is rather unexpected, because O'Brien's relationship makes the CVI proportional to $(\rho_p - \rho_m)$.

This discrepancy in the density contrast might be related to the initial definition of the inertial frame of reference. When sound is the driving force, (for either CVI or CVP), the correct inertial frame is a laboratory frame of reference because the acoustic wavelength is much shorter than the size of the sample chamber. Therefore, particles move with different phases inside the narrow sound beam. The chamber as an entity remains immobile.

The question of the frame of reference is more complicated in the ESA case where the electric field is the driving force. The wavelength of the electric field is much longer, and as a result, all the particles move in phase. This motion exerts a given force on the chamber. The motion of the chamber depends both on the mass of the chamber and mass of the sample. Depending on the construction of the instrument, the inertial system might be related either to the chamber, to the center of mass, or to some intermediate case (depending on the ratio of the masses of the chamber and the sample). This means that, for the ESA method, the final expression relating any measured ESA signal with the properties of the dispersed system might contain a multiplier that is dependent on the mass of the chamber.

REFERENCES.

1. Debye, P. "A method for the determination of the mass of electrolyte ions", J. Chem. Phys., 1,13-16, (1933)
2. Yeager, E., Bugosh, J., Hovorka, F. and McCarthy, J., "The application of ultrasonics to the study of electrolyte solutions. II. The detection of Debye effect", J. Chem. Phys., 17, 4, 411-415 (1949)
3. Yeager, E., and Hovorka, F., "The application of ultrasonics to the study of electrolyte solutions. III. The effect of acoustical waves on the hydrogen electrode", J. Chem. Phys., 17, 4, 416-417 (1949)
4. Yeager, E., and Hovorka, F., "Ultrasonic waves and electrochemistry. I. A survey of the electrochemical applications of ultrasonic waves", J. Acoust. Soc. Am., 25, 3, 443-455 (1953)
5. Yeager, E., Dietrich, H., and Hovorka, F., J. "Ultrasonic waves and electrochemistry. II. Colloidal and ionic vibration potentials", J. Acoust. Soc. Am., 25, 456 (1953)
6. Zana, R., and Yeager, E., "Determination of ionic partial molal volumes from Ionic Vibration Potentials", J. Phys. Chem, 70, 3, 954-956 (1966)
7. Zana, R., and Yeager, E., "Quantitative studies of Ultrasonic vibration potentials in Polyelectrolyte Solutions", J. Phys. Chem, 71, 11, 3502-3520 (1967)
8. Zana, R., and Yeager, E., "Ultrasonic vibration potentials in Tetraalkylammonium Halide Solutions", J.Phys.Chem, 71, 13, 4241-4244 (1967)
9. Zana, R., and Yeager, E., "Ultrasonic vibration potentials and their use in the determination of ionic partial molal volumes", J. Phys. Chem, 71, 13, 521-535 (1967)
10. Zana, R., and Yeager, E., "Ultrasonic vibration potentials", Mod. Aspects of Electrochemistry, 14, 3-60 (1982)
11. Bugosh, J., Yeager, E., and Hovorka, F., "The application of ultrasonic waves to the study of electrolytes. I.A modification of Debye's equation for the determination of the masses of electrolyte ions by means of ultrasonic waves", J. Chem. Phys. 15, 8, 592-597 (1947)
12. Derouet, B., and Denizot, F., C. R. Acad.Sci., Paris, 233, 368 (1951)

13. Dietrick, H., Yeager, E., Bugosh, J., Hovorka, F. "Ultrasonic waves and electrochemistry. III. An electrokinetic effect produced by ultrasonic waves", J. Acoust. Soc. Am., 25, 3, 461-465 (1953)

14. Hermans, J., Philos. Mag., 25, 426 (1938)

15. Rutgers, A.J. and Rigole, W. "Ultrasonic vibration potentials in colloid solutions, in solutions of electrolytes and pure liquids", Trans. Faraday Soc., 54, 139-143 (1958)

16. Enderby, J.A. "On Electrical Effects Due to Sound Waves in Colloidal Suspensions", Proc. Roy. Soc., London, A207, 329-342 (1951)

17. Booth, F. and Enderby, J. "On Electrical Effects due to Sound Waves in Colloidal Suspensions", Proc. of Amer. Phys. Soc., 208A, 32 (1952)

18. Oja, T., Petersen, G., and Cannon, D. "Measurement of Electric-Kinetic Properties of a Solution", US Patent 4,497,208, (1985)

19. Marlow, B.J., Oja, T. and Goetz, P.J., "Colloid Analyzer", US Patent 4,907,453 (1990)

20. Marlow, B.J., Fairhurst, D. and Pendse, H.P., "Colloid Vibration Potential and the Electrokinetic Characterization of Concentrated Colloids", Langmuir, 4,3, 611-626 (1983)

21. Levine, S. and Neale, G.H. "The Prediction of Electrokinetic Phenomena within Multiparticle Systems.1.Electrophoresis and Electroosmosis.", J. of Colloid and Interface Sci., 47, 520-532 (1974)

22. O'Brien, R.W. "Electro-acoustic Effects in a dilute Suspension of Spherical Particles", J. Fluid Mech., 190, 71-86 (1988)

23. Rider, P.F. and O'Brien, R.W., "The Dynamic Mobility of Particles in a Non-Dilute Suspension", J. Fluid. Mech., 257, 607-636 (1993)

24. Ohshima, H. "Dynamic Electrophoretic Mobility of Spherical Colloidal Particles in Concentrated Suspensions", J. of Colloid and Interface Sci., 195, 137-148 (1997)

25. Ennis, JP, Shugai, AA and Carnie, SL, "Dynamic mobility of two spherical particles with thick double layers"; Journal of Colloid and Interface Science, 223,21-36, (2000)

26. Ennis, JP, Shugai, AA and Carnie, SL, "Dynamic mobility of particles with thick double layers in a non-dilute suspension"; Journal of Colloid and Interface Science, 223, 37-53, (2000).

27. Dukhin, A. S., Shilov, V.N. and Borkovskaya Yu. "Dynamic Electrophoretic Mobility in Concentrated Dispersed Systems. Cell Model.", Langmuir, 15, 10, 3452-3457 (1999)

28. Dukhin, A.S., Shilov, V.N, Ohshima, H., Goetz, P.J "Electroacoustics Phenomena in Concentrated Dispersions. New Theory and CVI Experiment", Langmuir, 15, 20, 6692-6706, (1999)

29. Dukhin, A.S., Shilov, V.N, Ohshima, H., Goetz, P.J "Electroacoustics Phenomena in Concentrated Dispersions. Effect of the Surface Conductivity", Langmuir, 16, 2615-2620 (2000)

30. Dukhin, A.S., and Goetz, P.J. "Acoustic and Electroacoustic Spectroscopy for Characterizing Concentrated Dispersions and Emulsions", Adv. In Colloid and Interface Sci., 92, 73-132 (2001)

31. Colic, M. and Morse, D. "The elusive mechanism of the magnetic memory of water", Colloids and Surfaces, 154, 167-174 (1999)

32. Lubomska, M. and Chibowski, E. "Effect of Radio Frequency Electric Fields on the Surface Free Energy and Zeta Potential of Al_2O_3 ", Langmuir, 17, 4181-4188 (2001)

33. O'Brien, R.W. "Electro-acoustic effects in a dilute suspension of spherical particles", Preprint, School of Mathematics, The University of New South Wales, (1986)

34. O'Brien, R.W., Midmore, B.R., Lamb, A. and Hunter, R.J. "Electroacoustic studies of moderately concentrated colloidal suspensions", Faraday Discuss. Chem. Soc., 90, 1-11 (1990)

35. O'Brien R.W., Rowlands W.N. and Hunter R.J. "Determining charge and size with the Acoustosizer", in S.B. Malghan (Ed.) Electroacoustics for Characterization of Particulates and Suspensions, NIST, 1-21 (1993)

36. O'Brien, R.W. "Determination of Particle Size and Electric Charge", US Patent 5,059,909, Oct.22, (1991).

37. O'Brien, R.W. Particle size and charge measurement in multi-component colloids", US Patent 5,616,872, (1997).

38. O'Brien, R.W. "Electro-acoustic Effects in a dilute Suspension of Spherical Particles", J.Fluid Mech., 190, 71-86 (1988)

39. O'Brien, R.W., Wade, T.A., Carasso, M.L., Hunter, R.J., Rowlands, W.N. and Beattie, J.K. "Electroacoustic determination of droplet size and zeta potential in concentrated emulsions", JCIS, (1996)

40. O'Brien, R.W., Cannon, D.W. and Rowlands, W.N. "Electroacoustic determination of particle size and zeta potential", JCIS, 173, 406-418 (1995)

41. Hunter, R.J. "Review. Recent developments in the electroacoustic characterization of colloidal suspensions and emulsions", Colloids and Surfaces, 141, 37-65 (1998)

42. Lyklema, J., "Fundamentals of Interface and Colloid Science", vol. 1-3, Academic Press, London-NY, 1995-2000.

43. Bondarenko, M.P. and Shilov, V.N. "About some relations between kinetic coefficients and on the modeling of interactions in charged dispersed systems and membranes", JCIS, 181, 370-377 (1996)

44. Katchalsky, A. and Curran, P.F. "Nonequlibrium thermodynamics in biophysics", Harvard University press, Cambridge (1967)

45. Kruyt, H.R. "Colloid Science", Elsevier: Volume 1, Irreversible systems, (1952)

46. Dukhin, S.S. and Shilov V.N. "Dielectric phenomena and the double layer in dispersed systems and polyelectrolytes", John Wiley and Sons, NY, (1974).

47. Rider, P.F. and O'Brien, R.W., "The Dynamic Mobility of Particles in a Non-Dilute Suspension", J. Fluid. Mech., 257, 607-636 (1993)

48. Ohshima, H., Healy, T.W., White, L.R. and O'Brien, R.W. "Sedimentation velocity and potential in a dilute suspension of charged spherical colloidal particles", J. Chem. Soc. Faraday Trans., 2, 80, 1299-1377 (1984)

49. Shilov, V.N., Zharkih, N.I. and Borkovskaya, Yu.B. "Theory of Nonequilibrium Electrosurface Phenomena in Concentrated Disperse System.1.Application of Nonequilibrium Thermodynamics to Cell Model.", Kolloid. Zh., 43,3, 434-438 (1981)

50. Dukhin, A.S. and Goetz, P.J. "Acoustic Spectroscopy for Concentrated Polydisperse Colloids with High Density Contrast", Langmuir, 12 [21] 4987-4997 (1996)

51. Loewenberg, M. and O'Brien, R.W. "The dynamic mobility of nonspherical particles", JCIS, 150, 1, 159-168 (1992)

Chapter 6. EXPERIMENTAL VERIFICATION OF THE ACOUSTIC AND ELECTROACOUSTIC THEORIES

In the previous sections describing acoustic and electroacoustic theory, we tried to justify the superposition approach to the description of the interaction of ultrasound with colloids. Altogether six interaction mechanisms have been introduced. They are assumed to operate independently, with very little interaction in real-world practical colloids. In this chapter we demonstrate that it is, indeed, possible to prepare a real disperse system in which only one of the six mechanisms dominates the acoustic and electroacoustic properties of the colloid. This chapter is structured such that each sub-section corresponds to a different mechanism of ultrasound interaction. In each of these sub-sections we present experimental tests which verify the corresponding theory.

6.1. Viscous losses

According to our theory, it is the viscous loss that is dominant, compared to all other loss mechanisms, for those colloids composed of sub-micron sized rigid particles that have a substantial density contrast with respect to their suspension medium. It is obvious that a wide variety of such colloids exist. We will test our viscous loss theory by selecting two examples of stable colloids, performing a dilution experiment, and testing whether 1) the measured attenuation conforms to the theory 2) the calculated size using this theory is correct, and, importantly, 3) the calculated size remains constant with dilution. If these three criteria are met, then we have strong evidence that the viscous mechanism theory adequately describes the particle-particle interactions. We closely follow here published works by A. Dukhin and P. Goetz [1, 2] using a Dispersion Technology Acoustic Spectrometer DT-100, Mount Kisco, NY, USA.

The protocol for testing volume fraction dependence involves making an equilibrium dilution of an initially concentrated and stable system. The initial concentrate must itself be stable if we are to ensure that the actual particle size remains constant throughout the experiment. The dilution protocol requires that we obtain a suitable diluent that is chemically identical, in all respects, to the original suspending medium, in order to ensure stability as we dilute the sample. The chemical composition of the medium must remain the same for all volume fractions. In summary, if the initial sample is stable, and we retain the same chemical environment, then the particle size should remain constant for any volume fraction. In other words, the measured particle size should be independent of the volume fraction.

We performed this dilution test using dispersions of two different materials: colloidal silica (Ludox TM-50) and rutile (R-746), both produced by duPont. We selected the silica because of its small size; the nominal area-based size, reported by DuPont, based on titration measurement, is approximately 22 nm and it is provided as a 50%wt aqueous slurry. Acoustic measurement of the silica gave a weight-basis particle size of 28 nm. We selected the rutile as an example of enhanced density contrast. We took 100 ml of this dispersion and weighed it; it weighed 234 g, which yields a particle density of 3.9 g/cm^3. This density was somewhat lower than the density of regular rutile, perhaps because of the addition of stabilizing additives. The rutile is also supplied as an aqueous slurry, at a weight fraction of 76.8%; the nominal size is about 300 nm.

We used two different techniques to produce the equilibrium dispersion medium needed for dilution: centrifugation and dialysis. For the larger rutile particles, having a significant density contrast, we easily extracted the needed equilibrium solution by centrifugation. However, such gravimetric separation does not work for the very small silica Ludox particles; for this we employed dialysis.

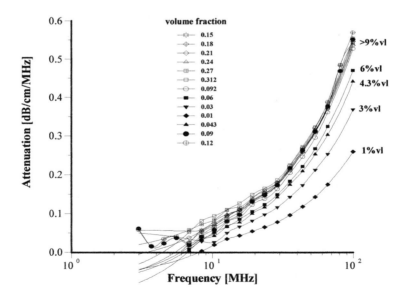

Figure 6.1 Attenuation spectra of 28 nm silica at various volume fractions.

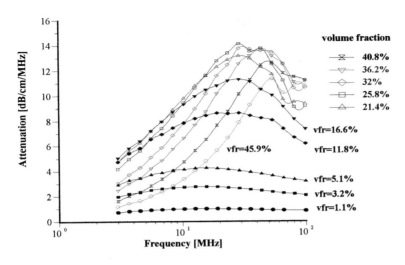

Figure 6.2 Attenuation spectra of 300 nm rutile at various concentrations.

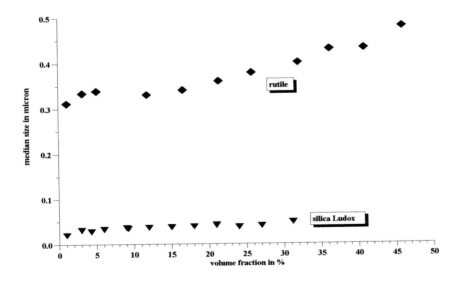

Figure 6.3 Median particle size of rutile and silica Ludox calculated from the attenuation spectra in Figures 6.1 and 6.2.

The measured attenuation spectra for the silica and rutile samples are shown in Figures 6.1 and 6.2, respectively. It is clear that the attenuation of the silica is much smaller than that of the rutile. This is because of the smaller size and lower density contrast of the silica. For the silica slurry the attenuation spectra are almost indistinguishable at concentrations above 9%vl. This reflects a nonlinear dependence of the attenuation with volume fraction. It arises because of particle-particle interactions.

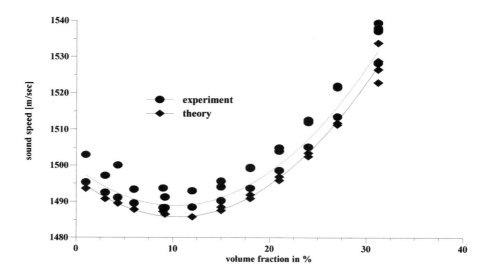

Figure 6.4 Sound speed at 10 MHz of silica Ludox at various volume fractions

Such non-linearity has been known for quite some time [1], and it is even more pronounced for the rutile (Figure 6.2). The interactions also shift the critical frequency to higher values. Hence the attenuation, at low frequency, decreases with increasing volume fraction above 16.6%. It is exactly the same effect that makes the attenuation constant with volume fraction for silica. Our new theory takes this non-linear effect into account. As a result the values for particle size, calculated from these attenuation spectra, become almost constant irrespective of volume fractions for silica (Figure 6.3). The increase in the particle size of the rutile slurry at high volume fraction cannot be accounted for by particle-particle interaction. It is plausible that some increase might be due to unavoidable aggregation, but it is also possible that the theory may yet need some further refinement. Nevertheless, using the simple dilute case theory would certainly yield particle sizes that would decrease very dramatically with increasing volume fraction. In the case of this rutile sample the dilute case theory would yield a size of about 100 nm at the highest volume fraction, giving an error exceeding 200%.

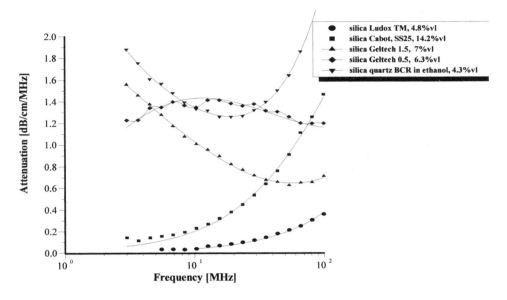

Figure 6.5 Experimental attenuation spectra compared with theoretical fit calculated for various silica suspensions.

In addition to attenuation, the viscous loss theory also yields an expression for sound speed, as the real part of a complex wavenumber (Equation. 2.24 and 4.37). The dilution experiment allows us to test this part of the theory as well. Figure 6.4 shows the experimental and theoretical sound speed for silica slurries at various volume fractions. The variation in sound speed is relatively small; only about 2% for a change in volume fraction from 1 to 31%. Nevertheless, this small variation is measurable, and is in good agreement with theoretical calculations.

Verifying the accuracy of the size measurement at high volume fraction requires standard materials with well-known particle size. Unfortunately, there are no accepted particle size standards for materials in the submicron range that have, in addition, a high density contrast. Traditional polymer latex standards do not exhibit the needed viscous loss because of their low density contrast. Accordingly, we show here data that we have collected over the last five years using ten commercially available slurries, namely: colloidal silica Ludox TM; silica Geltech 1.5 and 0.5; silica quartz BCR-70; silica Cabot SS25; rutile

Dupont R746; alumina Sumitomo AKP 15 and AKP30; zirconia TOSO TZ-3YS; and Silicon Nitride Ube E-10. The particle size of these materials, as determined by their respective manufacturers, is summarized in Table 6.1.and compared with the particle size determined using acoustic attenuation.

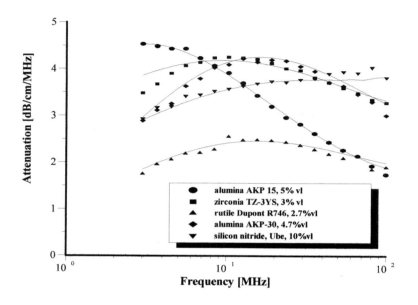

Figure 6.6 Experimental attenuation spectra and theoretical fit calculated for various solid materials.

Figure 6.5 and 6.6 show the attenuation spectra for all ten of these materials, at various volume fractions. All of the slurries are considered relatively stable. The solid lines represent the best theoretical fit to the experimental data that can be achieved with the new theory for viscous losses, using the particle size distribution as an adjustable parameter. The median particle size, corresponding to this best fit for each slurry, is given in Table 6.1. The median size measured using acoustic spectroscopy agrees very well with the independent data from the manufacturer's literature. Unfortunately, we do not know the measuring technique used by every manufacturer to determine the particle size of their material. Nevertheless, the general agreement for this wide variety of materials is compelling evidence for the accuracy of our viscous loss theory. The precision will be discussed later in Chapter 7.

Table 6.1 Median particle sizes reported by manufacturer and calculated from the acoustic attenuation spectra for various solid particle materials.

Sample Material	Median Size (manufacturer data), microns	Median Size (acoustics), microns
Silica Ludox TM	0.022 (a)	0.028
Silica Geltech 1.5	1.5	1.55
Silica Geltech 0.5	0.5	0.53
Silica quartz BCR 70	2.9 (b)	2.85
Silica Cabot SS25	0.1 (c)	0.087
Rutile Dupont R746	0.3	0.31
Alumina Sumitiomo AKP-15	0.7	0.67
Alumina Sumitomo AKP-30	0.3	0.33
Zirconia TOSO TZ-3YS	0.3	0.34
Silicon Nitride Ube E-10	0.5	0.47

(a) Using titration method, thus giving smaller area-weighted particle size.
(b) Using sedimentation method, thus giving a weight basis hydrodynamic size.
(c) Using Chromatographic HydroDynamic Fractionation (CHDF), thus emphasizing largest dimension for irregularly shaped particles.

6.2 Thermal losses

Allegra and Hawley [3] provided an early demonstration of the ability of Isacovich's theory to characterize moderately concentrated colloids. They observed almost perfect correlation between experiment and their dilute case ECAH theory for several systems: a 20%vl toluene emulsion, a 10%vl hexadecane emulsion, and a 10%vl polystyrene latex dispersion. This fortuitous result can be understood by noting that their ECAH dilute case theory reduces to Isacovich's theory for the special case of submicron soft particles. Studies by McClements and Povey [4-6], using emulsions, have provided similar results. Recent work by Holmes, Challis and Wedlock [7-9] also shows good agreement between ECAH theory and experiments on a 30%vl polystyrene latex suspension.

Dukhin and Goetz [10] have shown that particle-particle interactions can be neglected for small (160 nm) neoprene latex dispersions. In this case they found that the attenuation was a linear function of volume fraction up to 30%vl. (Figures 4.4 and 4.6). Such linear behavior is a strong indication that each particle fraction contributes to the total attenuation independently of other fractions, *i.e.* the total attenuation is a superposition of the individual contributions. Superposition works only when particle-particle interactions are insignificant.

Anson and Chivers have also shown [11] that thermal losses are important for soft particles up to *ka=0.5*. Above this limit, other mechanisms may dominate the attenuation. However, for those emulsion systems in which, of all the relevant physical properties, only the thermal ones differ substantially between the two phases, the thermal losses may be important for much larger values of ka. Their study involved 72 liquids. Water based emulsions are somewhat unique because of its unusual thermal properties.

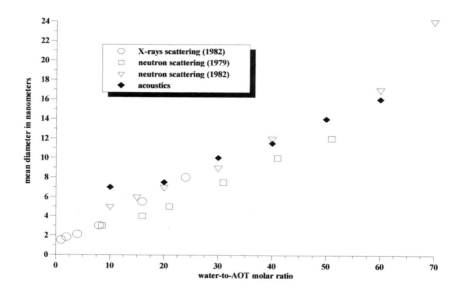

Figure 6.7 Comparison of the median particle size of a water in heptane microemulsion measured using several techniques.

Allison and Richardson [12] measured attenuation in several emulsions including: benzene in water, water in benzene, and benzene in glycerin-water mixtures. To interpret their experiments, they applied a combination of Urick's theory for the viscous losses, and the Rayleigh scattering theory. They completely neglected "thermal losses" and consequently observed large deviations between experiment and their theory.

One of the most convincing confirmations of Isacovich's theory was obtained by Wines et al [13] with a classical water-in-heptane microemulsion, stabilized with AOT surfactant. Independent information of the droplet size was obtained using both neutron scattering and X-ray scattering [14-19]. Figure 6.7 shows the data from these earlier experiments together with new data obtained from attenuation spectroscopy. The droplet size is seen to increase with increasing water content from just a few nanometers to about 20 nm, probably because of a deficit in the amount of surfactant (AOT) necessary to stabilize the additional surface. Notwithstanding this size increase, it is clear that the particle size calculated from the attenuation spectra using Isacovich's theory correlates well with the particle size from two totally different techniques.

Recently the focus of research in this area has shifted towards more concentrated systems where Isacovich's theory certainly fails. In particular are the experimental tests by McClements, Hemar *et al* [20-23] for verifying the "core-shell model" theory. These experiments were made using hexadecane-in-water emulsions and corn oil-in-water emulsions. The results from these studies show that deviation occurs (from Isacovich's dilute case theory) at volume concentrations exceeding approximately 30%. However, the new "core-shell model" provides a good fit with experimental data up to 50%vol.

In summary, we can conclude that multiple experiments performed by different independent groups have confirmed the validity both of Isacovich's theory, for dilute soft colloids, and the "core-shell model" theory for more concentrated soft colloids.

6.3 Structural losses

Here we use data from experiments performed at the National Institute for Resources and Environment, Tsukuba, Japan [46]. The authors used two alumina powders: Showa Denko AL-160SG-4 and Sumitomo Chemical Industry ALM-41-01. The median size of very dilute aqueous samples of each powder were measured using both laser diffraction and photo-centrifugation (Table 6.2).

Table 6.2 Measured particle size of two alumina samples, showing importance of structural losses.

Measurement Details	Median particle size, microns	
	ALM-41-01	AL-160SG-4
Photo-centrifugation, Horiba CAPA-700, very dilute sample	1.47	0.56
Laser diffraction, Sympatec Helos, very dilute sample	1.98	0.71
Acoustic spectroscopy, DT1200, < 20%vol	1.79	0.52
Acoustic spectroscopy, DT1200, 40%vol. Initial analysis ignoring structural losses	1.07 (fit error 19.2%)	0.8 (fit error 18.4%)
Acoustic spectroscopy, DT1200 40%vol. Subsequent analysis including structural losses	1.63 (fit error 6.1%)	0.77 (fit error 2.3%)

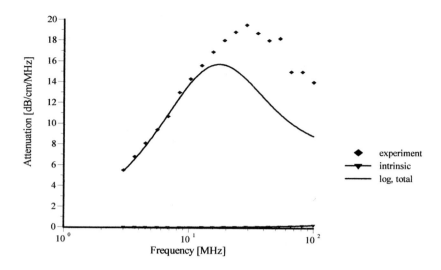

Figure 6.8 Experimental attenuation for alumina AL-160SG-4 and theoretical fit assuming only viscous losses.

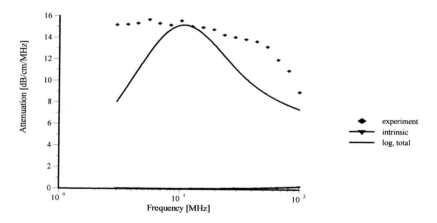

Figure 6.9 Experimental attenuation for alumina ALM-41-01, and theoretical fit assuming only viscous losses.

Slurries of both alumina powders, stabilized using a sodium polycalboxyl acid surfactant and ball milled for three days before measurement, were prepared at various concentrations from 1% to 40%vol. The acoustic attenuation spectrum was then measured for both slurries at various concentrations. For concentrations up to 20%vol, the particle size calculated from these attenuation spectra agreed quite well with independent data on dilute samples. However, for the highest volume fraction of 40%, the agreement was not as good.

The actual attenuation spectrum for both slurries at 40%vol is shown in Figure 6.8 and 6.9. In both cases it is clear that the experimental data exceeds the theory by a very substantial degree. The theory, however, is based only on viscous losses. Hence, based on this "excess attenuation", the authors concluded that there was an as yet unknown factor that becomes significant only at high volume fractions.

We suggest the possibility of "structural losses" to explain this excess attenuation. We have described such structural loss in some detail previously (Section 4.2.4). Specifically, equation 4.61 can be included in the analysis to compute any additional structural loss (to account for this excess attenuation). In doing this we assume that the first coefficient, H_1, is zero. The second coefficient, H_2, is then used as an adjustable parameter, in addition to the median size and standard deviation that are used to define the lognormal particle size distribution. In essence, the searching routine looks for that particle size distribution, and that value of H_2, that provide the best fit between theory and experiment.

The addition of this new adjustable parameter, H_2, allowed us to achieve a much better fit between theory and experiment for both alumina samples, as can be clearly seen in Figures 6.10 and 6.11. The resultant particle size and fitting errors, after allowance for these structural losses, are given in Table 6.2. Since the addition of structural losses leads to a dramatic improvement in the fitting error, this strongly suggests that such a mechanism does explain, at least in part, the observed excess attenuation.

The calculated particle size data confirms this conclusion, because the calculated values are then much closer to independent measurements on the dilute samples, and to acoustic measurements of the more dilute case.

It is interesting that the value of the adjustable parameter, coefficient H_2, turns out to be the same for both samples, i.e. 0.8. It is independent of the particle size, as one might expect, because it is a property of the polymer "springs". This parameter characterizes the flexibility of the bonds linking the particles together into a structure at high volume fractions. In the case of polymer bridging, it characterizes the rheology of the polymer chains. This rheological property might be very useful in optimizing the chemical formulation of concentrated systems, such as these ceramic slips, to obtain specific structural characteristics. Unfortunately, we do not know of any independent way to calculate or measure this parameter.

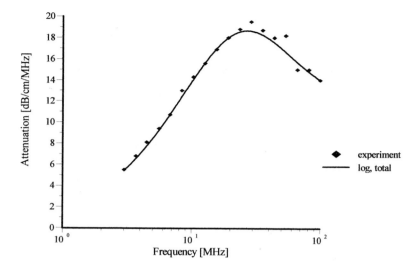

Figure 6.10 Experimental attenuation for alumina AL-160SG-4, and theoretical fit including both viscous and structural losses.

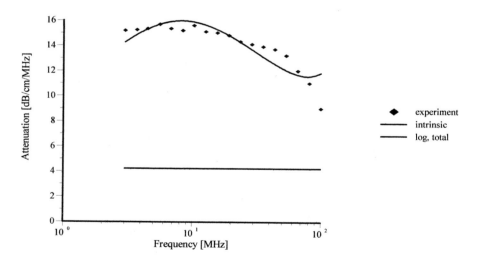

Figure 6.11 Experimental attenuation for alumina ALM-41-01, and theoretical fit including both viscous and structural losses.

The addition of these structural losses is only justified, however, when a simpler theory neglecting such losses fails, and the experiment shows such a characteristic excess attenuation. This excess attenuation then becomes a source of additional experimental information about the micro-rheological properties of concentrated systems.

6.4 Scattering losses

There appear to be only two experimental studies where the scattering losses are the dominant mechanism and all other losses can be neglected. The first was by Faran [24] who in 1951 measured the angular dependence of scattered ultrasound for large spheres and cylinders. The second was by Busby and Richrdson [25] who in 1956 measured the attenuation of large (95 micron) glass spheres. They observed a linear dependence of attenuation with respect to concentration, up to 13.6% vol.

There are many other studies in which the scattering was not dominant, but rather just one component of the total attenuation spectra. One example was shown previously in Figure 4.2, for the silica quartz BCR-70 particle size standard. The scattering losses for this sample dominate only the high frequency portion of the attenuation spectra, whereas at low frequency viscous dissipation is predominant. Ideally, to verify the scattering loss theory, the particle size must be sufficiently large to eliminate contributions by any absorption mechanism to the total attenuation spectrum.

We have made such a test using glass beads of different sizes produced by Sigmund Lindner GmbH. These particles are quite monodisperse and almost spherical. The median particle size varies from 25 microns to 200 microns (Table 6.3).

In addition, we tested alumina Sumitomo AA-18, having a particle size reported by the manufacturer of approximately 18 microns. These particles are also sufficiently large to eliminate ultrasound absorption over the frequency range of 3 MHz to 100 MHz.

Both the glass and alumina dispersions were prepared in distilled water; no stabilizer was used.

Attenuation spectra were measured using a Dispersion Technology DT-100 Acoustic Spectrometer. This instrument measures the attenuation of the transmitted sound pulse, as opposed to the sound scattered outside the incident beam path, which eliminates undesirable non-linear effects that result from multiple scattering [26]. To demonstrate the absence of such multiple scattering, we measured one of the glass samples at different concentrations, from 1 to 46 %vol. The measured attenuation is seen to depend linearly with concentration, within the statistical confidence range (Figure 6.12).

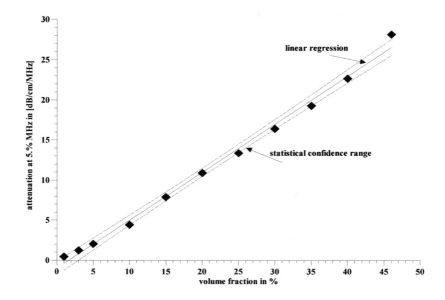

Figure 6.12 Attenuation of 120 micron glass spheres at 5 MHz as function of concentration.

Figures 6.13 and 6.14 show the attenuation spectra for various large particles dispersions. The concentration in each case was 4%vol. The samples were continuously pumped through the chamber using a peristaltic pump to prevent sedimentation.

As predicted by theory, the attenuation spectra are bell shaped. The position of the critical frequency shifts to lower frequency with increasing particle size, which is in agreement with the theoretical predictions of Morse [26].

According to Morse, there is a simple way to estimate the median particle size from these attenuation spectra. The critical frequency at which attenuation reaches a maximum can be defined with:

$$ka = \frac{\pi d}{\lambda} \approx 2 \qquad (6.1)$$

This simple equation allows us to calculate the median particle size as approximately 2/3 of the wavelength at the critical frequency. Table 6.3 lists the particle sizes calculated from the peak frequency from each curve in Figure 6.13. This calculated particle size is very close to the independent particle size data provided by manufacturer.

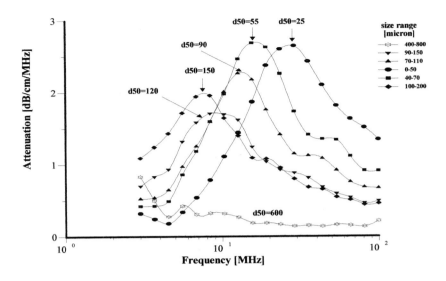

Figure 6.13 Attenuation spectra for glass spheres having different sizes.

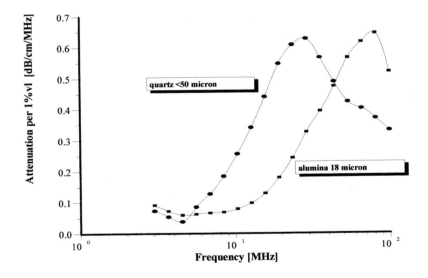

Figure 6.14 Attenuation spectra normalized by volume fraction for glass spheres and alumina.

Table 6.3. Particle size calculation using simple algorithm.

Sample ID	Size range, Manufacturer [μm]	Median size, Manufacturer [μm]	Median size, Acoustics [μm]	Critical Frequency, [MHz]	Wavelength, [μm]
S 5210	0-50	25	30	30	50
S 5211	40-70	55	54	18	83
S 5212	70-110	90	74	13	115
S 5213	90-150	120	107	9	166
S 5214	100-200	150	137	7	214
AKP-18		18	14	70	22

These results confirm the main conclusions of the Morse theory. There is, however, one unexplained feature of the experimental data; the variation of the amplitude of the attenuation spectra with particle size. It is not completely clear what causes this variation. It is not density variation; alumina and quartz with similar particle size range exhibit the same attenuation per 1%vl (Figure 6.14). There are two potential factors to explain this phenomenon: polydispersity and particle shape. The latter might be much more important for scattering losses compared to the various mechanisms of ultrasound absorption.

We conclude from this test that Morse's scattering theory allows the particle size range, which can be characterized using acoustics, to be expanded up to millimeters.

6.5. Electroacoustic phenomena

The first experimental confirmation of the electroacoustic phenomena in colloids was made by Hermans [27] in 1938. Then, in 1958 Rutgers and Rigole published data on the CVP of silver iodide sols [28]. At about the same time, Yeager's group made measurements of colloidal silica (Ludox) at various weight fractions [29]. They showed that the measured signal (CVP) increased with weight fraction, and they observed this increase for solutions having different ionic strengths. These first measurements were performed using an instrument employing a continuous wave technique. This creates complications when attempting to separate the desired electroacoustic response from other unwanted signals. Nevertheless, this first work did illustrate the existence of the electroacoustic effect in colloids, but did not yield sufficient precision for quantitative analysis.

The next step forward was made by the Yeager-Zana group when they switched to a pulse technique [30] in the early 1960's. Their new measurements with colloidal silica gave the first quantitative confirmation of the main conclusions of the Enderby-Booth theory.

Marlow, Fairhurst and Pendse [31] made CVP measurements on concentrated rutile dispersions and observed a very good correlation with independent microelectrophoretic measurements. Although they observed a nonlinear dependence in the CVP on concentration starting at about 5%vl, they managed to perform measurements up to 30%vl. These studies confirmed the volume fraction dependence predicted by the Levine cell model.

Marlow and Rowell [32] used CVP measurements to characterize the electric surface properties of coal slurries. They observed a deviation from linear dependence above 5%vol, but were able to perform measurements even at 40%vl. They tried to apply a "cell model" concept for describing the non-linearity of this volume fraction dependence.

This initial period of electroacoustic studies dealt only with CVP, using ultrasound as the driving force. At the end of the 1980's, the focus began to shift to the reverse electroacoustic effect, Electric Sonic Amplitude (ESA), using the electric field as the driving force. This new direction was initiated by Cannon, O'Brien, Hunter, and others from the international Australian-USA group associated with development of the first Matec instruments.

One of the first results, presented by Hunter [33] in 1988, concerned a quantitative comparison of mobility (of polymer latex particles), as estimated from the ESA effect (after correction for inertia using O'Brien's theory [34]) with microelectrophoresis results.

Shortly afterwards, Babchin, Chow, and Samatzky [35] compared electroacoustic dynamic mobility and electrophoretic mobility for a silver iodide sol, and showed very good correlation. They also measured the CVP in both bitumen-in-water emulsions and water-in-crude oil emulsions, and showed that it is possible to monitor the effect of a surfactant on emulsion stability.

The most vigorous test of O'Brien's theory for dilute systems has been carried out by O'Brien's own group, using monodisperse cobalt phosphate and titanium dioxide dispersions [36]. They observed very good correlation with microelectrophoresis data for different chemical compositions of the suspensions. They also observed a linear dependence of CVP for monodisperse latex of particle size 88 nm.

Later ESA investigations of dilute dispersions (below 5%vol) have been directed towards characterizing polydispersity, aggregation, and the DL model, for example the work by James et al [37] on alumina and silicon nitride. They confirmed that it is the mass-averaged particle size that should be used when estimating inertial effects in dilute colloids. Gunnison, Rasmussen, Wall, Dahlberg and Ennis [38] tested a Standard Stern model in electroacoustics using titration of dilute aqueous hematite dispersions. They obtained good agreement at high ionic strengths of 50 mM and 100 mM NaNO3. They found some deviation at lower ionic strengths, which they explained as being a consequence of aggregation of the hematite particles.

However, the most intriguing issue remained - the volume fraction dependence. Indeed, this was the strongest limitation on the early stages of electroacoustic theory. For instance, James, Texter and Scales [39] showed that the ESA remains a linear function of concentration only up to 5% vol. Their studies used polymer latexes, alumina (AKP), and colloidal silica (Ludox). Using an organic pigment, Texter [40] observed a linear dependence only up to 2%vol, and then a strongly nonlinear behavior. The overall conclusion formulated by Hunter is that the dilute case theory is valid only up to 2-5%vl. Beyond this concentration, particle-particle interactions result in non-linearity.

Marlow suggested using a cell model to incorporate particle-particle interactions, and derived a corrected volume fraction dependence [31]. Despite his initial success with rutile and coal dispersions [32], this path was not followed further until recently. O'Brien and Hunter even reported failure of the Levine cell model when describing ESA effects [41].

The development of the ESA theory for concentrates has proceeded along different lines, and there are experimental data confirming these new developments. For details we suggest that the interested reader go to the original publications [41, 42]. As an example, O'Brien, Hunter and others [43, 44] have tested ESA theory using dairy cream, intravenous fat emulsions and bitumen emulsions. Both particle size and zeta potential were reported at volume fractions up to 38%. The results suggest that electroacoustics can indeed be used to characterize concentrated emulsions.

Now, we would like to present an experimental verification of the new CVI theory that was described in Chapter 5. From our viewpoint this new theory has several advantages over the latest ESA theories. In particular, it does not require the use of a superposition principle to describe polydispersity. This and other advantages have already been described. The main goal of the studies described here is to test the validity of this new CVI theory in concentrated systems.

There are two aspects which are subjected to experimental verification. The first one is a comparison with independent microelectrophoretic results, and the second is a test of the volume fraction dependence.

We have successfully addressed the first aspect of this test many times. However, the most convincing results are from independent tests. As an example, we present results by Van Tassel and Randall [47], who studied the ζ-potential of both concentrated and dilute dispersions of alumina in ethanol. For the concentrated slurries they used the CVI measurements from the Dispersion Technology Model DT-1200. For the dilute samples they used the Coulter Delsa 440 microelectrophoretic measurements. They observed almost perfect correlation between the CVI based data and microelectrophoretic measurements. This confirms the new CVI theory since it is the basis of the Dispersion Technology DT-1200 software.

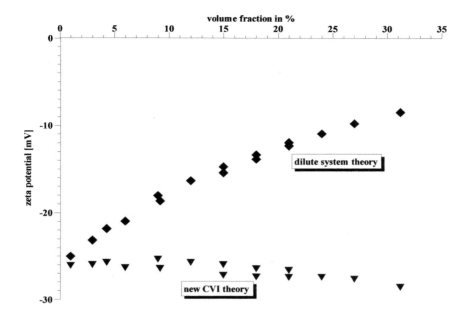

Figure 6.15 Comparison of the zeta potential of colloidal silica calculated using the dilute case theory, and the new CVI theory.

In regard to the volume fraction dependence test, the equilibrium dilution is again the logical experimental protocol, as with the case of viscous attenuation. Equilibrium dilution maintains the same chemical composition of the dispersion medium for all volume fractions, which means that the ζ-potential calculated from the CVI should remain constant for all volume fractions. Any variation of the ζ-potential with volume fraction is an indicator that a particular theory does not correctly reflect the particle-particle interactions that occur at high volume fraction.

Colloidal silica (Ludox TM-50) is ideal for these studies, because the small particle size allows us to eliminate the particle size dependence dictated by Equation 5.28, since the factor G is reduced to unity. Using small particles gives another simplifying advantage; it eliminates any contribution to the overall attenuation because small particles do not attenuate sound at low frequency. Thus the choice of small particles allows us to test only .the volume fraction dependence. This is critical because volume fraction dependence is the parameter that produces the most pronounced difference between the various theories. In addition, the small particle size would then reproduce a low frequency Smoluchowski limit, the important reference point in the electroacoustic theories.

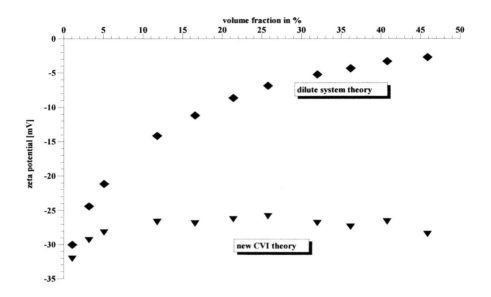

Figure 6.16 Comparison of the zeta potential of rutile using the dilute case theory and the new CVI theory.

With a nominal size of less than 30 nm, the colloidal silica satisfies all specified conditions of Equation 5.10 and 5.32. The measured size using acoustics agrees fairly well with the manufacturer's specifies value (Table 6.1). At the same time, the silica slurry also satisfies the thin double layer restriction (Equation. 5.32) because of its relatively high ionic strength, about 0.1 mol/l. Otherwise we would have to generalize the theory to remove this thin double layer restriction, following the Henry-Ohshima correction [48].

The selection of rutile as the second dispersion gives us an opportunity to test the particle size dependence, and enhance the density contrast contribution. An equilibrium dilution protocol requires a pure solvent that is identical to the medium of the given dispersed system. In principle, one can try to separate the dispersed phase from the dispersion medium using either sedimentation or centrifugation.

The dilution protocol used is the same as that described earlier (Section 6.1) for testing the theory for viscous losses; centrifugation for the rutile and dialysis for the colloidal; silica. We carried out our studies using a Dispersion Technology Acoustic and Electroacoustic Spectrometer DT-1200. Details concerning the hardware are described in Chapter 7.

Figures 6.15 and 6.16 summarize the results of the dilution experiment for the colloidal silica and rutile slurries respectively. It is seen that ζ-potential calculated from the measured CVI using the new theory yields a ζ-potential that remains almost constant over the volume fraction range studied. Variations do not exceed 10%. These experimental measurements strongly support the new CVI theory for concentrated colloids.

Figures 6.15 and 6.16 also show results using the dilute case theory. For the rutile the error exceeds 1000%. The volume fraction dependence is, however, less pronounced for the less dense silica particles, in agreement with the results obtained by Rasmussen [45] for latex particles using ESA. This confirms one of the main positions of the new CVI theory; the magnitude of the volume fraction dependence is significantly affected by the choice of inertial frame of reference. For the CVI this is reflected in the multiplier, $(\rho_p - \rho_s)/\rho_s$, and this is less volume fraction dependent for the lighter particles.

REFERENCES.

1. Dukhin, A.S. and Goetz, P.J. "Acoustic Spectroscopy for Concentrated Polydisperse Colloids with High Density Contrast", Langmuir, 12, 21, 4987-4997 (1996)

2. Dukhin, A.S., Shilov, V.N, Ohshima, H., Goetz, P.J "Electroacoustics Phenomena in Concentrated Dispersions. New Theory and CVI Experiment", Langmuir, 15, 20, 6692-6706, (1999)

3. Allegra, J.R. and Hawley, S.A. "Attenuation of Sound in Suspensions and Emulsions: Theory and Experiments", J. Acoust. Soc. Amer., 51, 1545-1564 (1972)

4. McClements, J.D. and Povey, M.J. "Ultrasonic velocity as a probe of emulsions and suspensions", Adv. Colloid Interface Sci., 27, 285-316 (1987)

5. McClements, D.J. "Ultrasonic Characterization of Emulsions and Suspensions", Adv. Colloid Int. .Sci., 37, 33-72 (1991)

6. McClements, D.J. and Povey, M.J.W. "Scattering of Ultrasound by Emulsions", J. Phys. D: Appl. Phys., 22, 38-47 (1989)

7. Holmes, A.K., Challis, R.E. and Wedlock, D.J. "A Wide-Bandwidth Study of Ultrasound Velocity and Attenuation in Suspensions: Comparison of Theory with Experimental Measurements", J. Colloid and Interface Sci., 156, 261-269 (1993)

8. Holmes, A.K., Challis, R.E. and Wedlock, D.J. "A Wide-Bandwidth Ultrasonic Study of Suspensions: The Variation of Velocity and Attenuation with Particle Size", J. Colloid and Interface Sci., 168, 339-348 (1994)

9. Holmes, A.K., and Challis, R.E "Ultrasonic scattering in concentrated colloidal suspensions", Colloids and Surfaces A, 77, 65-74 (1993)

10. Dukhin, A.S., Goetz., P.J. and Hamlet, C.W. "Acoustic Spectroscopy for Concentrated Polydisperse Colloids with Low Density Contrast", Langmuir, 12, 21, 4998-5004, (1996).

11. Anson, L.W. and Chivers, R.C. "Thermal effects in the attenuation of ultrasound in dilute suspensions for low values of acoustic radius", Ultrasonic, 28, 16-25 (1990)

12. Allison, P.A. and Richardson, E.G. "The Propagation of Ultrasonics in Suspensions of Liquid Globules in Another Liquid", Proc. Phys. Soc., 72, 833-840 (1958)

13. Wines, T.H., Dukhin A.S. and Somasundaran, P. "Acoustic spectroscopy for characterizing heptane/water/AOT reverse microemulsion", JCIS, 216, 303-308 (1999)

14. Fletcher, P.D.I., Robinson, B.H., Bermejo-Barrera, F., Oakenfull, D.G., Dore, J.C., and Steytler, D.C. *in* "Microemulsions" (I.D. Rob, Ed.), p. 221, Plenum Press, New York, (1982)

15. Cabos, P.C., and Delord, *P., J. App. Cryst.* 12, 502 (1979).

16. Radiman, S., Fountain, L.E., Toprakcioglu, C., de Vallera, A., and Chieux, P., *Progr. Coll. Polym. Sci.* 81, 54 (1990).

17. Huruguen, J.P., Zemb, T., and Pileni, M.P., Progr. Coll. Polym. Sci. 89, 39 (1992).

18. Pileni, M. P., Zemb, T. and Petit, C., *Chem. Phys. Lett*. 118, 4, 414 (1985).

19. Bedwell B., and Gulari, E., *in* "Solution Behavior of Surfactants" (K.L. Mittal, Ed.), Vol. 2., Plenum Press, New York, (1982)

20. Hemar, Y, Herrmann, N., Lemarechal, P., Hocquart, R. and Lequeux, F. "Effect medium model for ultrasonic attenuation due to the thermo-elastic effect in concentrated emulsions", J. Phys. II 7, 637-647 (1997)

21. McClements, J.D., Hermar, Y. and Herrmann, N. "Incorporation of thermal overlap effects into multiple scattering theory", J. Acous. Soc. Am., 105, 2, 915-918 (1999)

22. Chanamai, R., Coupland, J.N. and McClements, D.J. "Effect of Temperature on the Ultrasonic Properties of Oil-in-Water Emulsions", Colloids and Surfaces, 139, 241-250 (1998)

23. Chanamai, R., Hermann, N. and McClements, D.J. "Influence of thermal overlap effects on the ultrasonic attenuation spectra of polydisperse oil-in-water emulsions", Langmuir, 15, 3418-3423 (1999)

24. Faran, J.J. "Sound Scattering by Solid Cylinders and Spheres", J. Acoust. Soc. Amer., 23, 4, 405-418 (1951)

25. Busby, J. and Richrdson, E.G. "The Propagation of Ultrasonics in Suspensions of Particles in Liquid", Phys. Soc. Proc., v.67B, 193-202 (1956).

26. Morse, P.M. and Uno Ingard, K., "Theoretical Acoustics", 1968 McGraw-Hill, NY, 1968, Princeton University Press, NJ, 1986, 925 p.

27. Hermans, J., Philos. Mag., 25, 426 (1938)

28. Rutgers, A.J. and Rigole, W. "Ultrasonic vibration potentials in colloid solutions, in solutions of electrolytes and pure liquids", Trans. Faraday Soc., 54, 139-143 (1958)

29. Yeager, E., Dietrick, H., and Hovorka, F., J. "Ultrasonic waves and electrochemistry. II. Colloidal and ionic vibration potentials", J. Acoust. Soc. Am., 25, 456 (1953)

30. Zana, R., and Yeager, E., "Ultrasonic vibration potentials and their use in the determination of ionic partial molal volumes", J. Phys. Chem, 71, 13, 521-535 (1967)

31. Marlow, B.J., Fairhurst, D. and Pendse, H.P., "Colloid Vibration Potential and the Electrokinetic Characterization of Concentrated Colloids", Langmuir, 4,3, 611-626 (1983)

32. Marlow, B.J. and Rowell, R.L. "Acoustic and Electrophoretic mobilities of Coal Dispersions", J. of Energy and Fuel, 2, 125-131 (1988)

33. Hunter, R.J., Liversidge lecture, J. Proc. R. Soc. New South Wales, 121, 165-178 (1988)

34. O'Brien, R.W. "Electro-acoustic Effects in a dilute Suspension of Spherical Particles", J. Fluid Mech., 190, 71-86 (1988)

35. Babchin, A.J., Chow, R.S. and Sawatzky, R.P. "Electrokinetic measurements by electroacoustic methods", Adv. Colloid Interface Sci., 30, 111 (1989)

36. O'Brien, R.W., Midmore, B.R., Lamb, A. and Hunter, R.J. "Electroacoustic studies of Moderately concentrated colloidal suspensions", Faraday Discuss. Chem. Soc., 90, 1-11 (1990)

37. James, M., Hunter, R.J. and O'Brien, R.W. "Effect of particle size distribution and aggregation on electroacoustic measurements of Zeta potential", Langmuir, 8, 420-423 (1992)

38. Gunnarsson, M., Rasmusson, M., Wall, S., Ahlberg, E., Ennis, J. "Electroacoustic and potentiometric studies of the hematite/water interface", JCIS, 240, 448-458 (2001)

39. James, R.O., Texter, J. and Scales, P.J. "Frequency dependence of Electroacoustic (Electrophoretic) Mobilities", Langmuir, 7, 1993-1997 (1991)

40. Texter, J., "Electroacoustic characterization of electrokinetics in concentrated pigment dispersions", Langmuir, 8, 291-305 (1992)

41. O'Brien R.W., Rowlands W.N. and Hunter R.J. "Determining charge and size with the Acoustosizer", in S.B. Malghan (Ed.) Electroacoustics for Characterization of Particulates and Suspensions, NIST, 1-21 (1993)

42. Hunter, R.J. "Review. Recent developments in the electroacoustic characterization of colloidal suspensions and emulsions", Colloids and Surfaces, 141, 37-65 (1998)

43. O'Brien, R.W., Wade, T.A., Carasso, M.L., Hunter, R.J., Rowlands, W.N. and Beattie, J.K. "Electroacoustic determination of droplet size and zeta potential in concentrated emulsions", JCIS, (1996)

44. Kong, L., Beattie, J.K. and Hunter, R.J. "Electroacoustic study of concentrated oil-in-water emulsions", JCIS, 238, 70-79 (2001)

45. Rasmusson, M. "Volume fraction effects in Electroacoustic Measurements", JCIS, 240, 432-447 (2001)

46. Hayashi, T., Ohya, H., Suzuki, S. and Endoh, S. "errors in Size Distribution Measurement of Concentrated Alumina Slurry by Ultrasonic Attenuation Spectroscopy", J. Soc. Powder Technology Japan, 37, 498-504 (2000)

47. Van Tassel, J. and Randall, C.A. "Surface chemistry and surface charge formation for an alumina powder in ethanol with the addition HCl and KOH", JCIS 241, 302-316 (2001)

48. Ohshima, H. "A simple expression for Henry's function for retardation effect in electrophoresis of spherical colloidal particles", JCIS, 168, 269-271 (1994)

Chapter 7. ACOUSTIC AND ELECTROACOUSTIC MEASUREMENT TECHNIQUES

7.1 Historical Perspective

Ultrasound is a versatile tool for characterizing materials. The most widely used application is the nondestructive characterization (NDC) of defects in solid bodies; a review is given by Bhardwaj [1]. Ultrasound can also yield vital information on texture, microstructure, density, porosity, elastic constants, mechanical properties, corrosion, and residual stress. However, we do not cover these many applications concerning solids. Instead, we will describe the use of acoustic and electroacoustic measurements to characterize only liquid based systems.

Over the years, many useful fluid properties have been computed from acoustic measurements. For example, Hamstead [1] measured sound speed using the time of flight method to determine the density of fluid in a pipe. Dann and Halse [3] used acoustic impedance measurements to determine, not only the density of slurries, but also rheological properties. Kikuta [4] used attenuation measurements to characterize the composition of electro-deposition coating materials containing resin and pigment. Sowerby [5] used a combination of gamma-radiation and acoustic measurements to determine the volume fraction of solid particles in suspensions. Although a wide range of techniques have been used to measure fluids, many of these methods have not worked well for studying colloidal dispersions.

For example, one technique is based on the interaction between light and ultrasound. Variations in the density of the liquid, induced by the pressure variation within the sound wave, cause a periodic pattern in the refracted light wave. Although this interaction allows one to measure the sound speed of pure liquids with a precision of 0.01% [6], the method is not suitable for colloids because of the additional scattering of the sound waves caused by the colloidal particles.

Another technique is related to the use of "surface acoustic waves" (SAW). Although Ricco and Martin [10] describe construction of a simple acoustic viscosity-meter, and Kostial [11] used this to monitor the concentration of ions in water, the method has not been employed to characterize colloidal dispersions.

Finally, Apfel [12] suggested an exotic droplet levitation technique for the measurement of sound speed, density and compressibility.

None of the above methods have been employed successfully to characterize colloidal dispersions. However, there are a number of other techniques that have indeed been found useful for characterizing, not only pure liquids, but colloidal systems as well. For attenuation and sound speed, these techniques can be divided into two basic categories: "interferometric" and "transmission". The interferometric technique uses standing waves, whereas transmission utilizes transverse waves. This classification was initially suggested by Sette [6] and later adopted by McClements [7-9].

7.2 Difference between measurement and analysis

At this point it is important to stress the difference between the measured data and the desired output parameters. For acoustics, the measured data can include: the attenuation coefficient, the sound speed and the acoustic impedance. For electroacoustics, the measured data might include the Colloid Vibration Current (CVI), Colloid Vibration Potential (CVP) or ElectroSonic Amplitude (ESA). However, the investigator is typically not very interested in these measured properties; more commonly the particle size distribution, ζ potential, or rheological properties of the colloidal dispersion is of prime importance.

Obtaining these desired output parameters involves two steps. The first is to perform experiments on the disperse system to obtain a set of measured values for macroscopic properties such as temperature, pH, acoustic attenuation, Colloid Vibration Current, etc. The second step is to then analyze this measured data, and compute the desired microscopic properties such as particle size or ζ potential. Such an analysis requires three tools: a model dispersion, a prediction theory, and an analysis engine.

A "model dispersion" is an attempt to describe a real dispersion in terms of a set of model parameters including, of course, the desired microscopic characteristics (Chapter 2). The model, in effect, makes a set of assumptions about the real world in order to simplify the complexity of the dispersion and thereby also simplify the task of developing a suitable prediction theory. For example, most particle size measuring instruments make the assumption that the particles are spherical, which allows a complete geometrical description of the particle to be given by a single parameter, its diameter. Obviously such a model would not adequately describe a dispersion of high aspect ratio carpet fibers, and any theory based on this over-simplified model might well give incorrect results. The model dispersion may also attempt to limit the complexity of a particle size distribution by assuming that it can be described by certain conventional distribution functions, for example a lognormal distribution.

A "prediction theory" consists of a set of equations that describes some of the measurable macroscopic properties in terms of these microscopic properties of the model dispersion. For example, a prediction theory for Electroacoustics would attempt to describe a macroscopic property such as Colloid Vibration

Current in terms of microscopic properties such as the particle size distribution, zeta potential, viscosity, and dielectric permittivity.

An "analysis engine" is essentially a set of algorithms, implemented in a computer program, which calculate the desired microscopic properties from the measured macroscopic data, using the knowledge contained in the prediction theory. The analysis can be thought of as the opposite, or inverse, of prediction. Prediction describes some of the measurable macroscopic properties in terms of the model dispersion. Analysis, on the other hand, given perhaps only few values for the model parameters, attempts to calculate the remaining microscopic properties by an evaluation of the measured data. There are many well-documented approaches to this analysis task.

In the succeeding sections we shall first speak of details concerning the measuring technique for obtaining the required set of macroscopic properties, and then discuss the specific analysis methods that can be used to extract the desired microscopic properties.

7.3 Measurement of attenuation and sound speed using Interferometry.

We will consider interferometry just briefly before turning our full attention to the transmission method. Interferometry requires generating standing waves inside the sample chamber. This can be achieved by placing an acoustic reflector in front of the ultrasound transducer at some distance across the chamber. The incident wave generated by the transducer interferes with the reflected wave. This interference creates standing wave with a structure dependent on both the wavelength and the distance between transducer and reflector. In this setup the transmitting transducer works as the receiver as well. It monitors the intensity of the ultrasound at its surface. It is possible to change this amplitude by varying either wavelength (swept-frequency interferometry, Sihna [30]) or distance (swept-distance interferometry, McClements [7-9]).

The measured intensity of the ultrasound goes through a series of minima and maxima when either the wavelength or distance changes. The distance between successive maxima is equal to half of the ultrasound wavelength in the colloid. This allows a simple calculation of the sound speed using the following equation:

$$c_s = f\lambda_s \tag{7.1}$$

where c_s is the sound speed, f is frequency, and λ_s is the measured wavelength in the colloid.

This is a very accurate and precise method for sound speed measurement.

Attenuation measurement using interferometry is also possible. The amplitude of the maxima depends on the distance between transducer and receiver. The amplitude decays with increasing distance. It is possible to measure this maxima amplitude-distance dependence and to thereby derive an attenuation coefficient. However, Sette [6] suggested that the transmission technique is better suited for attenuation measurements because of its ability to handle a much larger dynamic range.

7.4. Measurement of attenuation and sound speed using the transmission technique

7.4.1 Historical development of the transmission technique

The first transmission measurements of ultrasound attenuation had nothing to do with colloids or particle sizing. The goal was only to measure the ultrasound attenuation of pure liquids. The general principles of this technique were initially formulated in the early 1940's by Pellam and Galt [13], and independently by Pinkerton [14]. Further developments continued through the middle 1960's. These early investigators were perhaps the first to apply two basic principles: the use of a pulse technique and the use of a sensor with a variable acoustic path length (variable gap). These early instruments worked only at a single frequency, using one transducer in a pulse-echo mode. However, they both managed to measure sound velocity with an accuracy of 0.05%, and sound attenuation with an accuracy of 5%.

In each case, the velocity measurement was made by determining the distance the transducer must be moved to delay the received echo by a specified increment [6-9, 22, 23, 27-29]. Similarly, the attenuation was measured by determining the attenuation that needed to be added or subtracted in order to keep the received signal constant as the transducer was moved. The velocity was obtained directly from the slope of the measured pulse delay versus the distance traveled. The attenuation was obtained from the slope of the compensating attenuation versus distance. In both cases there was no need to know the exact distance traveled by the sound because the measured parameters depend only on the difference in the acoustic path.

This approach was significantly improved by Andrea *et al* from 1958 to 1962 [15-17]. They expanded the Pellam-Galt approach to include measurements at multiple frequencies. Instead of using a single transducer, in a pulse-echo mode, they used separate transducers for transmitting and receiving. Like Galt, they also used a variable gap and measured the change in the pulse amplitude due to variation of the gap. Again, the final attenuation and sound speed are computed from the slope of the measured attenuation or time delay versus gap. These researchers made four key observations concerning this measurement technique that are still important today:

1. At any one frequency, there is a wide variation in the attenuation coefficient of different liquids, and this variation is even more pronounced for colloids. Different samples require different path lengths.

2. The attenuation is strongly frequency dependent. For pure liquids the attenuation (in db/cm) typically increases with the square of frequency. Again this makes different path lengths necessary for different frequencies.

3. At low frequencies diffraction effects caused by the transmitting transducer having a finite radius must be taken into account. The sound field can approximately be divided into the Fresnel region, near the transducer, and the more distant Fraunhofer region; the boundary between two is called the Rayleigh distance (Chapter 2 and 4). In the Fresnel region, the total ultrasonic energy is still confined to a narrow beam, but the pressure amplitude fluctuates rapidly across the beam and along its axis. If, however, the receiving transducer is made somewhat larger than the transmitting transducer, and thus includes the whole cross section of the beam, a constant signal is obtained throughout the Fresnel region (in the absence of attenuation). This is not exactly true for transmitting and receiving transducers of equal radius, but even then the diffraction effects are, in practice, usually negligible at frequencies above 5 MHz and with ultrasonic path lengths not exceeding the Rayleigh distance of $a_T^2 / 2\lambda$, where a_T is transducer radius. On the other hand, in the Fraunhofer region, a correction for beam divergence is required.

4. When the time taken for the ultrasonic pulse to traverse the liquid path is less than the pulse duration an additional time delay must be provided in the ultrasonic system. The necessary time delay may be provided conveniently by introducing a length of fused quartz rod into the ultrasonic path. Fused quartz possesses a low attenuation and gives a delay of about 1.7 μs/cm.

There are eight US patents dedicated to transmission attenuation measurements for the purpose of obtaining raw data from which particle size might be derived: Gushman *et al*, 1973 [18], Uusitalo *et al*, 1983 [19], Riebel 1987 [20], Alba 1992 [21], Dukhin and Goetz, 2000 [22,23,24,25]. However, from the standpoint of the actual measuring technique, these patents offer only a few improvements over the basic ideas suggested originally by Andrea.

The Uusitalo *et al* patent [19] suggests using the measurement of the incident beam attenuation in combination with a measurement of the sound scattered off-axis. The scattered part of the measurement provides complementary information only for large particles (> 3 microns) because smaller particles do not scatter sufficient ultrasound. This might be useful for monitoring aggregation, the presence of a few large particles on a background of a large number of small particles. Erwin and Dohner [26] similarly proposed monitoring colloid aggregation by using a single transducer equipped with a spherical quartz lens to monitor agglomerates passing the focal point. To our knowledge, this interesting idea has not been implemented in any commercial instrument.

Neither Riebel [20] nor Alba [21] brings new ideas to the acoustic measurement *per se*, but rather address methods of calculating the desired results. Riebel assumes that the particle size range is known in advance and selects a correspondingly appropriate frequency range. This makes it difficult to use as a general research tool but offers utility for dedicated online applications. Alba repeats some ideas from Andrea's work and then concentrates on the conversion of this attenuation data into particle size information using ECAH theory (see Chapter 4).

Perhaps the most significant modification to the original transmission method is called the "wide-bandwidth" method. It is closely associated with the names of Holmes, Challis and others from the UK [27, 28]. Whereas Andrea used an ultrasonic tone-burst containing a finite number of ultrasound cycles at a selected set of frequencies, the wide-bandwidth method instead applies a single sharp video pulse to the transmitting transducer. Such a sharp pulse contains a wide range of frequency components. After the pulse has traveled through a known path length of the sample, the frequency components of the received signal are extracted using a Fourier transformation of the received pulse.

Like all techniques, this wide-bandwidth technique has some advantages and disadvantages. The hardware to generate the video pulse is somewhat simpler than in the tone-burst method, but the computer computations are more difficult because of the necessity to maintain an extreme degree of precision. It does have the advantage that the entire spectrum is obtained during the same time interval which may be important in colloidal systems that are very unstable and varying with time. Among commercial instruments, Colloidal Dynamics and Dispersion Technology both use a tone-burst technique. There is no commercially available wide bandwidth instrument.

7.4.2 Detailed Description of the Dispersion Technology DT100 Acoustic Spectrometer

The Dispersion Technology DT100 is a modern example of an Acoustic Spectrometer that, in its basic design, follows the transmission tone-burst variable gap technique pioneered by Andrea [15-17]. A photograph of a typical system is shown in Figure 7.1.

The unit consists of two parts, the measuring unit and the associated electronics. The description here follows various patent disclosures by Dukhin and Goetz [22-25].

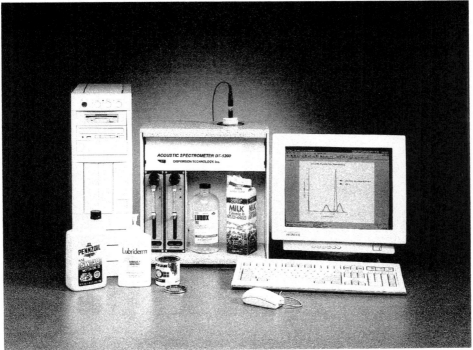

Figure 7.1 The Dispersion Technology DT100 Acoustic Spectrometer.

Acoustic sensor

The acoustic sensor is the portion of the measuring unit that contains the sample being measured and positions the piezoelectric transducers appropriately for the attenuation measurement (Figure 7.2). The sensor utilizes two identical piezoelectric transducers, separated by an adjustable gap that is controlled by a stepping motor. Each transducer incorporates a quartz delay rod that provides a delay of 2.5 usec. The transmitting transducer (on the right side of photo) is fixed, but has an adjustment accurately set and maintain the parallelism between the two transducers. The receiving transducer (on the left side) is mounted in a piston assembly whose position can be adjusted. The gap between the face of the transmitting and receiving transducers can be set from close to zero up to 20 mm. The moving piston is sealed against the sample by a wiper seal and supplemental O rings. The space between the seals is filled with fluid to keep the withdrawn portion of the piston wetted at all times. This prevents buildup of any dried deposit. Ports associated with this space between the seals can be used to periodically inspect the intra-seal fluid for any contamination. The presence of any contamination serves as advance warning of possible seal degradation when making continuous measurements of very aggressive samples such as concentrated Portland cement. The position of the receiving transducer piston is controlled by a stepping motor linear actuator that is, in turn, controlled by the electronics and software, with a precision of a few microns.

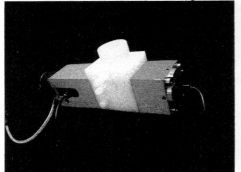

Figure 7.2 (a) The Acoustic sensor.   Figure 7.2 (b) Top view of acoustic sensor.

The sample is contained in the space between the transducers. Depending on the installation, the sample might be stirred with a magnetic stir bar, pumped through the sensor with a peristaltic pump, or in some cases not stirred at all. When the piston is fully withdrawn, the transducers are essentially flush with the walls of the sample compartment to facilitate cleaning. During operation, tonebursts at selected frequencies are applied to the transmit transducer. The amplitude of the received pulses, received with some delay, is then measured.

Electronics

A block diagram of the Electronics Unit is shown in Figure 7.3. A frequency synthesizer **30** generates a continuous wave (CW) RF signal **31**. A field programmable gate array 32 (FPGA) controlled by the host computer via digital interface **60** generates transmit gate **33**, attenuator control **34**, receive gate **35**, switch control **36**, and A/D strobe command **37**. The transmit gate **33** is used by gated amplifier **38** to form a pulsed RF signal from the synthesizer CW signal **31**. A power amplifier **39** increases the peak power of this RF pulse to approximately 1 watt. A directional coupler **40** provides a low-loss path for the forward signal from the power amplifier to the transmit switch **41** which then routes this 1 watt RF pulse **61** to any of three channels, labeled A, B, and C according to the FPGA switch control **36**.

Channel A is called the reference channel and in this case the output pulse is routed to a precision fixed 40 dB reference attenuator **42**. The output of this fixed attenuator is connected to channel A of the Receive Switch **43**, where it is then routed to the RF input signal **44**. The RF input signal is connected to a wide band RF amplifier **45** having a programmable gain set by FPGA attenuator control **34**, and thence to a gated amplifier **46** controlled by the FPGA receiver gate **35** which allows only pulses arriving within a predetermined time interval to pass, and thence to a quadrature mixer **47** keyed by the synthesizer output **31** which demodulates the received signal providing analog pulses **48** and **49** proportional to the in-phase and quadrature components of the received RF pulse respectively, thence to separate matched filters **50, 51** for the in-phase and quadrature components, thence to a dual a/d converter **52** which digitizes the peak amplitude of these quadrature signals upon an A/D strobe command **37**, thence to a digital signal processor (DSP) **53** chip that accumulates data from a large number of such pulses and generates statistical data on a given pulse set, and finally this statistical data is routed to the computer bus **54** so that the data can be analyzed to determine the amplitude and phase of the input signal level **44** for this reference condition. This reference signal level is computed in the following way. The statistical data from the two-channel A/D provides an accurate measure of the in-phase and quadrature components of the received signal at the A/D input. The magnitude is computed by calculating the square root of the sum of the squares of these two signals. The phase is computed from the inverse tangent of the ratio of the in-phase and quadrature components. The magnitude of the RF input signal **44** is then computed by taking into account the overall gain of the RF signal processor as modified by the gain control signal **34**.

The use of this reference channel with a known attenuation allows the attenuation to be measured precisely when using the remaining channels B and C. Since we know the input signal level for the 40 dB reference attenuator, any increase or decrease in this signal when using either of the other channels will be proportional to any increase or decrease with respect to this 40 dB reference value.

When the Transmit Switch and Receive Switch are set to Position C, the pulsed RF signal is routed to the Transmit transducer in the Acoustic Sensor where it is converted to an acoustic pulse and launched down the quartz delay rod. A portion of the pulse is transmitted into the colloidal dispersion, traverses this sample, is partially transmitted into the quartz delay rod of the receiving transducer, and after passing through this delay rod is converted back into an electrical pulse. The received pulse is then routed to port C of the Receive Switch for processing by the receiver.

Measurements can be performed for just one frequency, or for a chosen set of frequencies from 1 to 100 MHz. Typically the system averages at 800 pulses, but may accumulate as many as several million pulses if necessary to achieve a satisfactory signal to noise ratio. The number of pulses measured will depend on the properties of the colloid. For example, concentrated slurries at high frequency and large gaps may require a large number of pulses.

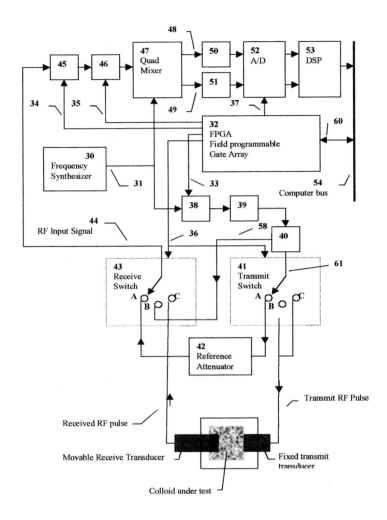

Figure 7.3 Block diagram of electronics.

Measurement procedure

Since the attenuation is determined by measuring the **relative** change in signal versus relative change in gap, it is not necessary to know precisely the

absolute value of the gap. The precision with which the relative gap can be measured is determined by the accuracy of the lead screw used in the linear displacement stepping motor. Nevertheless, the absolute position is calibrated to an accuracy of just a few microns the first time a measurement is made after starting the program. The calibration is performed by simply moving the receiving transducer towards the transmitting transducer for a sufficient time that it is certain that the two have contacted one another and the motor has stalled in that position. The gap position is then initialized and additional measurements are then relative to this known initial position.

Measurements are made using a defined grid consisting of a certain number of frequencies and gaps. In default, the system selects 18 frequencies between 1 and 100 MHz in logarithmic steps, and 21 gaps between 0.15 to 20 mm, also in logarithmic steps. These grid values may be modified automatically by the program, depending on *a priori* knowledge of the sample, or manually by an experienced user.

The Signal Processor generates a transmit gate which defines the one Watt pulse as well as the necessary signals to set the frequency for each point on the grid. At the beginning of each measurement the signal is routed through the reference channel attenuator as previously described to calibrate the attenuation measurement at each frequency.

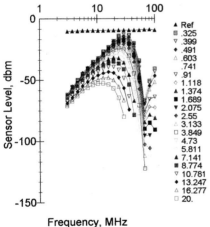
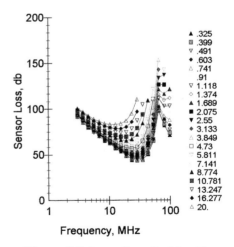

Figure 7.4 Sensor Level for 10 wt%. silica slurry.

Figure 7.5 Sensor Loss for 10 wt% silica slurry.

The next step in the measurement is to determine the losses in the acoustic sensor for each point on the measurement grid. The Signal Processor now substitutes the acoustic sensor for the reference channel attenuator. The one watt pulses are now sent to the transmitting transducer which converts these electric pulses into sound pulses. Each sound pulse propagates through the transducer's quartz delay rod, passes through the gap between the transducers (which is filled with the dispersion under test), enters an identical quartz rod associated with the receiving transducer, is converted back to an electrical signal in the receiving transducer and finally is routed through the interface to the input signal port on the Signal Processor, where the signal level of the acoustic sensor output is measured. For each point on the grid, the Signal Processor collects data for a minimum of 800 pulses. The number of pulses that are collected will automatically increase, as necessary, to obtain a target signal to noise ratio. Typically the target is set for a signal to noise ratio of 40 dB, which means that the received signal power will be 10,000 times that of the noise. This is usually more than sufficient to insure an acoustic spectrum of very high quality. However, an experienced user can tailor the measurement parameters to provide trade-offs between measuring speed and precision. A typical plot of the Signal Level for both the reference and acoustic sensor channel for all points on the grid for a 10 %wt dispersion of silica (Minusil 15) is shown in Figure 7.4.

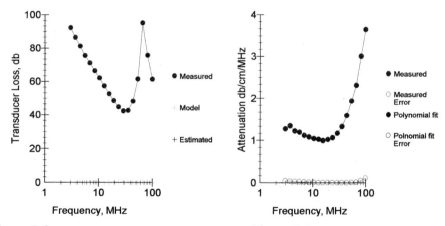

Figure 7.6 Transducer Loss computed while measuring a 10 wt% silica slurry.

Figure 7.7 Attenuation of 10 wt% silica slurry.

Comparison of the amplitude and phase of the acoustic sensor output pulse with that of the reference channel output pulse allows the software to calculate precisely the overall loss in the acoustic sensor at each frequency and gap. The overall Sensor Loss for this same sample is shown in Figure 7.5.

The overall Sensor Loss consists of the sum of two components: the "Colloid Loss" in passing through the gap between the transmitting and receiving transducers, and a "Transducer Loss". This Transducer Loss includes the loss in converting the electrical pulses to acoustic energy in the transmitting transducer, the reflection losses at the two interfaces between the transducers and the sample, and the loss in converting the acoustic energy received by the receiving transducer back to an electrical signal. Regression of the Sensor Loss with the gap yields two useful results. Since the loss in the sample, by definition, is zero for a gap of zero, the intercept of this regression analysis provides a measure of all other losses in the acoustic sensor, what we have defined as the Transducer Loss (Figure 7.6). Moreover, the slope of this regression analysis yields the attenuation per unit distance of the dispersion, expressed in dB/cm. Since, in general, the attenuation for most materials increases strongly with frequency, it is conventional to divide the above attenuation by the frequency in MHz so that the final output is expressed in dB/cm/MHz (Figure 7.7). Subtracting the Transducer Loss from the Sensor Loss now enables us to plot the Colloid Loss (Figure 7.8).

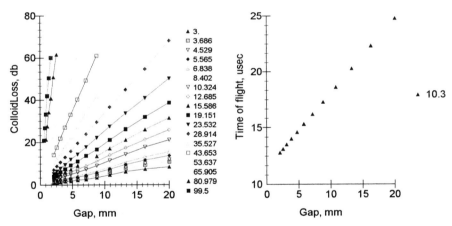

Figure 7.8 Colloid Loss for 10 wt% silica slurry.   Figure 7.9 Time of flight data for 10 wt% silica slurry.

Although the transducer loss is itself not necessary for the calculation of the attenuation, it does provide a good integrity check on the system performance and allows any degradation in the sensors to be detected long before any loss in system performance would be observable by the user. Inspection of Figure 7.6 shows a typical Transducer Loss. Both transducers are resonant at approximately 30 MHz and therefore have a maximum conversion efficiency (*i.e.* minimum loss) at 30 MHz. Inherently, such resonant transducers have a loss maxima at even multiples of their resonant frequency and a minima in their loss at odd multiples. Following this, we see a maximum in the transducer loss occurs at about 60 MHz and a subsequent minimum at about 90 MHz.

Accurate attenuation measurements require a reasonably precise knowledge of the sound speed in the sample. The transmitted pulse arrives at the receiving transducer with a time delay dependent on the width of the gap, as well as the sound speed in the sample. At the beginning of the measurement the expected sound speed computation is based on the recipe used to make the sample. However, this estimated velocity may not be exact, and the error becomes progressively more important as the gap is increased.

For this reason the sound speed is automatically determined during each measurement, at least at one frequency, typically 10 MHz. The instrument employs a certain synergism between the attenuation and sound speed measurements, using each to improve the performance of the other. Errors in sound speed lead to an apparent decrease in the signal level because the received pulse is sampled at the wrong moment in time. At the same time, the attenuation data is used to improve the accuracy of the sound speed measurements by accepting only data at those gaps for which the attenuation is not so great as to preclude an adequate signal to noise ratio.

Although the transmitted tone-burst pulse has a rectangular envelope, the received RF pulse processed by the Signal Processor exhibits a bell shaped response, primarily as a result of the matched filter used to optimize the received signal to noise ratio. For maximum accuracy the amplitude of this received pulse must be measured at the exact peak of this bell shaped curve. For small gaps, the pulses are sampled by the receiver at a time delay calculated using only the estimated sound speed and the present gap. For larger gaps, a more precise pulse arrival time is determined, for at least one selected frequency, by making a few additional measurements at time intervals slightly shorter and longer than the expected nominal value. The optimum arrival time, referred to as the "time of flight" (TOF), is then computed by fitting a polynomial curve to these few points measured on this bell shaped response. This procedure typically provides a precision of a few nanoseconds in the determination of the optimum arrival time. The measured error in the arrival time for this single frequency is accumulated with each increasing gap. For each subsequent point on the grid, the signal is sampled at a time interval which is the sum of the time calculated from the theoretical sound speed and the last updated value for the cumulative error in the arrival time.

When the measurement for the entire grid is complete, a regression analysis of the optimum arrival time with gap provides an accurate measure of the group sound speed. (Figure 7.9) This procedure of time of flight adjustment achieves two goals. First, it eliminates possible artifacts in the attenuation measurement. Second, it provides the group sound speed measurement which is of interest in its own right. This final sound speed measurement is much more reliable than that which might be obtained by measuring the TOF for just two gaps, because it reflects a regression analysis over data for many gaps. The attenuation of the sound pulse as it passes through the sample makes measurement of the time of flight possibly less accurate at larger gaps where the intensity of the arriving signal may be low. To insure the integrity of the sound speed measurement, the software selects only those gaps where the s/n ratio is sufficient to insure reliable data. This method makes it possible to measure the group sound speed over a much wider range of attenuation than would be possible with any fixed gap technique.

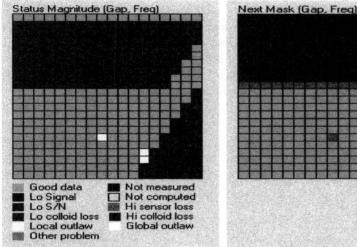
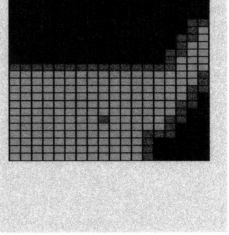

Figure 7.10 Status array for 10 wt% silica slurry.

Figure 7.11 Next Mask for 10 wt% silica slurry.

It is clear that taking data at all points on the grid provides some redundant information. In principle we need only two gaps for each frequency to determine the attenuation. The extra data provides a very robust system that can operate despite many difficult conditions. For example, when monitoring a flowing stream in a process control application, there may be occasional clumps of extraneous material, or perhaps a large air bubble, or maybe even a momentary loss of sample material altogether. In doing the regression at each frequency, the software automatically rejects any data points where that residual (to the best fit linear regression) differs by more than three sigma from the average. After any "outlaw" points are removed, it uses the remaining residuals to estimate the overall experimental error in the measured attenuation at each frequency. The final status of each point in the grid is then reflected in a Status

Array as shown in Figure 7.10. The legend gives the reason why any given point was rejected. In this example valid data at low frequency was obtained data only for larger gaps, whereas for high frequency valid data was obtained only for smaller gaps. Three points were discarded as being "outlaws".

Often it is desirable to make several measurements on the same sample. For example, if performing a titration experiment to test the effect of pH on particle size distribution, it will be necessary to measure the same sample after reagent addition. After the first measurement, the software constructs a "Next Mask" (Figure 7.11) that notes all of the points on the grid that produced useful results on the first measurement. Typically the number of such "good" points is only half those measured the first time the sample was measured. Deleting these non-productive points for succeeding measurements typically doubles the measurement speed after the first measurement. However, since the sample may change slowly from one measurement to the next, the Next Mask always includes additional points at the boundary of the acceptable points from the previous measurement. The Next Mask allows the system to adapt to samples that are changing from measurement to measurement yet not waste time collecting useless data.

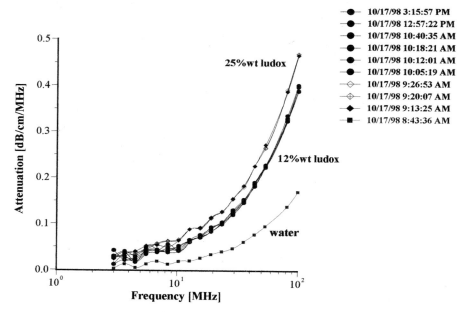

10/17/98 15:54:21 Graph1

Figure 7.12 Several attenuation spectra for colloidal silica slurries at two concentrations, with comparison to that of water.

7.5. Precision, accuracy, and dynamic range for transmission measurements

Precision is a measure of the reproducibility, whereas accuracy is a measure of agreement with a given standard. Quite often these two metrics of system performance are confused, or not clearly stated. For example, Andrea *et al* [15-17] reported that the accuracy of their attenuation measurement was 2% over the frequency range of 0.5 to 200 MHz, but they did not give any measure of the precision. Holmes and Challis [27-29] simply reported an error in sound speed measurement, using their "wide-bandwidth" technique, of 0.2 m/s. They characterized the error in the attenuation measurement for water by specifying only a signal-to-noise ratio of 66 dB at 5 MHz, after averaging 4000 pulses,

Table 7.1 Attenuation spectra for 12%wt silica, measured 9 times with 5 different loads.

Run#	\multicolumn{19}{c}{Frequency in MHz}

Run#	3	3.7	4.5	5.6	6.8	8.4	10.3	12.7	15.6	19.1	24	29	35.5	43.6	53.6	65.9	81	100
1	0.03	0.03	0.02	0.04	0.04	0.04	0.05	0.07	0.08	0.09	0.1	0.13	0.15	0.19	0.23	0.27	0.33	0.4
2	0.04	0.02	0.03	0.04	0.04	0.04	0.05	0.06	0.07	0.09	0.1	0.13	0.15	0.19	0.23	0.28	0.34	0.4
3	0.05	0.05	0.04	0.05	0.05	0.06	0.05	0.05	0.07	0.09	0.1	0.12	0.15	0.19	0.22	0.25	0.33	0.39
4	0	0.03	0.01	0.04	0.04	0.04	0.05	0.06	0.07	0.09	0.1	0.12	0.15	0.19	0.22	0.24	0.33	0.39
5	0.03	0.04	0.02	0.05	0.04	0.04	0.05	0.06	0.07	0.09	0.1	0.12	0.15	0.19	0.22	0.27	0.33	0.4
6	0.03	0.03	0.02	0.03	0.04	0.05	0.04	0.06	0.07	0.08	0.1	0.12	0.15	0.18	0.22	0.24	0.32	0.39
7	0.03	0.04	0.03	0.04	0.04	0.05	0.04	0.06	0.07	0.08	0.1	0.12	0.15	0.18	0.23	0.24	0.33	0.39
8	0.01	0.02	0.02	0.04	0.04	0.04	0.04	0.06	0.07	0.08	0.1	0.12	0.15	0.19	0.22	0.26	0.32	0.39
9	0.02	0.03	0	0.04	0.04	0.04	0.05	0.06	0.07	0.09	0.1	0.12	0.15	0.19	0.22	0.27	0.33	0.39
Std dev	0.01	0.007	0.008	0.003	0.001	0.004	0.003	0.002	0.001	0.003	0	0.002	0	0.002	0.003	0.013	0.003	0.003
Average	0.027	0.032	0.021	0.041	0.041	0.044	0.047	0.06	0.071	0.087	0.1	0.122	0.15	0.188	0.223	0.258	0.329	0.393

We demonstrate the precision of our technique using two test samples: a 10 %wt colloidal silica sample (Ludox TM50) and a 20 wt% slurry of much larger silica particles (Minusil 5). The colloidal silica sample provides a difficult test, because the attenuation of the slurry is only slightly greater than the intrinsic attenuation of the aqueous media. Figure 7.12 shows several attenuation spectra for two Ludox samples having different weight fractions, with a typical water spectrum included for comparison. Table 7.1 summarizes this numerical data. It is seen that precision is better than 0.01 dB/cm/MHz for all frequencies.

The precision of the attenuation measurement over a longer time period (30 hours) is shown in figure 7.13. The standard deviation at 100 MHz over this 30 hour period was approximately 0.002 db/cm/MHz.

Figure 7.14 shows the measured sound speed for the same colloidal silica. The standard deviation for these 43 measurements was 0.28 m/s. For comparison, Holmes and Challis [27-29] reported an error in sound speed measurement using their "wide-bandwidth" technique in "transmission mode" as 0.2 m/s.

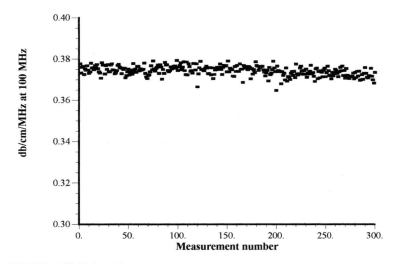

04/11/02 11:37:55 Graph1
Figure 7.13 Precision of Attenuation at 100 MHz for 10%wt silica Ludox over 30 Hours.

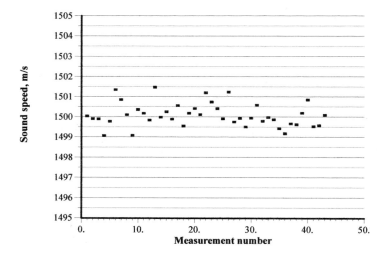

Figure 7.14 Precision of sound speed measurement for 10 wt% silica (Ludox TM-50).

Figure 7.15 shows the attenuation spectra for eight repeat measurements of a 10 wt% silica (Minusil 5). The precision of the attenuation measurement at mid-band (10MHz) was approximately 0.01dB/cm/MHz.

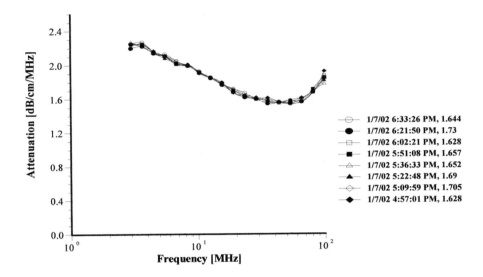

Figure 7.15 Several spectra for 20 wt% silica.

To be useful as a general purpose tool, an acoustic spectrometer must be able to accommodate a large dynamic range, i.e. work over a large range of attenuation. The instrument must be able to work for a low attenuation fluids such as the colloidal silica Ludox TM-50 already described as well as for high attenuation slurries such as concentrated rutile slurry. The Dispersion Technology database contains over 60,000 measurements, accumulated over the past eight years from many instruments around the world, for hundreds of different applications. This database was polled to find the sample that had a maximum attenuation at each of three frequencies, namely 3, 10, and 100 MHz. Several spectra for each of these three samples are shown in Figure 7.16. If we take the envelope of the spectra for these three samples, they provide a conservative estimate of the maximum attenuation that can be reliably measured with this technique. Considering these samples, the maximum attenuations at 3, 10, and 100 MHz is approximately 130, 60, and 18 dB/cm/MHz, respectively. Describing the attenuation in terms of dB/cm (by multiplying by frequency) we obtain values of 390, 600, and 1800 dB/cm.

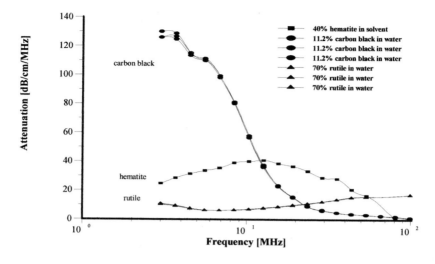

Figure 7.16 Spectra for several very highly attenuating slurries.

7.6 Analysis of Attenuation and Sound Speed to yield desired outputs

7.6.1 The ill-defined problem

It has been known for a long time that different particle sizing techniques can yield quite different particle size distributions. Prof. Leschonski wrote in 1984:

"...*Each group of measurement techniques based on the same physical principle provides, in general, different distribution curves. The reason for this must partly be attributed to the fact that, in particle size analysis, every possible physical principle has been used to determine these distributions. Due to many reasons, the results will, in general, not be the same .... It means that the instruments based on different physical principles yield different PSD. ...However, apart from this, the distributions differ due to two reasons. Firstly, different physical properties are used in characterizing the "size" of individual particles and, secondly, different types and different measures of quantity are chosen...*" [31].

We suggest that particle sizing methods can be broadly classified into three general categories: (1) counting methods, (2) fractionation techniques, and (3) macroscopic fitting techniques. Each category is closely linked to a most appropriate way to describe the particle size distribution. Particle counting yields a "full PSD" on a "number basis". This category includes various microscopic image analyzers, electro-zone methods such as the Coulter Counter, and light obscuration based counters.

Fractionation techniques also yield a "full PSD", but usually on "volume basis". They operate by physically separating the sample into fractions that contain different sized particles. This category includes sieving, sedimentation, and centrifugation methods.

Macroscopic techniques include methods that are related to the measurement of macroscopic properties that are particle size dependent. All macroscopic methods, in addition to a measurement step, require the step of calculating the particle size distribution from this measured macroscopic data. For attenuation spectroscopy, for example, measuring the spectra is only the first step in characterizing the particle size. The second step is to use a set of algorithms, implemented in a computer program, to calculate the desired microscopic properties from the measured macroscopic data, using the knowledge contained in the prediction theory. The analysis can be thought of as the opposite, or inverse, of prediction. Prediction describes some of the measurable macroscopic properties in terms of the model dispersion. Analysis, on the other hand, given perhaps only a few values for the model parameters, attempts to calculate the remaining microscopic properties by an evaluation of the measured data.

However, in attempting the analysis we often run into a problem. There may be more than one solution. Perhaps more than one size distribution yields the same attenuation spectrum, at least within the experimental uncertainties. This problem of "multiple solutions" is often termed the "ill-defined" problem. It is not unique to acoustics, but is common to all macroscopic methods, particularly light scattering methods.

Each category of measuring techniques has some advantages and disadvantages. No one method is able to address all particle sizing applications. The "counting method", for instance, has an advantage over "macroscopic techniques" in that it yields a "full PSD". Obtaining a statistically representative analysis, however, requires counting an enormous number of particles, especially for a polydisperse system, and this can be very time consuming. Also, sample preparation is rather complicated and may affect the results. Additionally, transformation of the number based PSD to a volume based PSD is not accurate in many practical systems.

Fractionation methods allow one to collect data on a very large numbers of particles and may more easily provide statistically significant results. However, fractionation methods are not very suitable for small particles. For example, when using an ultra-centrifuge it may take hours to perform a particle size analysis of a sample containing submicron particles. These methods also dramatically disturb the sample and consequently might affect the measured particle size distribution.

Macroscopic methods, in general, are fast, statistically representative, and easy to use. However, there is a price to pay. We must solve the ill-defined problem. Fortunately, in many case there is no need to know all of the details of the particle size distribution. A value for the median size, a measure of the width of the distribution, and perhaps some measure of bimodality may be a sufficient description for many practical colloidal dispersions. For this reason, macroscopic methods are very widely used in many research laboratories and manufacturing plants.

One of the first steps in selecting a particle sizing technique should be to decide how much detail one needs concerning the PSD. A "full PSD" is only needed in rare cases. Do you need this degree of detail? Ask yourself what you would actually do with such detail. If this elaborate PSD information is indeed necessary, you should consider some counting or fractionation techniques. If not, you can use macroscopic techniques, with the understanding you will not obtain the exposition of the PSD.

The nature of this ill-defined problem is well understood in the field of light scattering. For instance, Bohren and Huffman, in their well-known handbook on light scattering, [32] wrote:

*"...The **Direct Problem**. Given a particle of specified shape, size and composition, which is illuminated by a beam of specified irradiance, polarization, and frequency, determines the field everywhere. This is the "easy" problem. **The Inverse Problem**. By suitable analysis of the scattered field, describe the particle or particles that are responsible for the scattering. This is the "hard" problem...the ill-defined problem ...."*

"To understand why we have labeled this the "hard" problem, consider that information necessary to specify a particle uniquely is (1) the vector amplitude and phase of the field scattered in all directions, and (2) the field inside the particle. The field inside the particle is not usually accessible to direct measurements. The total amplitude and phase of the scattered light are rarely achieved in practice. The measurement usually available for analysis is the irradiance of the scattered light for a set of directions. We are therefore almost always faced with the task of trying to describe a particle (or worse yet, a collection of particles) with a less than theoretically ideal set of data in hand"

*"...Geometrical optics can often provide quantitative answers to small-particle problem which are sufficiently accurate for many purposes, particularly when one **soberly considers the accuracy with which many measurements can be made**...."*

This quote from a respected expert in the field of particle characterization suggests that we should be cautious in our expectations when any macroscopic method is employed. In general, we are able to measure some macroscopic property (scattered light, neutrons, X-rays or sound attenuation) with some experimental error, Δ_{exp}. In turn, this experimental error limits our ability to calculate an exact PSD from the measured experimental data. Figure 7.17, for instance, illustrates a hypothetical "full" size distribution as a histogram. One can see that there is certain irregular variation in the height of the columns. Is it possible to characterize this detailed variation with any macroscopic technique? Is it important to characterize this fine detail in order to formulate, manufacture, or use a colloidal product? The answer to both questions in many cases is "No".

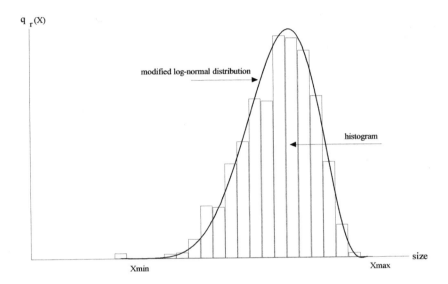

Figure 7.17 Example of the "full PSD" and fitting modified log-normal PSD.

These small variations of the PSD would create corresponding variations in the measured parameter, ΔM. If this variation is less than the experimental error (i.e. $\Delta M < \Delta_{exp}$), then the given macroscopic method will not be able to resolve this variation in PSD. The problem of "multiple solutions" prevents us from finding a unique PSD using the measured parameters. There may be several PSD that would each yield the same measured parameter within the limit of experimental errors. This rather simple discussion shows that all macroscopic techniques have some limits in the detail with which they can describe the actual PSD of a given colloidal dispersion.

There are many instances, however, when this limitation is simply ignored, and a "full PSD" is claimed as a valid output of some macroscopic measurement and analysis. For instance, US patent 6,119,510 by Carasso *et al*

[33] suggests an inversion procedure for calculating a "full PSD" from either acoustic or optical attenuation spectra. This invention presents the PSD as an abstract form, based on a selected number of parameters - perhaps 50 or even more. No means is suggested for determining this important number, which can be considered as a measure of the amount of information contained in the experimental data.

Is it possible to determine the limitations of a given "macroscopic technique" for characterizing the finer details of a PSD? We suggest a simple procedure, which follows the general scientific method. Let us assume that we have two PSD, each proposed as fitting the same set of experimental data. In the case when both PSD yield the same quality of fit, we suggest that the solution employing the fewer number of adjustable parameters should be selected. This common sense idea was formulated in the fourteenth century by William of Occam [34] as a general philosophical principle (Occam's razor), which briefly states that "given a set of otherwise **equivalent** explanations, the **simplest** is most likely to be correct".

In the case of fitting theory to experimental data, two solutions producing the **same fitting error are deemed equivalent**, whereas **the number of adjustable parameters is inversely related to simplicity**.

If we decide to stay with general scientific principles, we should start with minimum number of adjustable parameters, and increase it in steps till the fitting error no longer improves significantly relative to the experimental errors. This procedure then determines the maximum number of adjustable parameters which one can extract from the particular set of experimental data. This is just a general principle. There are many ways to implement it in calculating the PSD from the attenuation spectra.

The first step is obvious: we assume a monodisperse PSD with the size as the only adjustable parameter. It is possible to run a global search for the best monodisperse solution (the one that yields the best fit between the theoretical attenuation and the experimental data). In the case of very small particles, this first step will often be the last because the attenuation of the dispersion may be very low, and only slightly greater than the intrinsic attenuation of the sample. The added attenuation from the presence of the particles may be comparable to the experimental error which then limits the amount of particle size information that one can extract from the measured data. Our experience with DT1200 indicates that this often occurs when the particle size is 10 nanometers or less.

For particles larger than 10 nm there is usually sufficient information in the attenuation spectra for calculating additional adjustable parameters. The next step is to use two adjustable parameters, resulting in a "unimodal PSD". There are several ways to select these two parameters. By default we use a lognormal PSD, which is of course defined by two adjustable parameters: a median and standard deviation. It is also possible to use a modified lognormal PSD (see

Chapter 2) when the particle size range is known. This is helpful for characterizing non-symmetrical PSD, as was illustrated in Figure 7.17.

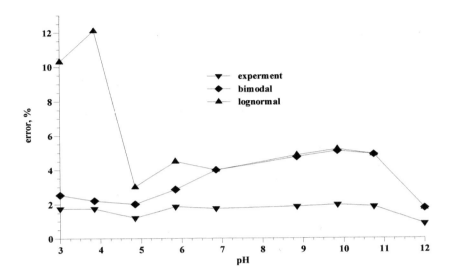

Figure 7.18 Experimental error and errors of theoretical fit based on lognormal and bimodal PSD for 7%vl rutile dispersion.

The next level of complexity brings us to a "bimodal PSD", which can be modeled as the sum of two lognormal PSD's. This increases the number of adjustable parameters to five (a median and standard deviation for each mode and the relative concentration of these two modes) or perhaps only four if we assume that the two modes have similar standard deviation.

Our experience with many thousands of samples is that it is seldom necessary, or allowable, to increase the complexity much beyond a bimodal PSD.

The analysis of an attenuation spectra proceeds by increasing the complexity (the number of adjustable parameters) in steps, and at each step computing the best fit PSD and the corresponding fitting error. We can illustrate how this process works by examining the results of a pH titration experiment on a 7%vl rutile dispersion. The rutile particles aggregate somewhat in the vicinity of the isoelectric pH, about 4.0. Thus, we would expect that the PSD becomes bimodal close to the isoelectric point, with one peak representing the primary particles, and the second peak comprising the aggregates. Figure 7.18 shows fitting errors for the lognormal and bimodal cases, as compared with the experimental errors. For pH values larger than 6, the lognormal and bimodal fitting errors are identical. Following Occam's principle, we should select then the simpler lognormal distribution. However, in the vicinity of the IEP, i.e. for pH < 6, the fitting error of the lognormal solution dramatically increases, an indication that the lognormal PSD fails to fit the experimental data in this pH range. In contrast, the fitting error for the bimodal solution remains small, indeed close to the experimental error. In light of this information, following Occam's principle allows us to claim that the PSD is bimodal in this pH range.

7.6.2 Precision, accuracy, and resolution of the analysis

The precision (not accuracy) of the PSD measurement is determined almost exclusively by the error in the attenuation measurement since the computations can be done with high precision. Figure 7.19 shows the mean size for a colloidal silica (Cabot SS25) measured over 18 hours. The standard deviation was less than 1 nm.

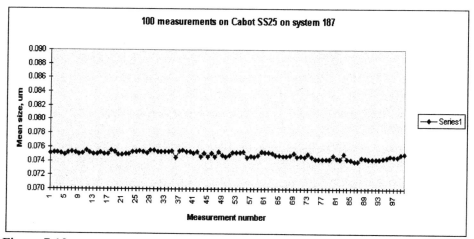

Figure 7.19 Precision of mean particle size for 100 measurements of 10 %wt silica slurry (Cabot SS25).

The absolute accuracy (as opposed to precision) depends on several factors, including:
1. The accuracy of the required physical properties of the sample.
2. How well the sample conforms to the limitation of the model.
3. The accuracy of the attenuation measurement.
4. The accuracy of the prediction theory.
5. The performance of the analysis engine in finding the best fit between theory and experiment.

The accuracy of attenuation spectroscopy for determining the PSD was briefly demonstrated in Chapter 6 using a glass (BCR-70) slurry. Here we show results for another sample, a 20 wt% silica (Minusil-5) slurry. The attenuation spectra for this sample were shown previously in Figure 7.15. The analysis of these spectra yields the corresponding cumulative size distributions shown in Figure 7.20. The eight separate measurements are virtually indistinguishable. Table 7.2 compares key points on this cumulative distribution with values provided by the manufacturer (U.S. Silica).

The term "resolution" is meant to characterize the ability of the acoustic technique to resolve sub-populations of particles having different sizes. An interesting resolution parameter is the amount of large particles that can be distinguished against a background of much smaller particles. Resolution, defined in this manner, is important for characterizing the aggregation in materials such as chemical mechanical polishing (CMP) slurries.

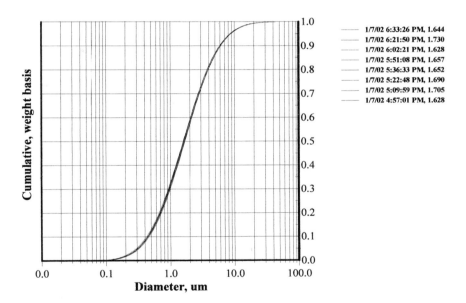

Figure 7.20 Precision of cumulative distribution for a 20 %wt silica (Minusil 5) dispersion.

One can estimate this resolution parameter by comparing the theoretical predictions for the attenuation spectra of a series of bimodal slurries, in which small amounts of large particles (representing the aggregates) are sequentially added to a base slurry of much smaller particles. Clearly the precision with which one can measure the attenuation spectra will determine the sensitivity with which one can detect any small changes due to the introduction of small amounts of larger particles. The precision in determining the attenuation of the base slurry was depicted earlier in Table 7.1, and was of the order of 0.01 dB/cm/MHz at low frequency. It follows that we should be able to detect a change from the addition of the larger particles, if this addition results in an increase in attenuation that is larger than this basic precision.

Table 7.2 Accuracy of PSD data for 20 %wt silica (Minusil 5).

Cumulative	Acoustic Measurement, microns	Manufacturer's data, microns
D_{10}	0.45	0.6
D_{50}	1.67	1.6
D_{90}	6.0	3.5

Figure 7.21 shows the increment in the theoretical attenuation of a bimodal slurry having 1% large particles, with respect to a unimodal distribution having no large particles. This increment is shown for three different sizes for the larger particles: 0.5, 1 and 3 microns. The precision of the attenuation measurement (from Table 7.1) is also shown for comparison. Agglomerates having a particle size of 0.5 micron would clearly be detectable, as the increment in attenuation would be considerably larger than the instrument error. On the other hand, we see that the incremental attenuation for 3 micron particles is somewhat less, which means that such particles would be more difficult to detect at this 1% level. As a result, a 1% addition of 3 micron particles would be detected only in the high frequency portion of the spectra.

Another conclusion follows from the theoretical curves of Figure 7.6 which is that the resolution depends on the particle size. We note that the addition of either the 0.5 or 1 micron particles contributes additional viscous loss and results primarily in an added attenuation at low frequencies, whereas the addition of the 3 micron particles contributes additional scattering losses, resulting primarily in added attenuation at higher frequencies. It is a fortunate coincidence that this acoustic technique is particularly sensitive for distinguishing between these "large" and relatively "larger" particles. Qualitatively, we can say that an increase in the attenuation at lower frequencies indicates that we are dealing with "large" particles of about 1 micron, whereas an increase in the attenuation spectra at higher frequencies is an indication that we have even "larger" particles exceeding 3 microns. Quantitatively, of course, these calculations are made more precise by the analysis software.

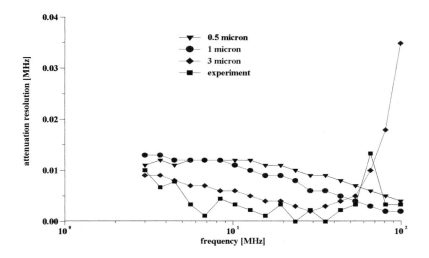

Figure 7.21 Increment in the predicted attenuation caused by agglomeration of 1% of the total particulates in a 12 wt% slurry of colloidal silica, for three sizes of agglomerates.

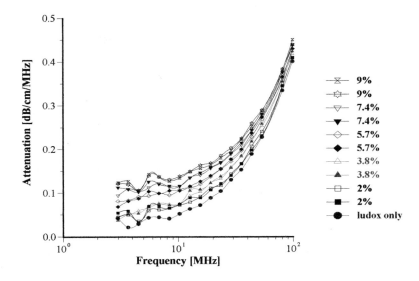

Figure 7.22 Attenuation spectra for a 12%wt slurry of colloidal silica (Ludox TM-50), with added amounts of larger silica (Geltech 0.5).

234

Figure 7.22 illustrates how large particles affect the total attenuation spectra. It is seen that 2% of larger particles induces visible changes in the attenuation spectra.

In conclusion, we can describe this level of resolution in the following terms. Acoustic spectroscopy is able to resolve a single large particle of 1 micron size against a background of 100,000 small 100 nm particles. This conclusion is rather approximate, it should be considered as reference point, but not as absolute number. It is obvious that the resolution might vary with particle size, weight fraction, and the properties of the particle and liquid media.

7.7. Measurement of Electroacoustic properties

7.7.1. Electroacoustic measurement of CVI

The Dispersion Technology DT300 is an example of a modern instrument for determining the zeta potential of colloidal dispersions by the measurement of electroacoustic properties and the subsequent analysis of this data. This instrument measures the Colloid Vibration Current (as opposed to ESA). Similar to the DT100, it consists of two parts, a CVI Probe and an Electronics Unit. A photo of a typical system is shown in Figure 7.23.

Figure 7.23 The probe for measuring colloid vibration current.

The electronics for the electroacoustic measurement are identical to that used for acoustic measurements. In fact, Dispersion Technology offers a single instrument which combines the two sensors (Acoustic and CVI) in a single instrument, the DT1200 for characterizing both particle size and zeta potential.

The CVI Probe consists of two parts: a transmitting and a receiving transducer. The transmitting transducer consists of a piezoelectric transducer to convert a radio frequency (RF) pulse into acoustic energy, and a delay rod that launches this acoustic pulse into the slurry after a suitable delay. This acoustic excitation of the slurry causes the particles to gain induced dipole moments. The dipole moments from the particles add up to create an electric field, which can be sensed by a receiver.

This receiver, in essence, consists of a two-element antenna immersed in the sample. The electric field changes the potential of one element with respect to the other. If the impedance of the measuring circuit associated with the antenna is relatively high with respect to the colloid, then the antenna senses an open circuit voltage, which is typically referred to as the Colloid Vibration Potential (CVP). Alternatively, in the preferred configuration, the impedance of the circuitry associated with the antenna circuit is low compared to that of the colloid, and therefore the electric field causes a current to flow in the antenna. This short circuit current is referred to as the Colloid Vibration Current (CVI). The probe can be dipped into a beaker for laboratory use, immersed in a storage tank, or inserted into a pipe for online monitoring of a flowing process stream.

Figure 7.24 shows a cross section of the probe. The transmitting portion utilizes an off-the-shelf transducer **3** in conjunction with a mating standard UHF cable connector **1** and input cable **2**. This transducer contains a cylindrical piezoelectric device **4** having a front electrode **5** and a back electrode **6** across which an RF pulse can be applied to generate an acoustic pulse. The front and back electrodes of the piezoelectric device are connected to corresponding terminals of the UHF connector by means of internal jumper wires **7** in a manner well known to those skilled in the art of such transducer design. The piezoelectric device in turn is bonded to a quartz delay rod **8** by means of a suitable adhesive. The resonant frequency of the piezoelectric device is selected depending on the frequency range for which electroacoustic data is required. Typically the resonant frequency is selected in the range of 2 to 10 MHz and the quartz delay rod has an acoustic delay of about 2.5 u sec. The quartz delay rod is extended by an additional buffer rod **9** having an acoustic impedance that is more closely matched to that of the slurry than is the quartz material.

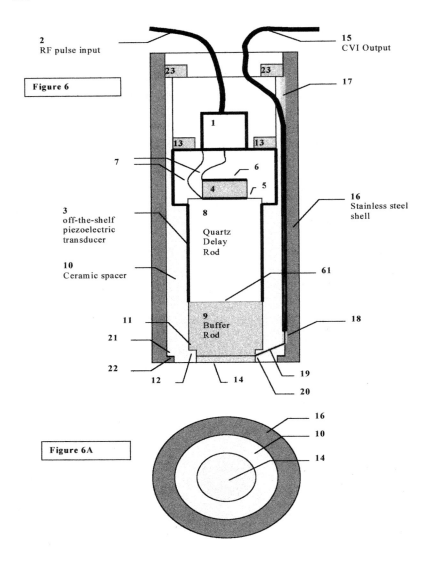

Figure 7.24 Diagram of probe for measuring colloid vibration current.

The plastic known commercially as Rexolite has been found to be quite suitable for this purpose. The length of this buffer rod is chosen to provide an additional time delay at least as long as the pulse length, typically 2.5 u sec. The interface between the delay rod and the buffer rod **62** provides a well characterized reflection and transmission coefficient for the incident acoustic pulse since the acoustic impedance of both materials is known precisely.

The acoustic transducer and the additional buffer rod are cemented together and inserted fully into a ceramic spacer **10** till the recess **11** on the buffer rod mates with the shoulder **12** on the spacer, whereupon it is fastened securely by means of brass retaining ring **13**.

The end of the buffer rod is coated with gold **14** in order to provide an electrode for measuring the electrical response of the colloid when excited acoustically. The gold coating wraps around the end of the buffer rod approximately 2 millimeters in order to provide a means for making an electrical connection to the gold electrode. Figure 6A provides an end view of the probe showing the relative location of the gold electrode **14**, the spacer **10**, and the stainless steel shell **16**.

A coax cable **15** carries the CVI signal, connecting to the two electrodes used to receive the colloid vibration signal. The center conductor of the coax connects to the gold electrode and the shield of the coax makes connection to the outer stainless steel shell **16** as will be explained shortly. A groove **17** is provided along the length of the spacer to provide a path for the coax cable. The outer insulation of the coax is removed from the cable over the length of this groove exposing the outer braid such that it will contact the stainless steel shell. The shield is removed from the final portion **18** of the coax cable center conductor, which is inserted in an access hole **19** in the spacer such that it makes contact with that part of the gold which wraps around the end of the buffer rod. The coax cable center conductor is permanently fastened to the gold by means of silver filled epoxy **20**.

The spacer, with its internal parts, is inserted into the stainless steel shell until a recess **21** at the periphery of the spacer mates with a flange **22** on the stainless steel shell, whereupon it is fastened securely by means of a retainer ring **23**. When the probe is immersed in a test sample, the colloid vibration signal between the gold central electrode and the surrounding stainless steel outer housing is thus available at the coax cable output.

7.7.2. CVI measurement using energy loss approach

The measured CVI signal reflects a series of energy losses that are experienced between the point the RF pulse is sent to the probe and the point at which it later returns. These losses include:
- the loss in converting the RF pulse to an acoustic pulse;
- the reflection losses at the quartz/ buffer rod interface;
- the reflection loss at the buffer rod/ colloid interface;
- and importantly the conversion from an acoustic pulse to a CVI pulse by virtue of the electroacoustic effect, which we are trying to characterize.

The following paragraphs describe how one calculates the CVI in terms of these effects.

The experimental output of the electroacoustic sensor, S_{exp}, is the ratio of the output electric pulse from the receiving antenna, I_{out}, to the intensity of the input RF pulse to the acoustic transmitter, I_{in}. It is different from the acoustic impedance measurement (Eq.7.11) due to the different nature of the output signal.

The intensity of the sound pulse in the delay rod, I_{rod}, is related to the intensity of this input electric pulse through some constant, C_{tr}, which is a measure of the transducer efficiency, and any other energy losses to this point. This relationship is given as Eq. 7.12. In the case of the electroacoustic measurement we are more interested in the sound pressure than the sound intensity. The sound pressure in the rod, according to Eqs.3.33 and 3.34 in Chapter 3, is given by:

$$P_{rod} = \sqrt{2 I_{rod} Z_r} = \sqrt{2 C_{tr} Z_r I_{in}} \qquad (7.2)$$

At the receiving transducer, we can define the electric pulse intensity as being proportional to the square of the colloid vibration current received at the antenna:

$$I_{out} = CVI^2 C_{ant} \qquad (7.3)$$

where the constant C_{ant} depends on some geometric factor of the CVI space distribution in the vicinity of the antenna, as well as the electric properties of the antenna.

Substituting Eq.7.2 into Eq.7.3 we obtain the following expression which relates CVI to the measured parameter S_{exp}:

$$\frac{CVI}{P_{rod}} = \sqrt{\frac{S_{exp}}{2 Z_r C_{tr} C_{ant}}} \qquad (7.4)$$

The magnitude of the measured CVI is proportional to the pressure near the antenna surface, P_{ant}, not in the rod itself. The pressure at the antenna is lower than the pressure in the rod, P_{rod}, because of a reflection loss at the rod-colloid interface, which gives us the following expression for P_{ant}:

$$P_{ant} = P_{rod} \frac{2 Z_s}{Z_s + Z_r} \qquad (7.5)$$

This correction leads to the following expression for CVI:

$$\frac{CVI}{P_{ant}} = \sqrt{\frac{S_{exp}}{2 Z_r C_{tr} C_{ant}}} \frac{Z_s + Z_r}{2 Z_s} \qquad (7.6)$$

The gradient of pressure ∇P in Eq.5.27 for CVI is equal to the gradient of pressure P_{ant}. Using this fact and substituting CVI from Eq.5.27, we obtain the following equation relating the properties of the dispersion with the measured parameter S_{exp}:

$$\frac{CVI}{\nabla P} = \sqrt{\frac{S_{exp}}{2Z_r C_{tr} C_{ant}} \frac{Z_s + Z_r}{2Z_s c_s}} \qquad (7.7)$$

where c_s is the sound speed in the dispersion.

For calculating ζ-potential or particle size from the measured CVI, one should use Eqs. 5.27-5.30. One can calculate ζ-potential for a known particle size, or calculate both ζ-potential and particle size. In order to use this relationship, we should define a calibration procedure for resolving the two calibration constants in Equation 7.7. There are two ways to eliminate them.

It is possible to combine them in one unknown calibration constant, C_{cal}, that is independent of the properties of the dispersion. This calibration constant can be determined using a colloid with a known zeta potential. For this purpose we use Ludox TM-50, a commercially available colloidal silica, which is diluted to 10%wt with 10^{-2} mol/l KCl. From independent measurement we have determined that this silica preparation has a ζ-potential of -38 mV at pH 9.3.

However, there is more efficient way. The conversion efficiency of the transducer, C_{tr}, is frequency dependent whereas the geometric constant for the CVI field distribution, C_{ant}, is not. Furthermore, the transducer efficiency may vary with the age of transducer, as well as the ambient temperature. It would be more accurate if we apply a calibration procedure for eliminating the calibration constant C_{tr}. As a result we obtain the following equation:

$$\frac{CVI}{\nabla P} = \sqrt{\frac{S_{exp} |Z_b - Z_r|}{2C_{ant} S_{cal} Z_r (Z_b + Z_r)} \frac{Z_s + Z_r}{2Z_s c_s}} \qquad (7.8)$$

Calibration constant C_{ant} requires calibration with the known colloid, as described above. The unknown impedance of the colloid can be either measured using acoustic impedance measurement (Section 7.8) or calculated if colloid density and sound speed are known.

7.8. Zeta potential calculation from the analysis of CVI

The analysis of the CVI signal to determine ζ potential is considerably simpler than the analysis of an attenuation spectrum to determine a PSD. Unlike the PSD calculation, which requires no calibration, the zeta potential is calibrated using a colloid of known electrokinetic properties as described above. Figure 7.25 shows the precision that is normally obtained following calibration. The standard deviation for the zeta potential measurement is typically 0.3 mV.

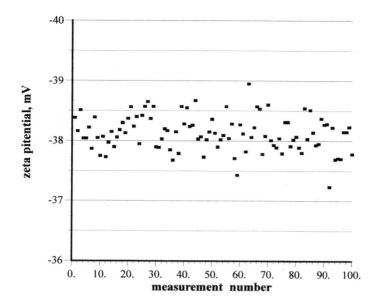

Figure 7.25 Zeta potential for 10 wt% colloidal Silica (Ludox TM-50).

7.9. Measurement of acoustic Impedance

Sound speed measurements, in principle, allow us to calculate the acoustic impedance of a sample if the density of the slurry is known. However, there is an independent way to measure acoustic impedance and thereby obtain additional data for the colloidal dispersion.

The acoustic impedance, Z, is introduced as a coefficient of proportion between pressure, P, and the velocity of the particles, v, in the sound wave. It directly follows from this (see Chapter 3) that the acoustic impedance equals:

$$Z = \rho c (1 - j \frac{\alpha \lambda}{2\pi}) \tag{7.9}$$

where the attenuation Z is in [nepers/m], the wavelength λ is in [m], the sound speed of the longitudinal wave c is in [m/sec], and the density ρ is in [kg/m³].

For most colloids, the contribution of attenuation to acoustic impedance, in the ultrasound range of 1 to 100 MHz, is negligible. Thus we can simplify Equation 7.9 to obtain:

$$Z_s = \rho_s c_s \tag{7.10}$$

If sound speed, c_s, is known, a measurement of the Acoustic impedance allows us to calculate an effective density of the slurry, which in turn can be used for monitoring the colloid composition.

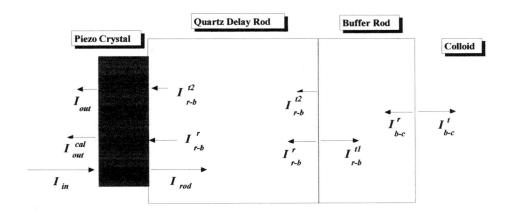

Figure 7.26 Transmission of an ultrasound pulse through the acoustic impedance probe.

According to the Equation 3.33 and 3.34, the acoustic impedance controls the propagation of the ultrasound wave through phase borders. The intensity of the reflected and transmitted waves depends on the relationship between the acoustic impedances of the various phases. These intensities are measurable, which opens up the way to characterize the otherwise unknown acoustic impedance of the colloid.

The design of the acoustic impedance probe is somewhat similar in design to that of the CVI probe previously described (Figure 7.24). The difference between this device and the CVI probe is the nature of the sound path and the reflections at various interfaces. These reflections and transmissions of the sound pulse are illustrated in Figure 7.26.

An electric pulse of intensity, I_{in}, is introduced to the piezoelectric crystal which converts it to an ultrasound pulse of intensity, I_{rod}. This pulse propagates through the delay rod, and eventually meets the border between the delay rod and the buffer rod. Part of the pulse, I^r_{r-b}, reflects back to the piezoelectric crystal, the other part, I^{t1}_{r-b} transmits further to the border between the buffer rod and the colloid. The last part of the pulse splits at this border, again into a reflected I^r_{b-c} pulse and a transmitted pulse. The. reflected part comes back towards the piezoelectric crystal, experiencing one more reflection-transmission split at the delay rod - buffer rod border. The last part of the initial acoustic pulse, I^{t2}_{r-b}, reaches the piezoelectric crystal, which converts it back to the electric pulse with intensity I_{out}.

The experimental output of the impedance probe, S_{exp}, is the ratio of intensities of the input and output pulses:

$$S_{exp} = \frac{I_{out}}{I_{in}} \tag{7.11}$$

We can express I_{out} through I_{in} by following the pulse path, and applying reflection-transmission laws (Eqs.3.33 and 3.34) at each border using the acoustic impedances of the delay rod, Z_r, buffer rod, Z_b, and of the colloid, Z_s. This gives us the following set of equations:

$$I_{rod} = C_{tr} I_{in} \qquad \text{:conversion electric-ultrasound} \tag{7.12}$$

$$I^{t1}_{r-b} = I_{rod} \frac{4Z_b Z_r}{(Z_b + Z_r)^2} = I_{in} C_{tr} \frac{4Z_b Z_r}{(Z_b + Z_r)^2} \qquad \text{:transmission delay rod-buffer} \tag{7.13}$$

$$I^r_{b-c} = I^{t1}_{r-b} \frac{(Z_b - Z_c)^2}{(Z_b + Z_c)^2} = I_{in} C_{tr} \frac{4Z_b Z_r}{(Z_b + Z_r)^2} \frac{(Z_b - Z_s)^2}{(Z_b + Z_s)^2} \qquad \text{:reflection buffer rod-colloid} \tag{7.14}$$

$$I^{t2}_{b-r} = I^r_{b-c} \frac{4Z_b Z_r}{(Z_b + Z_r)^2} = I_{in} C_{tr} \frac{16Z_b^2 Z_r^2}{(Z_b + Z_r)^4} \frac{(Z_b - Z_s)^2}{(Z_b + Z_s)^2} \qquad \text{:transmission buffer-delay rod} \tag{7.15}$$

$$I_{out} = C_{tr} I^{t2}_{b-r} = I_{in} C_{tr}^2 \frac{16Z_b^2 Z_r^2}{(Z_b + Z_r)^4} \frac{(Z_b - Z_s)^2}{(Z_b + Z_s)^2} \qquad \text{:conversion ultrasound-electric} \tag{7.16}$$

Substituting the final equation for the intensity of the output electric pulse into the definition of the experimental output of the impedance probe (Eq.7.11) we got the following result:

$$S_{exp} = \frac{I_{out}}{I_{in}} = C_{tr}^2 \frac{16Z_b^2 Z_r^2}{(Z_b + Z_r)^4} \frac{(Z_b - Z_s)^2}{(Z_b + Z_s)^2} \tag{7.17}$$

There are two unknown parameters in this equation: the acoustic impedance of the colloid, Z_c, and the efficiency of the piezoelectric crystal for conversion of electric energy to ultrasound, C_{tr}.

It turns out that we can eliminate the last parameter using pulse reflected from the delay rod – buffer interface during forward transmission [23]. The intensity of this pulse is marked as I^r_{r-b}. This pulse returns back to the piezo crystal before the pulse reflected from the colloid. The piezoelectric crystal converts it back to the electric pulse with energy I^{cal}_{out}. We can measure the ratio of this intensity to the intensity of the initial pulse:

$$S_{cal} = \frac{I^{cal}_{out}}{I_{in}} = \frac{C_{tr} I^r_{r-b}}{I_{in}} = \frac{C_{tr} I_{rod}}{I_{in}} \frac{(Z_b - Z_r)^2}{(Z_b + Z_r)^2} = C_{tr}^2 \frac{(Z_b - Z_r)^2}{(Z_b + Z_r)^2} \tag{7.18}$$

Substituting this Equation into Eq. 7.17, we obtain the following simple expression for calculating the acoustic impedance of the colloid:

$$Z_s = Z_b \frac{1 - S_{exp}^N}{1 + S_{exp}^N} \qquad (7.19)$$

where

$$S_{exp}^N = \sqrt{\frac{S_{exp}}{S_{cal}} \frac{|Z_b^2 - Z_r^2|}{4 Z_b Z_r}}$$

This calibration procedure minimizes the potential problem related to variations in the transducer conversion efficiency.

REFERENCES.

1. Bhardwaj, M.C. "Advances in ultrasound for materials characterization", Advanced Ceramic Materials, 2, 3A, 198-203 (1987)
2. Hamstead, P.J. "Apparatus for determining the time taken for sound energy to cross a body of fluid in a pipe", US Patent 5,359,897 (1994)
3. Dann, M.S., Hulse, N.D. "Method and apparatus for measuring or monitoring density or rheological properties of liquids or slurries", UK Patent GB, 2 181 243 A (1987)
4. Kikuta, M. "Method and apparatus for analyzing the composition of an electro-deposition coating material and method and apparatus for controlling said composition", US Patent 5,368,716 (1994)
5. Sowerby, B.D. "Method and apparatus for determining the particle size distribution, the solids content and the solute concentration of a suspension of solids in a solution bearing a solute", US Patent 5,569,844 (1996)
6. Sette, D. "Ultrasonic studies" in "Physics of simple liquids", North-Holland Publ., Co., Amsterdam, (1968)
7. McClements, J.D. and Powey, M.J.W., "Scattering of ultrasound by emulsions", J.Phys.D:Appl.Phys., 22, 38-47 (1989)
8. McClements, D.J. "Ultrasonic characterization of food emulsions", in Ultrasonic and Dielectric Characterization Techniques for Suspended Particulates, Ed. V.A. Hackley and J. Texter, Amer. Ceramic Soc., 305-317, (1998)
9. McClements, J.D. "Principles of ultrasonic droplet size determination in emulsions:, Langmuir, 12, 3454-3461 (1996)
10. Ricco, A.J., and Martin, S.J. "Acoustic wave viscosity sensor", Appl. Phys. Lett., 50, 21 1474-1476 (1987)
11. Kostial, P. "Surface Acoustic Wave Control of the Ion Concentration in Water", Applied Acoustics, 41, 187-193 (1994)
12. Apfel, R.E., "Technique for measuring the adiabatic compressibility, density and sound speed of sub-microliter liquid samples", J. Acoust. Soc. Amer., 59, 2, 339-343 (1976)

13. Pellam, J.R. and Galt, J.K. "Ultrasonic propagation in liquids: Application of pulse technique to velocity and absorption measurement at 15 Megacycles", J. of Chemical Physics, 14, 10 , 608-613 (1946)

14. Pinkerton, J.M.M. "A pulse method for measurement of ultrasonic absorption in liquids", Nature, 160, 128-129 (1947)

15. Andreae, J., Bass, R., Heasell, E. and Lamb, J. "Pulse Technique for Measuring Ultrasonic Absorption in Liquids", Acustica, v.8, p.131-142 (1958)

16. Andreae., J and Joyce, P. "30 to 230 Megacycle Pulse Technique for Ultrasonic Absorption Measurements in Liquids", Brit.J. Appl. Phys., v.13, p.462-467 (1962)

17. Edmonds, P.D., Pearce, V.F. and Andreae, J.H. "1.5 to 28.5 Mc/s Pulse Apparatus for Automatic Measurement of Sound Absorption in Liquids and Some Results for Aqueous and other Solutions", Brit.J. Appl. Phys., vol.13, pp. 551-560 (1962)

18. Cushman, et all. US Patent 3779070, (1973)

19. Uusitalo, S.J., von Alfthan, G.C., Andersson, T.S., Paukku, V.A., Kahara, L.S. and Kiuru, E.S. "Method and apparatus for determination of the average particle size in slurry", US Patent 4,412,451 (1983)

20. Riebel, U. "Method of and an apparatus for ultrasonic measuring of the solids concentration and particle size distribution in a suspension", US patent 4,706,509 (1987)

21. Alba, F. "Method and Apparatus for Determining Particle Size Distribution and Concentration in a Suspension Using Ultrasonics", US Patent No. 5,121,629 (1992)

22. Dukhin, A.S. and Goetz, P.J. "Method and device for characterizing particle size distribution and zeta potential in concentrated system by means of Acoustic and Electroacoustic Spectroscopy", patent USA, 09/108,072, (2000)

23. Dukhin, A.S. and Goetz, P.J. "Method and device for Determining Particle Size Distribution and Zeta Potential in Concentrated Dispersions", patent USA, pending

24. Dukhin, A.S. and Goetz, P.J. "Method for Determining Particle Size Distribution and Structural Properties of Concentrated Dispersions", patent USA, pending

25. Dukhin, A.S. and Goetz, P.J. "Method for Determining Particle Size Distribution and Mechanical Properties of Soft Particles in Liquids", patent USA, pending

26. Erwin, L. and Dohner, J.L. "On-line measurement of fluid mixtures", US Patent 4,509,360 (1985)

27. Holmes, A.K., Challis, R.E. and Wedlock, D.J. "A Wide-Bandwidth Study of Ultrasound Velocity and Attenuation in Suspensions: Comparison of Theory with Experimental Measurements", J. Colloid and Interface Sci., 156, 261-269 (1993)

28. Holmes, A.K., Challis, R.E. and Wedlock, D.J. "A Wide-Bandwidth Ultrasonic Study of Suspensions: The Variation of Velocity and Attenuation with Particle Size", J. Colloid and Interface Sci., 168, 339-348 (1994)

29. Holmes, A.K., and Challis, R.E "Ultrasonic scattering in concentrated colloidal suspensions", Colloids and Surfaces A, 77, 65-74 (1993)

30. Sinha, D.N. "Non-invasive identification of fluids by swept-frequency acoustic interferometry" US Patent, 5,767,407 (1998)
31. Kurt Leschonski, "Representation and Evaluation of Particle Size Analysis Data", Part. Char., 1, 89-95 (1984)
32. Bohren, C. and Huffman, D. "Absorption and Scattering of Light by Small Particles", J. Wiley & Sons, 1983, 530 p.
33. Carasso, M., Patel, S., Valdes, J., White, C.A. "Process for determining characteristics of suspended particles", US Patent 6,119,510 (2000)
34. Hyman, A. and Walsh, J.J. "Philosophy in the Middle Ages", Indianapolis, Hackett Publishing Co, (1973)
35. Cannon D.W. "New developments in electroacoustic method and instrumentation", in S.B. Malghan (Ed.) Electroacoustics for Characterization of Particulates and Suspensions, NIST, 40-66 (1993)
36. Lubomska, M. and Chibowski, E. "Effect of Radio Frequency Electric Fields on the Surface Free Energy and Zeta Potential of Al_2O_3 ", Langmuir, 17, 4181-4188 (2001)
37. Colic, M. and Morse, D. "The elusive mechanism of the magnetic memory of water", Colloids and Surfaces, 154, 167-174 (1999)
38. Rowlands, W.N., O'Brien, R.W., Hunter, R.J., Patrick, V. "Surface properties of aluminum hydroxide at high salt concentration", JCIS, 188, 325-335 (1997)
39. Cannon, D.W. and O'Brien, R.W. "Device for determining the size and charge of colloidal particles by measuring electroacoustic effect", US Patent 5,245,290 (1993)
40. O'Brien, R.W. "Determination of Particle Size and Electric Charge", US Patent 5,059,909, Oct.22, (1991).
41. O'Brien, R.W. Particle size and charge measurement in multi-component colloids", US Patent 5,616,872, (1997).
42. Hunter, R.J. "Review. Recent developments in the electroacoustic characterization of colloidal suspensions and emulsions", Colloids and Surfaces, 141, 37-65 (1998)

Chapter 8. APPLICATIONS OF ACOUSTICS FOR CHARACTERIZING PARTICULATE SYSTEMS

The Table of Applications at the end of this chapter summarizes all ultrasound experiments related to colloid characterization that are known to us. Here we describe in some detail only the most important of those experiments, applications for which ultrasound provides a "cutting edge" over more traditional colloidal-chemical methods. We emphasize results obtained with Dispersion Technology instruments, but make reference and comparisons with independent results where it seems appropriate.

8.1. Characterization of aggregation and flocculation

Determining the stability of dispersions is one of the central problems in Colloid Science. Accurate and non-destructive stability characterization is the key to the successful formulation and use of many colloids. Ultrasound based techniques offer several advantages over traditional means for stability characterization. Eliminating the dilution protocol is one obvious advantage. Another important advantage is the ability to distinguish between aggregation and flocculation.

Aggregation is a process that results in the build-up of rigid agglomerates of the initial primary particles. These agglomerates are essentially new, inherently larger particles, which move as rigid entities in the liquid under stress.

Flocculation, on the other hand, is a process in which the original particles retain, to some degree, their independent motion. The particle size distribution remains the same as in the original stable system, but the particles are now weakly linked together by specific colloidal-chemical forces. The nature of these forces might arise from various sources, including perhaps the creation of polymer chains that form bridges between the particles, or electrodynamic forces corresponding to particles being held in the secondary minima.

Figure 8.1 illustrates these conceptual differences between the concepts of aggregation and flocculation. One might expect that the measured acoustic properties somehow reflect this difference between aggregation and flocculation, and in fact we have been able to observe such differences for several colloids. Here we will show results for rutile and alumina that can be interpreted as aggregation, and data for anatase that can be seen as flocculation. The aggregation data were previously published in Langmuir [1], whereas the flocculation data for anatase are presented here for the first time.

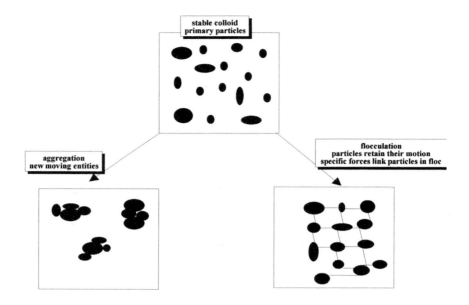

Figure 8.1 Illustration of the difference between aggregation and flocculation.

We present here data for alumina and rutile slurries that illustrate how attenuation spectra can be utilized to characterize an agglomeration process. We purposely chose two materials having quite different iso-electric points; pH 4 for the rutile, and pH 9 for the alumina. The rutile sample was prepared from commercially available aqueous rutile slurry, R746 from duPont. This slurry is supplied at 76.5%wt (44.5% by volume), but was diluted to 7%vl using distilled water. The particle density was 4.06 g/cm^3, which is slightly less than for some other rutile because of various surface modifiers used to stabilize the slurry. The manufacturer reported a median particle size of about 0.3 micron. The alumina slurry was prepared at a concentration of 11.6%wt (4%vol) from dry powder. The initial pH was 4. The manufacturer reported a median particle size of 100 nm; acoustically we measured 85 nm for the initial stable sample.

Titration of these alumina and rutile slurries and the corresponding measurements of the ζ-potential confirmed the expected iso-electric points for each material, as shown in Figure 8.2.

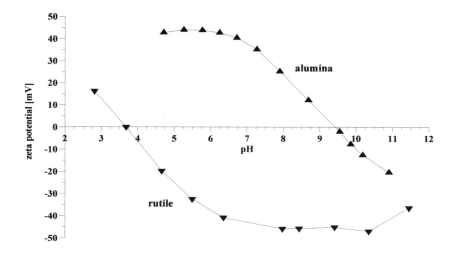

Figure 8.2 Electroacoustic pH titrations of rutile (7%vol) and alumina (4%vol).

According to general colloid chemistry principles, a dispersed system typically loses stability when the magnitude of the zeta potential decreases to less than approximately 30 mV. As a result, there will be some region surrounding a given isoelectric pH for which the system is not particularly stable. Within this unstable region, the particles may perhaps agglomerate, thereby increasing the particle size, and resulting in a corresponding change in the attenuation spectra. Indeed, this titration experiment showed that the attenuation spectra of both the alumina and rutile slurry does change with pH (Figure 8.3). For the rutile slurry, the attenuation spectra remained unchanged as the pH was decreased, until the zeta potential reached approximately -30 mV, at which point the spectra rapidly changed to a new state entirely. Similarly, for the alumina slurry, the attenuation spectra remained unchanged as the pH was increased, until the zeta potential reached +30 mV, at which point the spectra also changed dramatically. In both cases, that pH corresponding to the point where the magnitude of the zeta potential decreased to 30 mV became a critical pH, beyond which the pH could not be changed without degrading the stability of the slurry. In both cases, the spectra assumed a quite different, although more or less constant, shape within this unstable region: one spectra characteristic of the stable system, a second spectra characteristic of the unstable system.

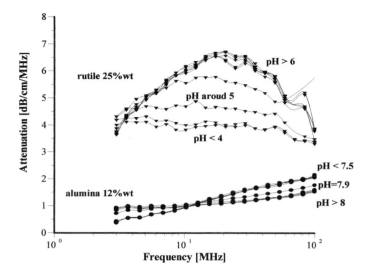

Figure 8.3 Attenuation spectra for 7%vl rutile (300nm) and 4%vl alumina (85nm), at different pH.

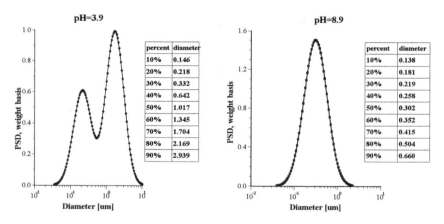

Figure 8.4 Particle size distribution of stable and unstable rutile at 7%vl.

The particle size distributions for the stable and unstable conditions for the rutile and alumina slurries are shown in Figure 8.4 and Figure 8.5 respectively. An error analysis presented in the following section proves that distribution becomes bimodal beyond the critical pH. It is interesting that the size of the aggregates do not exceed several microns. Apparently, there is some factor which prevents particles from collapsing into very large aggregates. We believe that these bimodal particle size distributions are stable with time because of stirring. The titration experiment requires the sample to be stirred to provide a rapid and homogeneous mixing of the added reagents. The shear stress caused by this stirring might break the larger aggregates. This shear can be a factor which restricts the maximum size that any aggregate might achieve. This idea can be tested in the future by making measurements at different stirring rates.

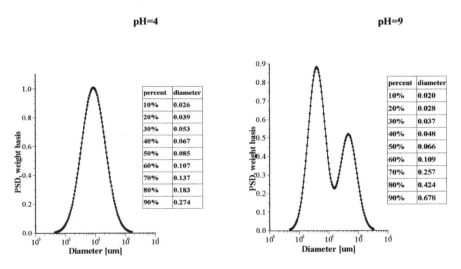

Figure 8.5 Particle size distribution of stable and unstable alumina at 4%vl

There remains another unresolved question. We don't know if the aggregation that we observe is reversible. Neither the alumina nor the rutile samples allow us to reach a ζ-potential above this critical 30mV value after crossing the iso-electric point. For both the alumina and rutile samples, the magnitude of the ζ-potential value is limited by the increased ionic strength at the extreme pH values. We will need another material with an iso-electric point at an intermediary pH in order to answer this question.

We can summarize this experiment on rutile and alumina slurries with the conclusion that the combination of Acoustics and Electroacoustics allows us to characterize the aggregation of concentrated colloids. Electroacoustics yields information about electric surface properties such as ζ-potential, while acoustic attenuation spectra provide data for particle size characterization.

In Chapter 3 we noted that the acoustic attenuation can sometimes be interpreted in rheological terms. Stokes, for example, 150 years ago showed that the viscosity of pure Newtonian liquids is proportional to its acoustic attenuation. Consider the rutile slurry in this light. The attenuation decreases near the isoelectric point. If we followed Stokes interpretation, this would suggest that the viscosity of the colloid decreases near the IEP, yet this contradicts our experience. Of course the rutile slurry is not a pure liquid, and the attenuation changes can readily be explained by changes in particle size resulting from aggregation of the primary particles. The rheological interpretation fails in this case because aggregation is essentially a heterogeneous effect. It needs to be interpreted using the notion of a "particle size distribution", which itself is heterogeneous in nature. A rheological interpretation assumes a homogeneous model for the colloid, which is apparently not valid for describing aggregation.

As for flocculation, it turns out that the relationship between these two possible interpretations of the acoustic attenuation spectra is reversed. We will demonstrate that such an aggregation based interpretation of the variation in the attenuation spectra of an anatase dispersion due to flocculation leads to meaningless results. At the same time, a rheological interpretation makes perfect sense, suggests an additional structural loss contribution to the total attenuation.

Whereas the rutile and alumina slurries exemplified an agglomeration process, we next present data for an anatase slurry, which illustrates how the attenuation spectra can be used to characterize flocculation as well. For this purpose, we selected a pure anatase supplied by Aldrich, following the suggestion of Rosenholm, Kosmulski *et al*. [2-6], who published several papers about the electrokinetic properties of this material at high ionic strength. We describe such an application later in this chapter. The anatase is easily dispersed at 2%vol in a 0.01 M KCl solution, resulting in an initial pH value between 2 and 3. Figure 8.6 shows the ζ-potential of this anatase during three back-and-forth titration sweeps over a wide pH range. During each pH sweep, the measured isoelectric pH was about 6.3. Following our earlier results on rutile and alumina, we might again expect a significant loss in stability when the magnitude of the zeta potential was reduced to less than 30 mV, corresponding in this case to an unstable pH range of 5-8. This loss in stability might then be expected to produce some change in the particle size distribution, along with a concomitant change in the attenuation spectra, in much the same way as we observed with the rutile and alumina slurries.

Indeed, measurements of the attenuation versus pH during these three pH sweeps did confirm that the attenuation varied with pH. However, this variation was quite different from that observed during the aggregation of the rutile or alumina. Figure 8.7 shows selected attenuation curves illustrating the changes in the attenuation spectra resulting from changes in pH.

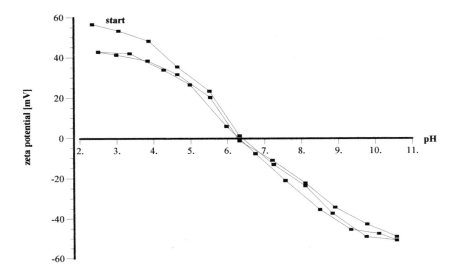

Figure 8.6 Three sweep pH titration of anatase at 2%vl in 0.01 M KCl solution.

After an initial adjustment caused by pumping, the attenuation reached the lowest level. This happens at pH 4.6, which is still in the stability region. Approaching the IEP more closely caused an increase in the attenuation, not a decrease as was the case for the rutile and alumina aggregation. In order to illustrate this behavior more clearly, we present in Figure 8.8 a plot of the attenuation, at only a single frequency (35 MHz), versus pH. It can be seen that this attenuation reaches a maximum value at the IEP.

There are two ways to evaluate this anatase experiment. The first approach is to assume that this sample is aggregating in the same manner as the rutile and alumina samples described above. We can calculate a new PSD from the attenuation spectra measured at each pH. The corresponding mean particle size and ζ-potential computed in such a way, is shown in Figure 8.9. We see that there is a dramatic correlation between the mean particle size and ζ-potential - but this correlation makes no sense. If we assume that an aggregation mechanism explains the changes in the colloid behavior, then the mean particle size should increase with decreasing ζ-potential, as we observed for the rutile and alumina slurries. For the anatase slurry, however, the size decreases with decreasing ζ-potential. Clearly our assumption of an aggregation mechanism must be incorrect. We need to look for some other mechanism to explain the attenuation behavior near the isoelectric pH.

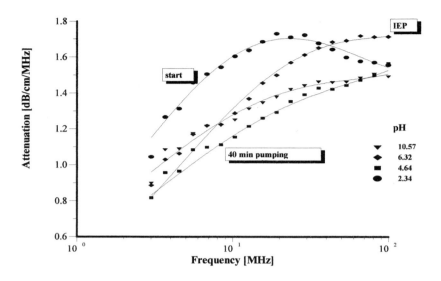

Figure 8.7 Attenuation spectra of 2%vl anatase in 0.01 M KCl, for different pH.

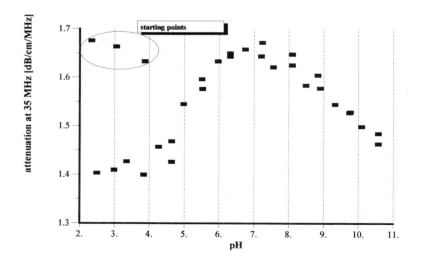

Figure 8.8 Attenuation of 2%vol anatase at 35 MHz versus pH.

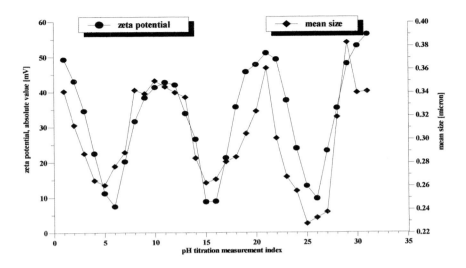

Figure 8.9 Mean particle size and ζ-potential of anatase assuming aggregation at IEP.

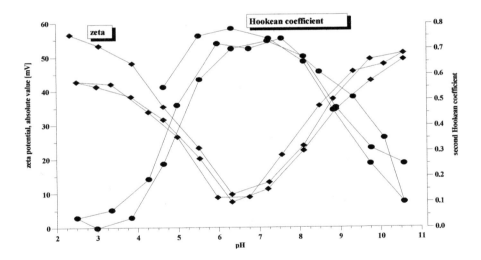

Figure 8.10 ζ-potential and Hookean coefficient H_2 for anatase at 2%vl versus pH, assuming flocculation at IEP.

For this anatase sample, let us consider replacing the initially proposed aggregation mechanism with an alternative flocculation mechanism. Rather than the creation of new large aggregates in the unstable pH region, instead we postulate some flocculation structure developing as was described in Figure 8.1. In Chapter 4 we introduced the concept of structural losses that can be described by two Hookean coefficients H_1 and H_2. We can now include the possibility of such structural losses, adding them to the viscous and other losses, when calculating the particle size from the attenuation spectra for this anatase sample. If we do this, we find that the PSD remains practically constant over the whole pH range, whereas the second Hookean coefficient, H_2, changes with pH as shown in Figure 8.10. H_2 reflects the variation with pH of the strength of the inter-particle bonds. These bonds become stronger closer to IEP, as one might expect. This might happen because of weakening electrostatic repulsion due to the decrease in ζ-potential decay in the vicinity of the IEP

In conclusion, we can formulate several general statements as follows:
- Electroacoustics provides a means to determine the IEP and the expected stability range of concentrated colloids.
- Acoustics allows us to distinguish between aggregation and flocculation
- In the case of aggregation, acoustic attenuation measurements yield information about variations in particle size.
- In the case of flocculation, acoustic measurements allow us to characterize not only the particle size distribution, but also the strength of the inter-particle bounds and the micro-viscosity of the colloid. To do this, the structural losses must be taken into account.

8.2. Stability of emulsions and microemulsions

The previous section discussed only solid rigid particles. In this section we will discuss means to characterize the stability of colloids having soft particles, such as latices, emulsions and microemulsions. Many papers amongst existing colloid science literature describe Acoustic and Electroacoustic techniques for characterizing such soft particles. Here we briefly present some of them, describing our own results in more detail.

The application of Acoustics to emulsions is associated to a great extent with names of McClements [7-15] and Povey [16,17,18]. Each leads a group that makes "Acoustics for Emulsions" a central theme of their research. Some of these papers are listed in the Applications Table at the end of this chapter. We mention here, in particular, experiments by McClements' group with water-in-hexadecane emulsions made in the early 90's [8,11,12], the patent by Povey's group for characterizing asphaltines in oil [20], and recent papers by McClements' group for testing a new theory for thermal losses in concentrates (see Chapter 4) using corn oil-in-water emulsions [21,22]. There are several reviews by these authors that summarize these results. In general, these studies showed that acoustic methods can provide droplet size distribution, and are suitable for monitoring surfactant effects, flocculation and coalescence, as well as monitoring phase transitions [23].

There is an interesting feature of these works: the raw data is most often a velocity measurement, whereas we prefer to use the measured attenuation. We suggest that readers interested in sound speed related results read the original papers of these authors. We illustrate such features of acoustic methods further down in this Chapter, by describing our own attenuation based experiments.

The first application of Electroacoustics to emulsions, as far as we know, is the CVP experiment made by Babchin, Chow and Samatzky [24] using relatively dilute (<10%vl) bitumen-in-water and water-in-crude oil emulsions. They showed that it was possible to monitor the effectiveness of a surfactant in achieving emulsion stability. Similar experiments by Isaacs, Huang, Babchin and Chow [25] showed that one could monitor the effectiveness of demulsifiers in water-in-crude oil emulsions, and that the change in CVP correlated well with the results from centrifugation and photomicrography.

Later, Goetz and El-Aasser [26] measured ESA and CVP using the Matec 8000 for toluene-in-water microemulsions that were stabilized by cetyl alcohol and sodium laurel sulfate. Although the trends were similar, whether measured electroacoustically or using microelectrophoresis, the amplitude of the ζ-potential measured electroacoustically was much lower than that measured by microelectrophoresis. This discrepancy was probably a result of using the O'Brien relationship as well as his dilute case theory for dynamic mobility.

The most recent work in the electroacoustics of emulsions is associated with O'Brien. For instance, O'Brien et al [27] have tested O'Brien's ESA theory for emulsions and other particles having low density contrast [28]. Particle size and zeta potential data were reported using the Acoustosizer for dairy cream, intravenous fat emulsions and bitumen emulsions at volume fractions up to 38%. Their results supported the possibility of characterizing concentrated emulsions using electroacoustics. These investigations have been continued by Kong, Beattie, and Hunter [29]. They applied ESA for characterizing the zeta potential and particle size of sunflower-in-water emulsions. Using O'Brien's theory for their calculations, they obtained meaningful results, including variations with temperature and surfactant content, and at volume concentrations up to 50%.

Our view on the best way to use ultrasound for characterizing soft particles such as emulsion droplets is quite different. For obtaining information on droplet size distribution and other related properties, we think that acoustic attenuation measurements are much more accurate, robust and reliable than electroacoustics measurements. We will support this statement with several experiments which follow.

One of our first studies with soft particles was made using a 5%vl water-in-car oil emulsion. We prepared this emulsion by simply adding water to the car oil, and then sonicating the mixture for two minutes. The emulsion remains stable for at least several hours. The two topmost curves of Figure 8.11 show the attenuation spectra for this emulsion; the bottom most curve shows the intrinsic attenuation of the pure car oil, which plays the role of background attenuation for the emulsion.

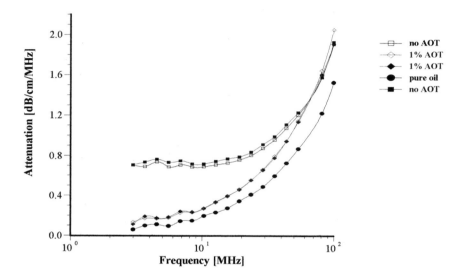

Figure 8.11 Attenuation spectra of the 5% water-in-car oil emulsion and microemulsion.

The right most curve in Figure 8.12 shows the droplet size distribution calculated from the attenuation spectra of the initial emulsion. The median particle size is about 1 micron, which correlates with our visual observation; the pure car oil is transparent, but becomes white after adding water and sonicating. This whitish appearance results from the light scattered by the relatively large water droplets.

In order to prove that the increase in the measured attenuation comes from the water droplets, we added AOT surfactant to reduce this emulsion to a microemulsion. The addition of just 1% AOT dramatically changed the attenuation spectra, as can be seen from the middle curves in Figure 8.11. At the same time, the whitish emulsion became completely transparent again, just like the initial untainted car oil. This visual appearance also proves that the droplet size became very small, indeed much smaller than the wavelength of light. The left most PSD curves on Figure 8.12 show that the addition of just 1% AOT reduced the droplet size more than two orders of magnitude, from 1 micron down to approximately 30 nm. The initial emulsion had become a microemulsion.

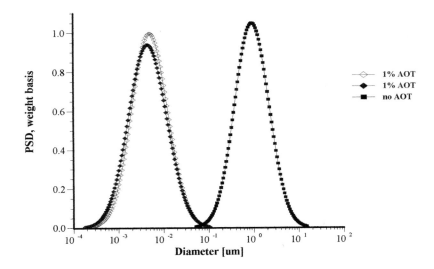

Figure 8.12 Droplet size distributions of 5% water-in-car oil emulsion and microemulsion.

We continued our study of surfactant related effects with classical water-in-heptane microemulsions using AOT as an emulsifier [30]. The idea of these experiments was to keep the surfactant content fixed, while increasing the amount of water in controlled steps. As the water content is increased, the limit of the available surfactant causes the droplet size to grow larger, based on the amount of available surfactant for stabilizing the droplet surface.

Figure 8.13 shows the attenuation spectra for these water-in-heptane microemulsions. It is seen that the attenuation grows with increasing amounts of water. This extra attenuation is certainly related to heterogeneity because the intrinsic attenuation of the heptane and water are much lower. Their superposition in the mixtures is much lower than the measured attenuation.

Figure 8.14 shows the droplet size distributions calculated from the attenuation spectra of Figure 8.13. The parameter R represents the ratio of water to AOT. The droplet sizes do indeed increase with increasing water content relative to the amount of AOT. The droplet size distribution becomes bimodal when the water content ratio R exceeds 50.

Acoustic spectroscopy is suitable for characterizing not only stable emulsions, but unstable ones as well. For instance, it is possible to monitor the changes in the attenuation spectra with time. Figure 8.15 shows an example of such time dependence for a 10%wt orange oil-in-water emulsion.

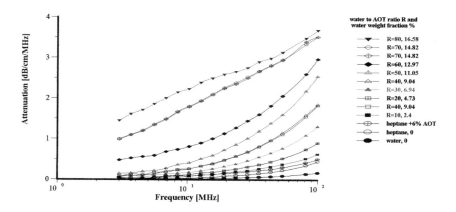

Figure 8.13 Attenuation spectra of water-in-heptane microemulsion. The AOT content is constant, but the water content varies.

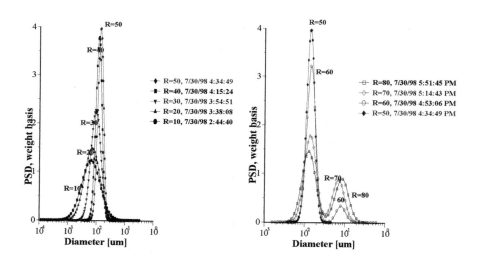

Figure 8.14 Droplet size distributions calculated from the attenuation spectra in Figure 8.13.

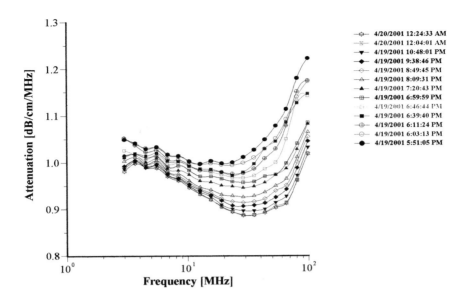

Figure 8.15 Attenuation spectra of the unstable 10%wt orange oil-in-water emulsion.

It is seen that the attenuation changes continuously over the course of the 18 hour experiment. There are two effects occurring simultaneously. First, the general attenuation decreases during the experiment; this decrease reflects the creaming of the orange oil, and the resultant decrease in the weight fraction of the oil phase which is still within the path of the sound beam. Second, in addition to this obvious creaming effect, the attenuation spectra changes shape, a reflection of the evolution of the droplet size, due to coalescence. Figure 8.16 shows the time dependence of the median droplet size and its standard deviation, as calculated from these attenuation spectra depicted in Figure 8.15. It is seen that the median size grows almost twofold, from 400 nm to 700 nm, in the first six hours. It is interesting that during this time, the width of the droplet size distribution becomes narrower, which implies that smaller droplets coalesce at a faster rate than the larger ones.

In general, the stability of an emulsion depends not only on the surfactant content, but also on any applied shear stress. By applying a shear stress to the emulsion, we might cause two opposing effects: coalescence of smaller droplets into larger ones, and a disruption of larger droplets into smaller ones. Quite often it is not clear which process might predominate. Acoustic spectroscopy offers a useful way to answer this question.

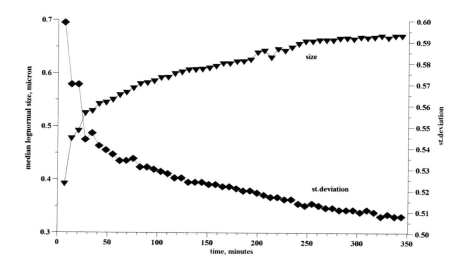

Figure 8.16 Time variation of orange oil droplet mean size and standard deviation.

To address the question concerning the effect of shear, the attenuation spectra for a 19%wt anionic oil-in-water emulsion was continuously measured, while the emulsion was pumped continuously for 20 hours with a peristaltic pump. The topmost curves in Figure 8.17 show the first and last attenuation spectrum from this long experiment. The gradual time dependence during this experiment is illustrated in the left hand plot in Figure 8.18, which shows the attenuation at only two selected frequencies (4 MHz and 100 MHz), versus time. It is seen that the shear-induced attenuation changes quite smoothly from the first to the last measurement. Even more significantly, however, we see that there are two quite different characteristic time constants associated with this shear-dependent change. The attenuation at 4 MHz changes rather quickly during the first 200 minutes, and then remains more or less constant. In contrast, the attenuation at 100 MHz changes exponentially over the 20 hour experiment, and apparently has still has not reached a steady state value at the end. We think that the variation of the low frequency attenuation reflects the breakdown of larger particles into smaller ones, whereas the change in the high frequency attenuations reflects coalescence of smaller particles into larger droplets. The right hand graph in Figure 8.18 illustrates the droplet size distribution at the beginning and end of the experiment.

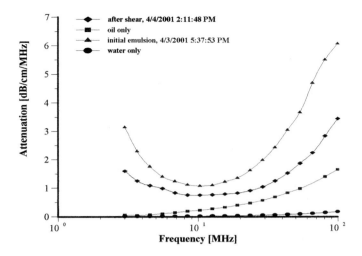

Figure 8.17 Attenuation of the 19%wt oil-in-water emulsion under continuous shear.

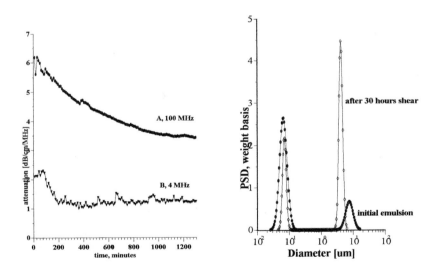

Figure 8.18 Variation of the attenuation and droplet size of the 19%wt oil-in-water emulsion, due to shear.

In general, the calculation of any PSD from a measured attenuation spectra requires knowledge of a certain set of input parameters. This problem has been discussed in some detail in Section 4.4. For submicron soft particles, such as emulsion or microemulsion droplets, we require information about the thermodynamic properties: the thermal conductivity τ, heat capacity C_p and thermal expansion β. Figures 2.1 and 2.2 previously illustrated the variation of these parameters for about 100 liquids. Fortunately, it turns out that τ and C_p are almost the same for all liquids, except water. In practice, this reduces the number of required thermal parameters to just one, the thermal expansion. This thermal expansion parameter plays the same role with respect to calculating "thermal losses" for soft particles, as does the density contrast in computing the "viscous losses" for hard, dense particles.

The thermal expansion is known for many liquids, but might present a particular problem for latices. The chemical composition of latex, and its preparation process, may dramatically affect the resultant thermal expansion coefficient. Initially, this was a serious obstacle for the development of acoustics as a characterization tool for latexes, but this situation has recently changed. It turns out that in many cases the thermal expansion coefficient, although unknown *a priori*, can actually be calculated from the measured attenuation spectra. Instead of being considered a required input parameter, the thermal expansion can now be regarded as an output parameter.

This section covered particle sizing only for simple colloids, with just one type of soft particles. Many real applications, however, are much more complex, and often consist of several different types of particles suspended in the liquid. The application of Acoustics to these more complex colloids is described in the following section.

8.3. Particle sizing in mixed colloids with several dispersed phases.

There are many important natural and man-made dispersed systems containing a high concentration of more than just one dispersed phase. For instance, whole blood contains many different types of cells, paint usually consists of latex with added pigment to provide color, and sunscreen preparations include an emulsion as well as ultraviolet light-absorbing particles. In many such systems there is a practical need to determine the particle size distribution of more than just one ingredient. In general, light-based techniques are not well suited for these applications, because most optical methods require the sample to be diluted prior to measurement, thereby distorting, or destroying altogether, the particle size information being sought. Furthermore, most light-based systems cannot handle multiple disperse phases, even in the most dilute case.

In contrast, acoustic attenuation spectroscopy enables us to eliminate this undesirable step of sample dilution. It is now well known that acoustic spectroscopy can characterize particles size at concentrations up to 50% by volume. Furthermore, acoustic attenuation spectroscopy can characterize the particle size distribution of concentrated dispersions having more than one dispersed phase. These two unique features make acoustic spectroscopy very attractive for characterizing the particle size distribution of many real-world dispersions.

There are at least three quite different philosophical approaches for interpreting the acoustic spectra in complex mixed colloids.

In the simplest "*empirical*" approach, we forego any size analysis *per se* and simply observe the measured acoustic attenuation spectra to learn whether, for example, the sample changes with time or if "good" or "bad" samples differ in some significant respect. Importantly, this empirical approach provides useful engineering solutions even in cases where nothing is known about the physical properties of the sample or whether, indeed, the sample is adequately described by our theoretical model.

In a more subtle "*validation*" approach, we assume in advance that we know the correct particle size distribution, and furthermore assume the real dispersion conforms to some well defined model. We then use some predictive theory based on this model, as well as the assumed size distribution, to test whether the predicted attenuation matches that actually measured. If the validation fails, it is a very strong indication that the model is inadequate to describe the system at hand.

As an example of this validation approach, consider the case where we construct a mixed system by simply blending two single-component slurries. The PSD of each single-component slurry can be measured prior to blending the mixed system. Since we have control of the blending operation, we know precisely how much of each component is added. If we claim that the combined PSD is simply a weighted average of the individual PSD for each component, we are in effect assuming that there is no interaction between these components. In this case the prediction theory allows us to compute the theoretical attenuation for this mixed system.

If the experimental attenuation spectrum matches the predicted spectrum then the assumption that the particles did not interact is confirmed. However, if the match is poor, it is likely that the mixing of the two components caused some changes in the aggregative behavior of the system. Perhaps new composite particles were formed by interaction of the two species. Or perhaps some chemical component in one sample interacted with the surface of another. Many interaction possibilities exist. Nevertheless, it seems appropriate to conclude that a necessary condition for ruling out aggregation in mixed systems is that the experimental and predicted attenuation curves match. In addition, we can probably also conclude that an error between theory and experiment is sufficient to imply that some form of aggregation or dis-aggregation has occurred within the mixed system. We will show that such prediction arguments are indeed able to monitor such aggregation phenomena.

Finally, we can take the ultimate leap of faith and use an "*analysis*" algorithm to search for that particle size distribution which, in accordance with the model and the predictive theory, best matches the experimental data. Whereas there are several papers [29-32] which demonstrate that acoustic spectroscopy is able to characterize bimodal distributions in dispersions when both modes are chemically identical, it is less well known that acoustic spectroscopy is also suitable for characterizing mixed dispersions when each mode is chemically quite different.

Here, we will discuss two models that can be particularly helpful for describing such mixed dispersions.

The first, "multi-phase", model assumes that we can represent the PSD of a real-world dispersion as a sum of separate lognormal distributions, one for each component in the mixed system. We assume that there are only two components, which then reduces the overall PSD to a simple bimodal distribution. When we calculate the attenuation of such a multi-phase system, we take into account the individual density, amongst other particle properties, for each component in the mixture. For a bimodal case, the multi-phase approach would typically require the analysis algorithm to fit five adjustable parameters: the median size and standard deviation of both modes, and the relative mass fraction of each mode. In this work we will assume that the weight fraction of each mode is known in advance. Furthermore, in an effort to avoid the well-known problem of multiple solutions, we will further assume that both modes have the same standard deviation. Altogether, these simplifications reduce the number of adjustable parameters to just three: the median size of each mode, and the shared standard deviation. The implications of these simplifications will be discussed later.

The second, "effective medium", model further assumes that one needs to determine the PSD of just one component in a multi-component mixed system. All other disperse phases are lumped together into an effective homogeneous medium, characterized by some composite density, viscosity, and acoustic parameters. By adopting this viewpoint, we significantly reduce a complex real-world mixture to a simpler dispersion of a single pre-selected dispersed phase, in a newly defined "effective medium". We need not even define the exact nature and composition of this new medium since we can simply measure, or perhaps calculate, the required composite density, viscosity, attenuation, and sound speed. If we assume that the key disperse phase can be described by a lognormal distribution, then we have reduced the degree of freedom to just two adjustable parameters, a median size and standard deviation.

Here we initially evaluate the effectiveness of both the "multi-phase" and the "effective medium" model for rigid particles with high density contrast.

These might be ceramic slurries, paints, and other oxides and minerals. Then we illustrate how this method works with the example of real cosmetic products, such as sun screens.

8.3.1. High density contrast - Ceramics, oxides, minerals, pigments.

We used three pigments from Sumitomo Corporation: AKP-30 alumina (nominal size 0.3 micron), AA-2 alumina (2 microns), and TZ-3YS zirconia (0.3 micron). A mixture of such oxides plays an important role in many ceramic composites [36]. In addition we used precipitated calcium carbonate (PCC) (0.7 micron) and Geltech silica (1 micron). Results of these tests are published in the paper [35].

Slurries of the AA-2 alumina and the zirconia were prepared in such a manner as to have quite good aggregative stability. Each slurry was prepared at 3%vl. by adding the powder to a 10^{-2} mol/l KCl solution, adjusted initially to pH 4 in order to provide a significant ζ-potential. Although the alumina showed very quick equilibration, the zirconia required about two hours for the zeta potential and pH to equilibrate. Electroacoustics allows us to establish this equilibration time by making continuous ζ-potential measurements over an extended time. This will be discussed later in Section 8.5.

Preparation of a 3 %vl PCC slurry was more problematic, since the ζ-potential right after dispergating was very low (1.3 mV). Control of pH alone was insufficient, and we therefore used sodium hexametaphosphate in order to increase the surface charge and thereby improve the aggregative stability of this slurry. In order to determine the optimum dose we ran a ζ-potential titration as described later (Section 8.5). The ζ-potential reaches a maximum at a hexametaphosphate concentration of about 0.5% by weight, relative to the weight of the PCC solid phase. The Geltech silica and the AKP-30 alumina were used only as dry powders, being added to the PCC slurry as needed.

The goals of the experiment were met in the following steps.

Step 1. A single component slurry was prepared for the alumina, zirconia, and PCC materials as described above.

Step 2. The attenuation spectra of these single component slurries was measured, and the particle size distribution for each was calculated.

Step 3. Three mixed alumina/zirconia slurries were prepared by blending the above slurries in different proportions, and the attenuation spectrum for each mixture was measured.

Step 4. Geltech silica powder was added to the initial PCC slurry, and the attenuation spectrum was measured for this mixed system.

Step 5. AKP-30 alumina powder was added to the initial PCC slurry, and the attenuation spectrum for this mixed system was measured.

Step 6. The particle size distribution was calculated for all of the mixed systems using the "multi-phase model".

Step 7. The properties of the "effective medium" were calculated for all mixtures.

Step 8. The particle size distribution for each of these mixed systems was calculated using the "effective medium" model.

Step 9. The results of the particle size calculation using two different approaches were compared.

Step 10. The validation approach was used to test for possible particle interactions in the mixed systems.

The experimental attenuation spectra for the three single component slurries and five mixtures are shown in Figures 8.19 and 8.20. In order to demonstrate reproducibility, each sample shown in Figure 8.19 was measured at least three times. Mixture 1, in fact was measured yet a fourth time after a fresh sample was loaded just to show that sample handling was not a detrimental factor. It is clear that the reproducibility is sufficient for resolving the relatively large differences in attenuation between the different samples.

The attenuation spectrum for the single component slurries of the alumina, the zirconia and the PCC allow us to calculate the particle size distribution for each material. The calculated sizes are given in the Table 8.1 and Table 8.2, and it can be seen that these acoustically defined sizes agree quite well with the nominal sizes given by the producers of these materials.

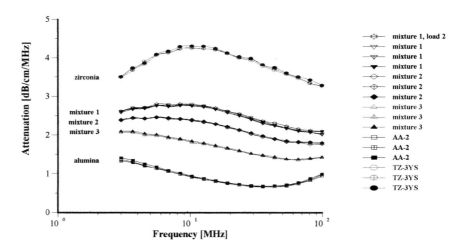

Figure 8.19 Experimental attenuation spectra for initial alumina AA-2 and zirconia TZ-3YS from Sumitomo and their mixtures with weight fractions given in the Table 8.1. This figure illustrates reproducibility by showing repeat measurements, including filling mixture 1 a second time.

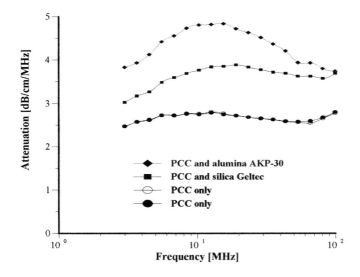

Figure 8.20 Experimental attenuation spectra for initial PCC slurry and its mixture with the added silica and alumina powders. Weight fractions are given in the Table 8.2.

As shown in figures 8.19 and 8.20, the attenuation spectra of the mixtures differ significantly from the attenuation spectra of the single component slurries. This difference in the attenuation spectra reflects the differences in both the particle size distributions and the density of the constituent components in the mixtures.

We want to compare the effectiveness of the "multi-phase" and the "effective medium" approach in calculating the PSD of these five different mixed systems.

First, let us consider the more or less straightforward "multi-phase" model. To use this approach we need only know the weight fraction and density of both disperse materials. The present DT-1200 software implementation assumes that the total particle size distribution is bimodal, and that each mode corresponds to one disperse phase material. In the alumina/zirconia mixture, for example, the smaller mode corresponds to the zirconia, and the larger mode corresponds to the alumina.

The software takes into account the difference in densities between materials of the first and the second modes. The PSD of each mode is itself assumed to be lognormal. In order to reduce the number of adjustable parameters, and in an effort to reduce the likelihood of multiple solutions, the present software implementation assumes that both modes have the same standard deviation. The software searches for some combination of the three adjustable parameters (two median sizes and their common standard deviation) that provide the best fit to the experimental attenuation spectra. It assumes the relative content of the modes to be known.

Table 8.1. Characteristics of alumina AA-2 and zirconia TZ-3YS slurries and their mixtures.

	Initial		Mixture 1		Mixture 2		Mixture 3	
	alumina	zirconia	alumina	zirconia	alumina	zirconia	alumina	zirconia
volume fraction,%	3	3	1.55	1.45	1.85	1.15	2.28	0.72
weight fraction, %	10.96	15.91	5.5	7.9	6.6	6.3	8.2	4
eff. viscosity [cp]			0.92		0.93		0.94	
eff. density [g/cm^3]			1.04		1.05		1.06	
att M0	1.593		1.21		0.982		0.823	
att M1	0.0845		0.0642		0.0521		0.0437	
att M2	-1.251		-0.95		-0.771		-0.646	
att M3	0.528		0.401		0.326		0.273	
Parameters of the particle size distributions, effective medium approach								
median lognormal [micron]	2.15 ±0.02	0.33 ±0.006	0.293 ±0.006		0.303 ±0.005		0.317 ±0.003	
st. deviation	0.26	0.43	0.38		0.378		0.372	
fitting error, %	6.6	1.9	1.4		1.2		0.95	
Parameters of the particle size distributions, two dispersed phases approach								
median size [micron]			0.565 ±0.002	0.558 ±0.001	2.922 ±0.088	0.352 ±0.005	3.582 ±0.182	0.303 ±0.003
st.deviation			0.53		0.3		0.21	
fitting error, %			5		7.6		4.4	

The corresponding PSD for these five mixed systems are shown in Figures 8.21 and 8.22. The parameters of these PSD are given in Tables 8.1 and 8.2. It is seen that in some cases this "multi-phase" approach yields approximately the correct size. For instance, the two zirconia/alumina mixtures with a lower zirconia content (mixtures 2 and 3) have almost the correct size combination. The size of the alumina particles is somewhat higher than expected (2.15 microns), but is still fairly acceptable. We can say the same about the PCC/alumina mixture from Table 8.2. The difference of the sizes relative to the nominal values does not exceed 10%.

Table 8.2. Characteristics of PCC slurry and its mixtures with alumina AKP-30 and silica Geltech.

	Initial PCC	Initial silica powder	PCC and silica		PCC and alumina	
			PCC	silica	PCC	alumina
Volume fraction,%	10.55		9.19	6.29	10.27	2.52
weight fraction, %	23.53		19.6	11.3	21.6	8.1
eff. viscosity [cp]	1.125		1.094		1.118	
eff. density [g/cm^3]	1.17		1.13		1.15	
att M0	1.053					
att M1	4.431					
att M2	-3.648					
att M3	0.9296					
parameters of the particle size distributions, effective medium approach						
median lognormal [micron]	0.684	1.26		0.454		0.325
st. deviation	0.31	0.35		0.015		0.015
fitting error, %	1.1	1.3		7.5		2.4
parameters of the particle size distributions, two dispersed phases approach						
median size [micron]			0.449	0.681	0.798	0.2715
st.deviation			0.16		0.19	
fitting error, %			8		1.9	

The multi-phase model, on the other hand, appears to be a complete failure for the alumina/zirconia mixture #1 as well as the PCC/silica mixture. It is not clear yet why this "multi-phase model" works for some systems and not for others. We think it probably is related to the fact that the present software assumes that both particle size modes have the same width. It is seen that the single component zirconia slurry has a PSD that is much broader (st.dev = 0.43) than the PSD of the AA-2 alumina (st.dev = 0.26). The bimodal searching routine finds the correct intermediate value for the standard deviation (0.3) only for mixture 2. It is interesting that this PSD solution is the closest match to the superposition of the initial PSD. The standard deviations for the other two mixtures are out of range completely, and the corresponding PSD also deviate from the expected superposition.

This observation allows us to conclude that our restriction that the standard deviation be the same for both modes might itself create a wrong solution. It is easy to eliminate this restriction, but in general as one adds additional degrees of freedom it is not uncommon to be faced with the problem of multiple solutions.

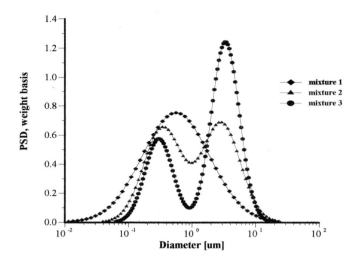

Figure 8.21 Particle size distributions calculated for alumina-zirconia mixtures using the "multi-phase model". The smaller size mode corresponds to zirconia, the larger size mode is alumina AA-2. The weight fraction and PSD parameters are given in Table 8.1.

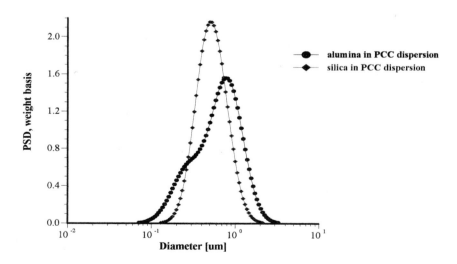

Figure 8.22 Particle size distributions calculated for PCC-alumina and PCC-silica mixtures using "multi-phase model". Weight fraction and PSD parameters are given in Table 8.2.

This multiple solution problem appears when the error function (difference between experimental and theoretical attenuations) has several local minimums resulting from different combinations of the adjustable parameters. In general, the problem of multiple solutions increases as the number of adjustable parameters increases. It seems clear that the maximum number of adjustable parameters that will avoid multiple solutions is not a fixed number, but rather depends on a combination of factors: the accuracy and amount of experimental data points, the degree to which the real world sample is described by the model, and how accurately the key parameters of the colloid such as weight fraction, density, etc are known. Our experience is that even bimodal PSD with only four adjustable parameters sometimes exhibit multiple solutions. We have found ways to resolve these multiple solutions in the case of single component dispersions, however the situation is more complicated for mixed dispersions with two or more chemically different components. For this reason, we restricted the number of the adjustable parameters to only 3 for this work.

These results indicate that the "multi-phase" model might sometimes lead to wrong solutions and it is unclear at this point how to completely eliminate the problem.

In contrast, the "effective medium" approach circumvents this problem by addressing only the question of determining a simple lognormal distribution, which describes only one disperse phase, in an otherwise complex mixture. Since we are then dealing only with two adjustable parameters (median size and standard deviation), the possibility for multiple solutions is diminished. On the downside, when using the "effective medium" approach we need to perform an additional experiment to measure the properties of this "effective medium", and this may not always be possible, or without other difficulties.

In the case of the PCC mixtures with the added alumina or silica, the original PCC slurry itself serves as the "effective medium". We need just three parameters to characterize this "effective medium" namely: density, viscosity, and attenuation. Importantly, all three parameters can be measured directly if we have access to this medium. The attenuation is the most important of these three required parameters. It is also the most challenging to characterize, because we need the attenuation of this medium as a function of frequency from 3 to 100 MHz. The current version of the DT 1200 software allows us to define the attenuation of the effective medium much the same way we would normally define the "intrinsic attenuation" of a pure liquid medium. This intrinsic attenuation as measured in dB/cm/MHz can be described in terms of a polynomial function:

$$\alpha(f) = M0 + f\,M1 + f^2\,M2 + f^3\,M3 \tag{8.1}$$

where f is frequency in MHz, and $M0$, $M1$, $M2$ and $M3$ are the polynomial coefficients.

For example, in the simplest case we can say that our effective medium is just water. Water has an attenuation that for practical purposes can be said to simply increase as a linear function of frequency, if attenuation is expressed in dB/cm/MHz. Thus *M0, M2*, and *M3* are zero and *M1* represents this linear dependence.

To use the effective medium approach for mixed systems, we simply need to define new coefficients which describe the intrinsic attenuation of this new medium. In the case of the alumina/zirconia mixtures we use the alumina slurry as the "effective medium". The coefficients for the alumina slurry can be calculated by fitting a polynomial to the attenuation data, as shown in Figure 8.23. These coefficients are also given in Table 8.1. Similarly, the coefficients for the PCC "effective medium" can be calculated from a polynomial fit of the attenuation data for that material, as shown in Figure 8.23. Likewise, these coefficients are given in Table 8.2.

We should keep in mind that the initial alumina slurry is diluted when we mix it with increasing amounts of the zirconia slurry. As a result, we need to recalculate the attenuation coefficients for each mixture, taking into account the reduced volume fraction of the alumina in each mixture. The suitably modified values for the attenuation coefficients of the effective medium for all three alumina/zirconia slurries are also given in Table 8.1. We avoided the need for making these additional calculations in the case of the PCC mixtures, by simply adding dry silica or alumina powder to the PCC effective medium, and therefore the coefficients for the PCC effective medium are the same for both mixtures.

For an aqueous medium, the software automatically calculates the intrinsic attenuation of water and subtracts this from the measured attenuation, in order to deduce the attenuation caused solely by the presence of the disperse particles. When using the "effective medium" model, the software actually works in the same way, except that the intrinsic attenuation of water medium is replaced by the attenuation of this new effective medium. For instance, in the case of the PCC/alumina mixture, the software calculates the attenuation due to the PCC contribution, and subtracts it from the total attenuation of the mixture. The residual part corresponds to the attenuation due to the alumina particles, and is the source of the particle size information for the alumina component. The software assumes a lognormal PSD, and fits this residual attenuation using the median size and standard deviation as adjustable parameters.

This effective medium approach allows us to calculate the particle size distribution of the zirconia in the alumina/zirconia mixtures, and of the silica or the alumina in the case of PCC mixtures. The corresponding values are shown in Tables 8.1 and 8.2. Figure 8.24 illustrates the corresponding PSD for each case.

In the case of zirconia we have almost the same PSD for all three mixtures. This PSD agrees well with the initial slurry. The fitting error is much smaller than in the "multi-phase model" which is an additional indication of the consistency of this method.

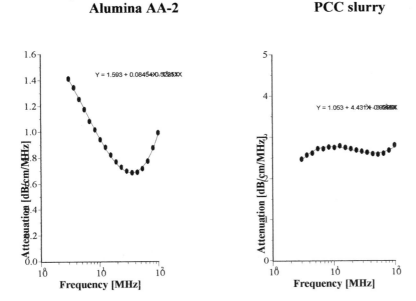

Figure 8.23 Experimental attenuation spectra measured for individual alumina AA-2 slurry and PCC slurry with polynomial fit.

In the case of PCC mixtures the situation is more complicated. We have a very good correlation with the nominal size of the AKP-30 alumina in the case of the PCC-alumina mixture, and a good fitting error.

However, the other PCC based mixture gives a particle size which is half that expected. Table 8.2 shows that the calculated size of the silica Geltech in this mixture is only 0.454 microns, whereas the nominal size is at least 1 micron. Acoustically we measured an even larger 1.26 microns for this silica: perhaps the size was slightly larger than reported by manufacturer because problems in dispersing this sample. We have found that this silica is difficult to disperse even at high pH and at high zeta potential. For instance, we measured a ζ-potential of -66 mV for this silica at pH 11, but even this was apparently not sufficient to disperse it completely.

Summarizing the Analysis results for these five mixtures, we conclude that in the case of the three mixed dispersions (alumina-zirconia mixtures 2 and 3, and the PCC-alumina mixture), the "multi-phase model" and the "effective medium model" gave similar results and reasonable PSD. For the other two mixtures, the results are more confusing. We suspect that the failure of the "multi-phase model" for the alumina-zirconia mixture 1 is related to the restriction on the PSD width, but particle aggregation is still a candidate as well. In the case of the PCC-silica mixture, a double failure of both models certainly points towards particle aggregation.

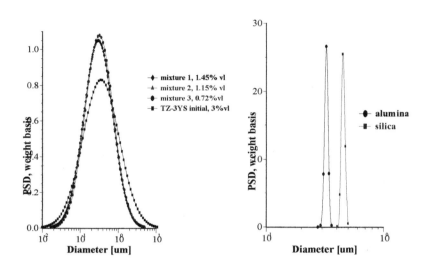

Figure 8.24 Particle size distribution calculated using "effective medium" model. In the case of zirconia the alumina AA-2 dispersion is the effective medium. Attenuation of the alumina is reduced according to volume fractions from Table 8.1. The density and viscosity are adjusted for the effective medium. In the case of alumina AKP-30 and silica the PCC dispersion acts as the effective medium.

We can evaluate these ideas about aggregation of the two difficult mixtures using the "validation" approach. To do this we must first compute the total PSD, using the known PSD of the individual single component dispersions. Next, we calculate the predicted attenuation for this combined PSD. This predicted attenuation should agree with the experimental spectrum for the mixed system, if there is no particle interaction between the species.

Figure 8.25 illustrates the predicted and experimental attenuation spectrum for the Zirconia-alumina mixture 1 and the PCC - silica Geltech mixture. For both mixtures we have also added the predicted attenuation corresponding to the best PSD calculated using the "multi-phase model" analysis.

It is seen that in the case of the zirconia-alumina mixture, a superposition PSD generates an attenuation spectrum that fits the experimental spectra much better than the best "multi-phase model" PSD. The fitting error has improved from 5% to 2.3%, and becomes comparable with the best fitting errors of the "effective medium" model. This correlation between Prediction and Experiment proves that our concern about using a common standard deviation for both modes was well founded. The Prediction program allows us to apply independent standard deviation for each mode of the PSD, and as a result we achieve much better fitting than in the case of the Analysis "multi-phase" model that uses the same standard deviation for both modes.

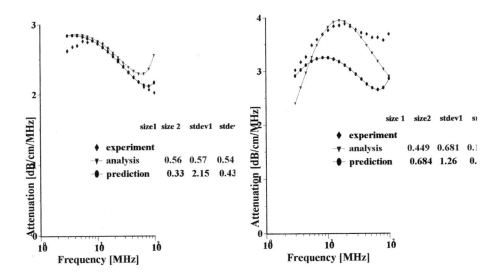

Figure 8.25 Experimental and theoretical attenuation for zirconia-alumina mixture #1 and the PCC-silica mixture. The theoretical attenuations are calculated for the best Analysis result, and for the combined PSD built from the individual distributions, assuming no particle aggregation.

In addition, we conclude that there is no aggregation between the alumina and zirconia particles in this mixed dispersion. Otherwise, the theoretical attenuation based on the superposition assumption would not fit the experimental data.

The situation with the second mixture (PCC-silica) is very different. In this case the predicted attenuation provides a much worse fit than the best "multi-phase" model analysis. The fitting error degrades from 8% to 17.2%.

This means that the superposition assumption is not valid. In this case there is apparently some aggregation between the PCC and silica particles.

We can conclude after these tests that Acoustic Spectroscopy is able to characterize the particle size distribution of dispersions with more than one dispersed phases. There are two models we can use to describe the dispersion: a "multi-phase" model and an "effective medium" model.

The "multi-phase" model describes the total distribution as a number of separate lognormal distributions, one for each disperse phase. Although this provides a complete description of the system, it also entails certain risks of multiple solutions because of the large number of adjustable parameters. For this reason we assumed here that the width for each mode of our binary mixtures was the same. This assumption might lead in some cases to an incorrect solution, especially when dealing with mixture of dispersed phases with widely different standard deviations.

The "effective medium model" is not complicated by multiple solution problems. It requires only two adjustable parameters for characterizing the lognormal distribution of the selected dispersed phase. All other components of the mixed dispersion are considered as a new homogeneous "effective medium". We assume that the attenuation of this "effective medium" is the same as measured for the mixed materials separately. In those cases where this assumption is valid, the "effective medium" model yields robust and reliable PSD for the selected dispersed phase. In the opposite case, when the assumption is not valid, we obtain a PSD that differs from the expected particle size of this dispersed phase. Observation of this difference can be used as an indication of particle-particle interactions between the particles from the "effective medium" and the selected dispersed phase.

In addition to the particle size distribution, acoustic attenuation spectroscopy is able to indicate the presence of particle aggregation. To perform this test, we need to calculate the attenuation spectra for the PSD that is built from the individual particle size distributions using the superposition assumption. Failure of this theoretical attenuation to fit experimental data is an indication of the particle aggregation.

8.3.2. Cosmetics - Sunscreen

Sun screen cosmetics are a very interesting and complex application. Sunscreens can be described as an oil-in-water emulsion with solid light absorbing particles mixed in the oil phase. The most widely used solid particles are rutile and zinc oxide. For this application we can speak of two different "continuous phases": a water phase for the oil droplets, and an oil phase as the dispersion medium for the solid particles. There are also two dispersed phases: the oil droplets in the water and the solid particles in the oil.

Correct characterization of the size distribution of the solid particles size is critical to insure the performance of sunscreen preparations. The efficiency of these sun screens in adsorbing ultra violet radiation, thereby preventing it from penetrating to the skin, depends strongly on the particle size. This particle size characterization must be done in the final sunscreen preparation. Any dilution of the final product might well affect the particle size distribution and yield misleading and possible dangerous results.

The fact that the attenuation spectra of the complete sun screen moisturizer is sensitive to particle size follows from the data presented in Figure 8.26 for moisturizers with two different ZnO particles: Microfine ZnO with a size 165 nm, and USP ZnO with a size of about 500 nm. There are two ways to extract this information.

First of all, we can measure the attenuation spectra of the oil phase containing the solid active ingredient. This is a simple two phase system. In order to calculate particle size from the total attenuation curve, the intrinsic attenuation of the pure oil phase is required. The attenuation spectra corresponding to the pure oil, and this oil with added 14%wt USP ZnO are shown in Figure 8.27.

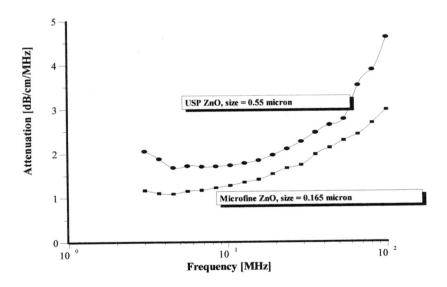

Figure 8.26. Attenuation of two sunscreens with identical chemical composition, but having ZnO particles of different size.

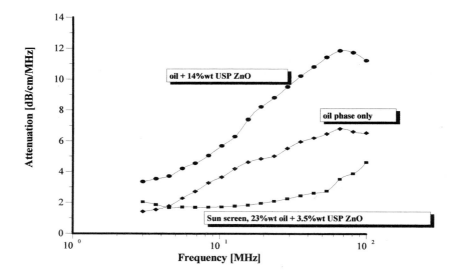

Figure 8.27 Attenuation spectra of the complete sunscreen and its components.

The attenuation of the pure oil allows us to estimate the viscosity of the oil phase. This is important because this phase often includes various additives. We can make an estimate of the viscosity using Stokes law, which relates the viscosity to the measured attenuation (see Chapter 3). The attenuation of the oil is about 35 times higher than for water, which means that we might estimate the viscosity of the oil phase as about 35 cp. The particle size calculated from the attenuation spectra using this viscosity and intrinsic attenuation is shown in Figure 8.28. It is very close to the expected value provided by the manufacturer.

A second way to characterize the particle size using the attenuation spectra can be accomplished by measuring the oil-in-water emulsion alone, without particles. This emulsion would represent the background for the particles in the attenuation spectra of the complete sun screen moisturizer. It would allow us to calculate the intrinsic attenuation, and extract the particles contribution to the total attenuation spectra. Figure 8.29 illustrates this approach, providing attenuation curves for a complete moisturizer prepared with Microfine ZnO and the emulsion only. The corresponding particle size is shown in Figure 8.30. It is very close to the value reported by the manufacturer for this ZnO.

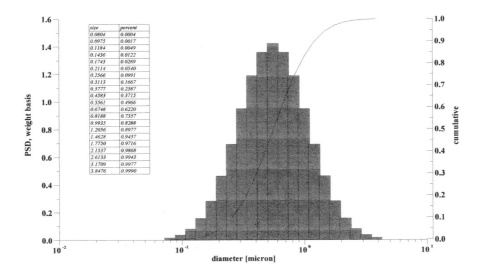

Figure 8.28 Particle size distribution of the ZnO particles.

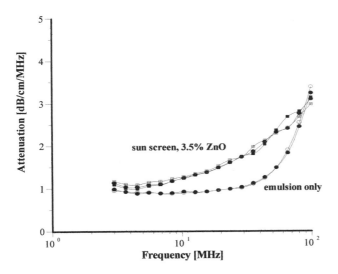

Figure 8.29. Attenuation spectra of the complete sunscreen, and of the basic emulsion.

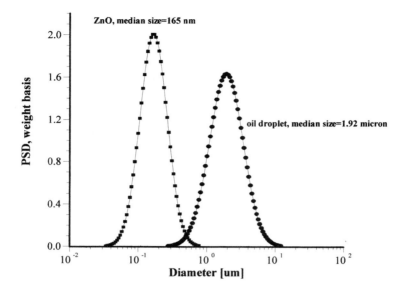

Figure 8.30. Particle size distribution of the ZnO particles and oil droplets.

It is important to point out that in this second approach we can calculate the size of the oil droplets. The attenuation for the emulsion consists of the intrinsic attenuation of water and oil, plus the thermal losses caused by the added oil droplets. Figure 8.30 shows the droplet size distribution for this emulsion.

In summary, we can conclude that acoustic spectroscopy is able to provide information about size of both the solid active ingredient, as well as the size of the emulsion droplets of sun screen moisturizers. In addition, it yields information about the rheological properties of these sun screens.

8.3.3. Composition of mixtures

Acoustic spectroscopy provides a means to determine the relative amount of a substance in a mixed dispersion comprised of several different materials. In the case of rigid particles, this opportunity is related primarily to the density differences between the various dispersed phase materials. A variation in the relative amount of each material causes a variation in the total volume fraction, and consequently the attenuation spectra. Experimental monitoring of these variations in the attenuation spectra allows one to characterize the corresponding variation in the composition of the dispersed phases. This idea was suggested by Kikuta in a US patent [138].

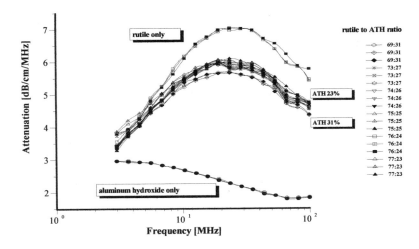

Figure 8.31 Attenuation spectra measured for rutile, aluminum hydroxide and their mixtures at 25%wt total solid.

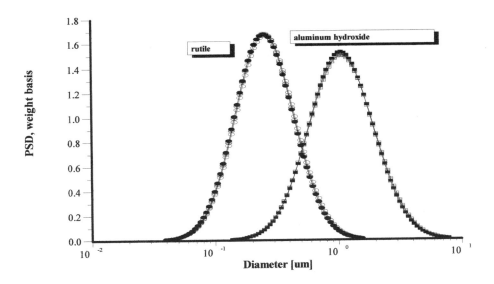

Figure 8.32 Particle size distributions of rutile and ATH calculated from the attenuation spectra in Figure 8.31.

We illustrate this application with an example of a mixture of rutile and aluminum hydroxide. Figure 8.31 shows the measured attenuation spectra for the initial single phase dispersions of rutile and aluminum hydroxide (ATH) as well as their mixtures. The content of ATH in these mixtures varies from 23% to 31%. Each sample was measured three times, in order to demonstrate reproducibility of the measured attenuation spectra, which in turn indicates the reliability of this method for characterizing ATH content.

The variation in the measured attenuation spectra reflects not only changes in the volume fraction and density, but the particle size as well. The attenuation of the mixture depends on the particle size distributions of the components. This fact is used in previous sections for characterizing particle size distributions of mixed dispersions. Particle size dependence is a potential obstacle for characterizing mixture composition. Fortunately, it is possible to minimize this factor's influence by proper selection of the frequency range. Figures 8.32 and 8.33 illustrate this procedure. Figure 8.32 shows particle size distributions calculated from the attenuation spectra measured for rutile only and ATH only. Note that the rutile particles are much smaller than ATH particles.

We can assume that the particle size distribution of the each material remains the same in the mixture, because the stability ranges for these two materials are about the same. Using this assumption, we can calculate theoretical attenuation spectra for mixtures with different ATH. content. These theoretical attenuation spectra are shown in the Figure 8.33. It can be seen that the level of ultrasound attenuation is almost proportional to the ATH content. The variation is relatively small at low frequency, and reaches a maximum at the mid frequency range. This means that the mid frequency range (from 20 to 40 MHz) is the most desirable for characterizing ATH content. Attenuation of ultrasound is the most sensitive to the ATH particles in this frequency range.

It turns out that the second factor determining ultrasound attenuation, particle size distribution, is also pointing to the mid frequency range as the most suitable for calculating ATH content. Ultrasound attenuation is least sensitive to particle size within the mid frequency range. Figure 8.34 illustrates this statement. The attenuation spectra shown in this figure are calculated for various particle size distributions. The particle size of rutile varies from 0.25 to 0.35 micron, and the particle size of ATH varies from 0.7 to 2 microns.

There are two conclusions which follow from the attenuation spectra in Figure 8.34. First of all, variation of the ATH particle size affects the attenuation spectra much less than variation of the rutile particle size. These variations become minimal in the mid frequency range between 20 and 40 MHz. The values of ultrasound attenuation given in the Table 8.3 indicate that the attenuation varies as little as 0.1 dB/cm/MHz for the rather wide variation of the particle sizes mentioned above.

Table 8.3. Attenuation of 75:25 rutile-aluminum hydroxide mixture at 25% total weight fraction for selected frequencies and particle size distributions. Standard deviation of each fraction is assumed to be 0.32.

frequency [MHz]	experiment	d1=0.3 d2=1.46	d1=0.3 d2=1	d1=0.3 d2=0.7	d1=0.3 d2=2	d1=0.25 d2=1	d1=0.28 d2=1	d1=0.33 d2=1	d1=0.35 d2=1
19.2	5.718398	5.600715	5.645591	5.679868	5.56728	5.541092	5.604283	5.66003	5.647861
23.5	5.651757	5.612663	5.658526	5.697022	5.582969	5.619873	5.651227	5.636786	5.594701
28.9	5.656034	5.587781	5.631235	5.67256	5.565599	5.659421	5.658074	5.574058	5.503365
35.5	5.506359	5.527928	5.568101	5.608497	5.517046	5.658066	5.624988	5.478223	5.381812

This short analysis indicates that the frequency range between 20 and 40 MHz is optimal for characterizing ATH content in rutile-ATH mixtures. The attenuation spectra is the most sensitive to the ATH particles in this frequency range, and the least affected by the possible variation of the particle size distribution of the components.

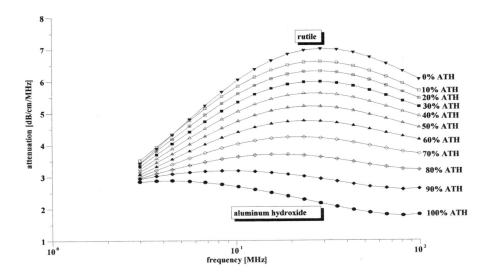

Figure 8.33 Attenuation spectra calculated for rutile-ATH mixture at 25%wt assuming particle size distributions from Figure 8.32.

The precision of the measurement of ATH content is related to the precision of the attenuation measurement. The total difference in attenuation between pure rutile and pure ATH is about 5 dB/cm/Mhz in this frequency range, as follows from the curves in Figure 8.31. Attenuation varies with ATH content almost as a linear function, according to Figure 8.33. This means that a variation of ATH content of 1% causes a variation of attenuation of about 0.05

dB/cm/MHz. It turns out that we can measure attenuation within this frequency range with such a high precision. Figure 8.35 presents results of multiple measurements performed with rutile-ATH mixtures with different component ratios. The frequency of the measurement is 40 MHz. The step size of the component ratio is about 1.1% of ATH. Eleven measurements were made for the each value of the component ratio. There is a clearly reproducible difference between the attenuations corresponding to the different component ratios. This means that this instrument allows us to characterize the content of ATH in the rutile-ATH mixtures with a precision of less than 1%.

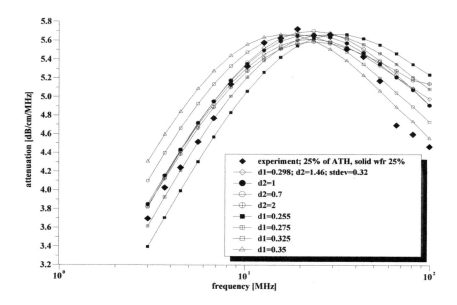

Figure 8.34. Theoretical attenuation spectra for rutile-ATH mixtures with different individual sizes.

This high level of precision is achievable in the particular case of the rutile-ATH mixtures. However, we cannot generalize this precision claim for every mixed dispersion. A similar analysis would be required in each particular case.

The suggested acoustic method of determining component ratios requires information about the total solid content. There are two ways to gain this information. The first one is a simple drying procedure, which is widely used in many industrial applications.

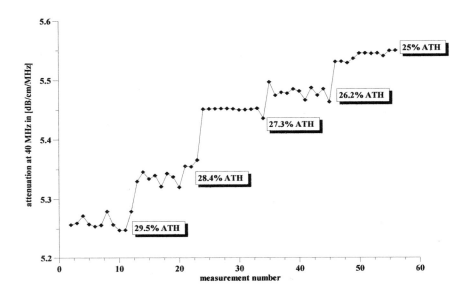

Figure 8.35. Attenuation of rutile-ATH mixture at 40 MHz for different values of ATH content.

The other method is to use density measurements. The density of the dispersion, ρ_s depends on the total weight fraction of solids, w_s, as well as on the ratio of the partial weight fraction of the components, w_{p1} and w_{p2}. The following formula relates these parameters:

$$w_{p1} = \frac{\rho_{p1}\rho_{p2}(\rho_m - \rho_s) + w_s\rho_s\rho_{p1}(\rho_{p2} - \rho_m)}{(\rho_{p2} - \rho_{p1})\rho_s\rho_m} \qquad (8.2)$$

$$w_{p2} = \frac{\rho_{p1}\rho_{p2}(\rho_m - \rho_s) + w_s\rho_s\rho_{p2}(\rho_{p1} - \rho_m)}{(\rho_{p1} - \rho_{p2})\rho_s\rho_m} \qquad (8.3)$$

where index 1 and 2 correspond to the first and the second solid phases.

In principle, density measurements, combined with a drying method for determining the total solid content, can provide information about the ratio of component weight fractions.

8.4. Chemical-mechanical polishing - Large particle resolution

Modern chemical-mechanical polishing (CMP) materials present a new challenge for measuring techniques. One can find extensive investigations of various characterization techniques for CMP slurries in publication by Anthony et al [37,38], as well as our own [34].

Three aspects of this application cause difficulties when using instruments based on traditional techniques. First, the particle size of a typical CMP slurry is too small for sedimentation based instruments or electric zone instruments. Typically, the mean size of CMP materials is approximately 100 nm, with no particles, or preferably just a few, larger than 500 nm. The desired range over which particle characterization is desired is very large, which eliminates most classical techniques. Third, CMP systems are typically shear sensitive. Shear caused by the delivery system or the polishing process itself may cause an unpredictable assembly of the smaller particles into larger aggregates. However, these aggregates may be weakly formed, and easily destroyed by subsequent sonication, high shear, or dilution. Therefore, any technique which requires dilution or other sample preparation steps may in fact destroy the very aggregates that one is attempting to quantify by measurement. We suggest that CMP systems must be characterized as is, without any dilution or sample preparation.

Acoustic spectroscopy provides an exciting alternative to the more classical methods; this technique resolves all three issues mentioned above. This is the conclusion of the Bell Labs tests [37,38] and we completely agree with it after our own extensive work using Dispersion Technology instruments.

In addition to particle sizing, ultrasound offers a simple and reliable way to characterize ς-potential using electroacoustics. The importance of electrokinetic characterization of CMP slurries are stressed by Osseo-Asare [39].

In this section, we present results of our work with CMP slurries using Dispersion Technology instruments for characterizing both particle size using Acoustics and zeta potential using Electroacoustics.

We emphasize here one feature of Acoustic Spectroscopy that thus far has not been described sufficiently in the literature: namely the ability to characterize a bimodal PSD. Although Takeda *et al* [29-31] demonstrated that Acoustic Spectroscopy is able to characterize bimodal distributions of mixed alumina particles, the ultimate sensitivity in detecting one very small sub-population in combination with another dominant mode was not studied. Yet, it is just this feature, the ability to recognize a small sub-population, which is most critical for CMP studies. This section addresses this important issue.

Unfortunately, there is no agreement in the literature as to the number of larger particles which might be acceptable in a CMP slurry. In Chapter 7, we have shown that the DT-100 Acoustic Spectrometer has resolution of bout one large particle of 1 micron size per 100,000 100 nm small particles. This sensitivity corresponds to large particles amounting to about 1% of the total weight of all particulates. This level of particle size resolution requires a precision in the attenuation measurement of roughly 0.01 dB/cm/MHz. We have shown experimentally in Chapter 7 that the Dispersion Technology instruments indeed meet this target requirement.

A set of experiments was made to test whether the attenuation spectra changed reproducibly, when a small amount of larger particles was added to a single component slurry of smaller particles. Two slurries were used for the small particles: Ludox-TM and Cabot SS25. Two Geltech silica with nominal sizes 0.5 and 1.5 micron were used as the model large particles. It was shown that the change in the attenuation spectra was statistically significant when the large particles amounted to at least 2 % of the total weight of all particulates. Expressed another way, the detection limit for this 12 wt % slurry corresponded to a sub-population which was only 0.24 wt % in terms of the total sample weight, or 0.24 g of large particles per 100 g of the slurry.

We used four silica materials altogether. Two small size particles were used, namely duPont Ludox-TM (50%wt) and Cabot SS25 (25%wt). Two large size particles were employed, namely Geltech 0.5 micron and Geltech 1.5 micron. We assumed a density of 2.1 g/cm^3 for all silica particles.

Slurries were prepared at 12 weight % for each material as follows.

The Ludox-TM was diluted to 12 wt % with 0.01 M KCl solution, resulting in a sample pH of 9.3.

The Geltech samples were supplied as a dry powder, which was dispersed in 0.01 M KCl solution, and adjusted to pH 9.6 with KOH. The dispersion was repeatedly sonicated, stirred, and allowed to equilibrate for 5 hours before being measured.

The Cabot SS25 silica was diluted to 12 wt % with 0.01 M KCl.

Table 8.4 presents particle size data provided by the manufacturer for each of these samples.

The goals of the experiment were met in three steps.

Step 1 - Reproducibility: The objective was to prove that the DT-1200 Acoustic Spectrometer is able to measure the attenuation spectra with the required precision of 0.01 dB/cm/MHz. To prove this, we measured five different fillings of the same 12%wt silica Ludox-TM. Each filling was measured nine times. A statistical analysis of the measured attenuation spectra yielded an average variation of the attenuation measurement, over the frequency range from 3 to 100 MHz. The details were given in Chapter 7.

Step 2 - Accuracy: The goal here was to prove that acoustic spectroscopy can characterize the particle size with sufficient accuracy. To achieve this goal all five silica slurries were measured individually. The measured particle size was then compared with independent information provided by the manufacturer.

Step 3 - Bimodal Sensitivity: This is the key step in the investigation. Model systems with bimodal PSD were prepared by mixing a small particle slurry (Ludox-TM or Cabot SS12) with increasing doses of the larger particle slurries (Geltech 0.5 or Geltech 1.5). The objective was to evaluate the accuracy of the PSD calculated by the DT-1200 software for these known systems.

The attenuation spectra for all five single component silica slurries is shown in Figure 8.36 and the corresponding particle size distributions for these same samples is shown in Figure 8.37. These tests allowed us to compare the particle size determined by acoustic spectroscopy for all five 12 wt % test slurries with independent data from the manufacturers. The values of the median size for each case is shown in Table 8.4. It is interesting to note that there is some difference between the acoustically measured data and that provided by the manufacturer. In large part this is related to differences in the characterization technique.

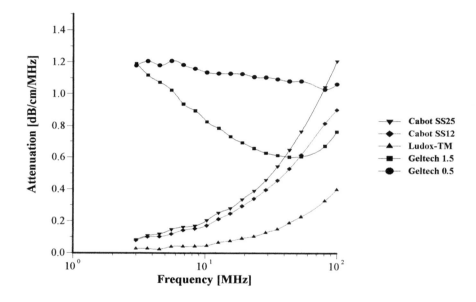

Figure 8.36 Attenuation spectra measured for Ludox TM silica, Geltech 0.5 and 1.5 silica, Cabot SS25 and SS12 silica. Total solid content is 12wt% except for SS25 which is 25wt%.

For instance, the size of the Ludox-TM slurry is determined by duPont using a titration method. This method yields an average size on an area basis. Acoustic spectroscopy gives us a size on a weight basis, which for a polydisperse system will always be somewhat larger than an area based size. In addition, acoustic spectroscopy implies some assumption about the real dispersed system when particle size is being calculated from the attenuation spectra. Any measuring technique does the same. These assumptions, and variations in the physical properties which are involved in the calculation can cause some variation in calculated size as well.

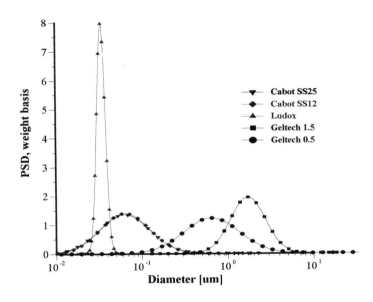

Figure 8.37 Particle size distributions calculated from the attenuation spectra in Figure 8.36.

Table 8.4 Particle size of the initial silica samples, expected and measured acoustically.

	Manufacturer	Acoustics
Ludox –TM	22 nm (area basis)	30 nm (weight basis)
Geltech 0.5	0.5 micron	0.65 micron
Geltech 1.5	1.5 micron	1.72 micron
Cabot SS12		63 nm
Cabot SS25		62 nm

Successful reproducibility and reasonable agreement with other techniques encouraged us to move to the third step, which is a test of the ability to correctly determine the bimodal PSD. We used Ludox TM and CMP SS12 as the major component of two test slurries. The Geltech 0.5 or Geltech 1.5 were added as "large" and "larger" particles to the above test slurries to form various mixed slurries. In each case the minor fraction was added to the Ludox-TM or the CMP SS12 systems in steps. Each addition increased the relative amount of the larger particles by 2%. The attenuation spectrum was measured twice for each mixed system in order to demonstrate reproducibility.

Figures 8.38-8.40 give the results of these mixed system tests. It cab be seen that attenuation increases with increasing amounts of the "large" or "larger" particles. The increase in the attenuation with increasing doses of the Geltech content is in all cases significantly larger than precision of the instrument. This demonstrates that the DT-1200 data contains significant information about small amount of the large particles. The final question is to determine whether the calculated PSD calculates a correct bimodal distribution for these mixed model systems.

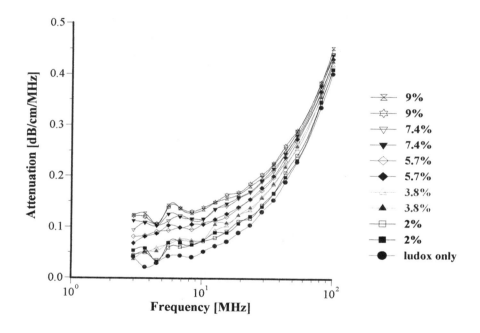

Figure 8.38. Attenuation spectra measured for Ludox TM-50 silica with various additions of Geltech 0.5 silica. Total solid content is 12% wt. The legend shows the fraction of the total solid content corresponding to the Geltech silica.

Tables 8.5 gives an answer to this question. The DT-1200 always calculates a lognormal and a bimodal distribution that best fits the experimental data. These two PSD are best in the sense that the fitting error between the theoretical attenuation (calculated for the best PSD) and the experimental attenuation is minimized. These fitting errors are important criteria for deciding whether the lognormal or bimodal PSD is more appropriate for describing a particular sample. For instance, the PSD is judged to be bimodal only if the bimodal fit yields substantially smaller fitting error than a lognormal PSD. The value for the lognormal and bimodal fitting errors are given in Table 8.5. It can be seen that fitting errors for the bimodal PSD are better than the lognormal fits for all of the mixed systems over the whole concentration range and for both the large and larger sized particles. According to the fitting errors, all PSD in the mixed Ludox-Geltech systems are bimodal, which of course is correct for these known mixed systems.

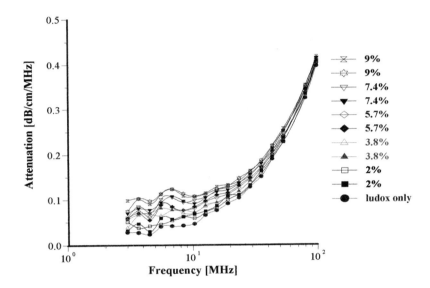

Figure 8.39. Attenuation spectra measured for Ludox TM-50 silica with various additions of Geltech 1.5 silica. Total solid content is 12% wt. Legend shows the fraction of the total solid content corresponding to the silica Geltech.

At the same time our experience tells us that this fitting error criterion alone is not a sufficient test for the bimodality of the PSD. It is always possible to obtain a better fit to a data set by allowing more degrees of freedom in the solution. A bimodal PSD provides at least 4 adjustable parameters for fitting the experimental attenuation curve, whereas a simple lognormal PSD provides only 2 parameters.

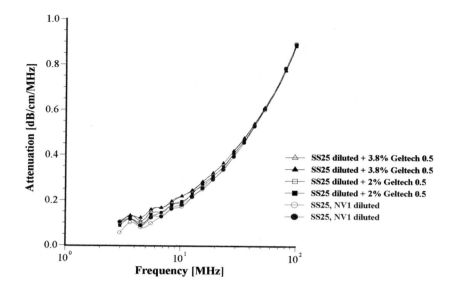

Figure 8.40. Attenuation spectra measured for Cabot SS25 silica diluted down to 12wt% with various additions of Geltech 0.5 silica. Total solid content is 12wt%. Legend shows the fraction of the total solid content corresponding to the silica Geltech.

The value of the larger particle content is another important parameter which must be taken into account. We have demonstrated that the precision of the DT-1200 is sufficient to detect larger particles present at concentrations larger than 2%. Therefore the software will not claim a bimodal PSD if the content of the large particles less than 2%, even if this yields a marginally better fit. Referring to Table 3, we see that the calculated large particle content in all of the mixed Ludox-TM/ Geltech systems is well above this 2% threshold and gives therefore an additional argument to claim a bimodal PSD for these systems. It is important that the calculated content of the large particles increases with the actual content of the added Geltech particles which confirms consistency of the PSD analysis.

Figures 8.40 and 8.41 illustrate a similar test performed with actual Cabot SS CMP slurry. It is seen that the DT-1200 is again able to resolve the presence of the small added amount of Geltech particles.

Table 8.5 Characteristics of the larger particles (Geltech silica) calculated from the attenuation spectra of the Figures 8.38 and 8.39.

actual Geltech content, %	calculated for Geltech 0.5				calculated for Geltech 1.5			
	content %	larger size micron	lognormal fitting error %	bimodal fitting error %	content %	larger size micron	lognormal fitting error %	bimodal fitting error %
9	14	0.7	14.1	7.3	11	1.6	24.5	4.4
9	15	0.9	10.9	8.7	12	1.7	17.7	5.5
7.4	8	0.9	13.6	4.5	10	1.9	17.8	6
7.4	12	1	15.4	7.1	10	1.9	17.3	5.4
5.7	7	0.9	13.8	4.3	7	1.9	17.2	4.1
5.7	10	1.3	12.9	7.2	6	1.6	18.5	4.7
3.8	4	0.9	9.5	3.7	5	1.3	18.1	3.9
3.8	4	0.7	12	4.3	6	1.6	16.9	3.8
2	4	0.6	10.8	3.6	3	1.9	12.2	2.6
2	5	1.2	12.2	3.7	4	1.6	12.3	3.9

In addition to particle sizing, ultrasound based techniques can be used for characterizing electrokinetic properties of the CMP slurries. Figure 8.42 illustrates typical two sweep pH titration curve of a CMP slurry that is manufactured by ECC. This technique is currently used by many CMP slurry manufacturers and consumers for quality and process control.

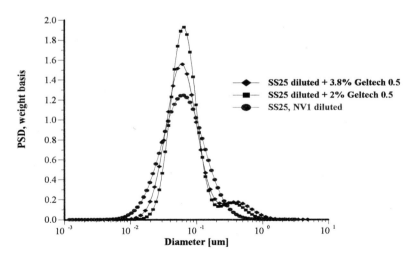

10/18/98 13:40:24 Graph1

Figure 8.41 Particle size distribution of Cabot SS25 silica diluted down to 12wt% with various additions of Geltech 0.5 silica. Total solid content is 12% wt. The legend shows the fraction of the total solid content corresponding to the silica Geltech.

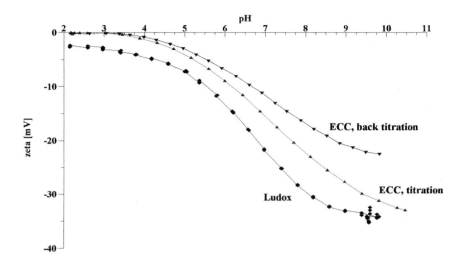

Figure 8.42. A one sweep titration of silica Ludox and a two sweeps titration of ECC CMP slurry.

8.5 Titration using Electroacoustics

Webster's dictionary [40] gives a definition of the word "titration" as it relates to general chemistry as "...the process of finding out how much of a certain substance is contained in a solution by measuring how much of another substance it is necessary to add to the solution in order to produce a given reaction". This term has a slightly different meaning in Colloid Chemistry. It emphasizes the second part of Webster's definition: observing the reaction caused by adding a certain substance to the solution. Colloid Chemistry deals with interfaces, and consequently the reaction of interest must somehow be related to the surface properties. An obvious "reaction" is the change in ς-potential due to additions to the chemical composition. We use this modified definition of the word "titration" later in this chapter.

8.5.1 pH titration

We start with the most widely known version of colloidal titration: "pH titration". In this case, acid and/or base are the chemicals used to affect the chemical composition of the colloid, namely the pH value. Measuring simply the variation of pH with changes in the acid-base content itself can yield useful information. However, this type of simple pH titration is not yet widely accepted in the field of colloid science. The more popular version of a pH titration is the simultaneous measurement of both pH and ς-potential, after the addition of either acid or base. In this case, the pH plays the role of a liquid solution chemistry indicator, whereas ς-potential characterizes the surface properties. The output ς-potential versus pH relationship characterizes the chemical balance between surface and bulk chemistry, and this determines its value for researcher.

Ultrasound is an important tool for pH titration because it provides a means for measuring the ζ-potential without dilution. Traditional microelectrophoresis techniques requires extreme dilution, which significantly affects the surface-bulk chemical balance. The Electroacoustic method eliminates the need for dilution, and allows us to obtain more meaningful and useful titration data.

We have already presented several examples of pH titration using the DT-300 Electroacoustic Probe, including: rutile (Figure 8.2), alumina (Figure 8.2), anatase (Figure 8.6), and silica (Figure 8.42). One more example is shown in Figure 8.43, which shows three titration sweeps of a 40%wt kaolin slurry.

The most important point on these pH titration curves is that pH value at which the ζ-potential equals zero. This is called the "iso-electric point" (IEP). If pH is the controlling factor, then we call this the iso-electric pH. The value of the IEP depends on the chemistry of surface and the chemistry of the solution. The IEP defines the range of instability of the sample being tested. As a general rule, a colloid is unstable if the magnitude of the zeta potential is less than approximately 30 mV. Generally this unstable region may extend over a range of ±2 pH units in the vicinity of the IEP, but obviously this depends upon the slope of the ς-potential vs. pH curve about the iso-electric point. In the examples mentioned above, the iso-electric pH of rutile is about 4, alumina about 9, anatase 6.3, and silica possibly about 2. The kaolin shown in Figure 8.42, under these conditions, does not have an iso-electric pH.

The iso-electric point is important and useful, because it is independent of the instrument calibration in the sense that it does not depend on the absolute values of the computed ς-potential. Consequently, the IEP can be measured with fairly high accuracy and precision. An international project is now underway by groups of scientists from NIST in the USA, the Federal Institute for Material Research in Germany, and the Japan Fine Ceramics Center for setting IEP standards using electroacoustics [41]. Each group uses the same alumina Sumitomo AKP-30 prepared in the same aqueous solution. Statistical analysis of multiple measurements made with the Matec ESA-8000 and Dispersion Technology DT-1200 lead them to conclude that the IEP can be measured electroacoustically with a precision of 0.1 pH unit.

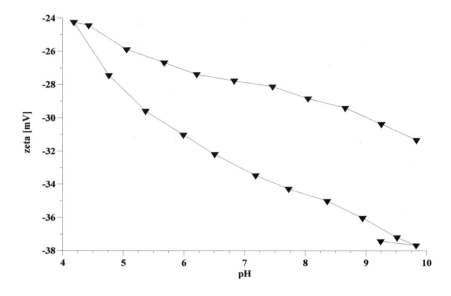

Figure 8.43 pH titration of a 40%wt kaolin slurry.

Such pH titrations might be useful not only in water but in other liquids as well. For example, Van Tassel and Randall applied this titration to alumina in ethanol. They used two methods to measure ς-potential: microelectrophoresis and CVI. The results were practically identical, making suitable Henry correction.

8.5.2 Time titration, kinetic of the surface-bulk equilibration

There is one important requirement for pH titration: it must go through points where the surface chemistry is in equilibrium with the liquid bulk chemistry. This is the only case where pH titration is meaningful and reproducible. It is in this situation that the IEP demarcates the instability range of the system. This implies that a true pH titration is actually an "equilibrium pH titration". This means that in order to run pH titration properly we should have some idea about chemical kinetics, and about the time required for reaching surface-bulk equilibrium after the addition of chemicals. We have observed the situation when investigators neglected this kinetic aspect of the pH titration and therefore measured completely meaningless, non-reproducible titration curves. One of the most striking examples is zirconia.

Figure 8.44 illustrates variations in the ς-potential and pH of a zirconia slurry with time following a step change of pH. We call this a "time titration" in the Dispersion Technology software. It is seen that equilibration takes more than half an hour. In contrast, the equilibration of a 5%vl alumina slurry in water takes less than 10 seconds. This difference between alumina and zirconia requires a large difference in the respective titration protocol used to obtain the data. Automated equilibrium titration of zirconia will take hundreds of times longer than that of alumina. It might require days or even weeks, which makes it, at the very , inconvenient or impractical.

Fortunately, zirconia represents a rather extreme case. Usually the surface-bulk chemistry balance reaches equilibrium in seconds, as with alumina. However, we strongly recommend to first perform a "time titration" when presented with a new material with unknown properties. This would allow one to determine the equilibration time and make some allowance in the titration protocol if necessary. However, the equilibration time often is itself a function of pH. At some pH values the surface may equilibrate quickly, at other pH it may equilibrate very slowly. For this reason, the Dispersion Technology software automatically adjusts the time allowed for equilibration depending on the dynamic response of the sample to each increment of added reagent.

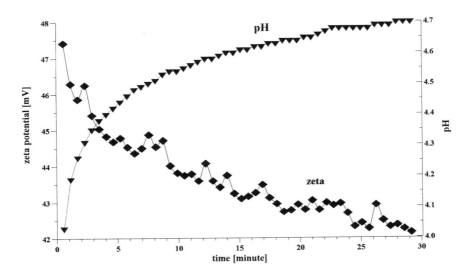

Figure 8.44. Equilibration of 3%vl zirconia slurry prepared in the KCl 10^{-2} with pH adjusted initially to 4.

There is one more factor that might affect the quality of titration data, the degree of agitation or mixing of the sample. According to our experience, the vast majority of problems with titration are related to how well the added reagent is mixed with the sample. This is also the conclusion of the international consortium that we mentioned above. There are several reasons for the importance of this agitation.
- It provides a homogeneous spreading of the reagent injected by the burettes into the sample.
- It prevents sedimentation, especially in the vicinity of the IEP.
- It prevents the build up of particle deposit on the sensor surface.
- It helps to speed up the surface-bulk equilibration.
- It breaks up aggregates and flocs when the chemical composition allows for colloid stability.

8.5.3. Surfactant titration

Figure 8.45 illustrates this last statement regarding the importance of agitation. The incremental addition of hexametaphosphate (HMPH) induces a negative electric charge on the kaolin particle surfaces. It can be seen that curves break at a HMPH to kaolin ratio of 0.06%wt. This happens due to sonication of the sample at this point. Sonication breaks up the aggregates of kaolin particles. This increases the total area of the kaolin-water interface, which in turn amplifies the CVI signal. This increase of CVI signal will be interpreted as increasing the ς-potential if one did not make a correction for the increasing surface area caused by the smaller particle size. It is interesting that after several more injections, the curve looks like it is recovering back to the initial slope. Still, this experiment confirms the importance of agitation.

Automatic "surfactant titration" is a simple way for determining the optimum dose of surfactant required to stabilize the system. In this case a surfactant solution of specific concentration is automatically dispensed by the burette. The software protocol specifies number of points to be measured and the total amount of surfactant to be added. The equilibration time is important here, as well as for pH titration. This is the time required for the injected surfactant molecules to reach the particle surface and establish surface-bulk equilibrium. Titration curves for PCC and for kaolin concentrated slurries are shown in Figures 8.46-8.47.

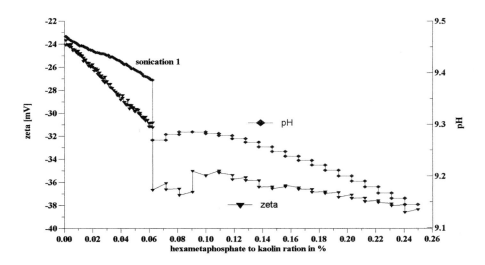

Figure 8.45 Effects of sonication on a hexametaphosphate titration of a 40%wt kaolin slurry.

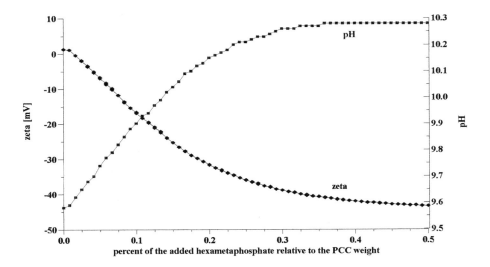

Figure 8.46 Titration of a PCC slurry with 0.1 g/g hexametaphosphate solution.

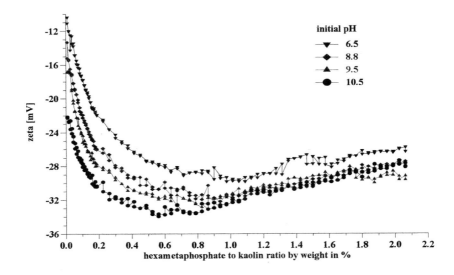

Figure 8.47 Titration of a 40%wt kaolin slurry using hexametaphospate with different starting pH.

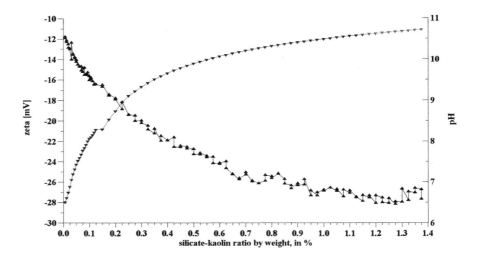

Figure 8.48 Titration of 40%wt kaolin slurry using silicate.

This method allows one to compare the efficiency of different surfactants. For instance, Figures 8.47 and 8.48 illustrate a "surfactant titration" of the same 40%wt kaolin slurry with two different agents: hexametaphosphate and silicate. It can be seen that hexametaphosphate is almost three times more efficient than silicate; it reaches a maximum ς-potential at 0.5%wt relative to the kaolin, whereas silicate requires 1.4%wt.

In addition, Figure 8.47 stresses the fact that the efficiency of a surfactant often depends on pH. In this particular case, for instance, hexametaphosphate efficiency improves with increasing initial pH. At the same time, the addition of hexametaphosphate changes the pH value as it is shown in Figure 8.49. This leads us to the conclusion that "surfactant titration" and "pH titration" are in many cases related. In order to represent the surface modification properly, we should use three dimensions, plotting ς-potential as a function of pH and surfactant content. Figure 8.50 represents such a three dimensional fingerprint that has been constructed from the data shown on Figures 8.47 and 8.49.

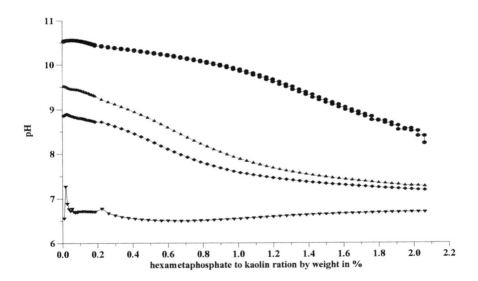

Figure 8.49 Variation of pH during hexametaphospate titration of 40%wt kaolin slurry at different starting pH.

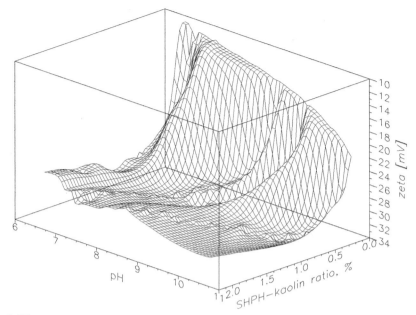

Figure 8.50 Fingerprint of the ζ-pH-hexametaphosphate titration of the 40%wt kaolin slurry.

Such a fingerprint presentation allows one to determine the optimum chemical composition for stabilizing a particular slurry. Electroacoustic fingerprints might have different axes, not necessarily pH and surfactant content. Conductivity, for example, is another important parameter, which characterizes the total ionic strength.

It is known that ς-potential depends on ionic strength; it generally becomes smaller with increasing ionic strength. This factor affects ς-potential values, because ionic strength changes during a titration. For instance, this factor is responsible for the difference in kaolin ς-potential at pH 9.25 between the three sweeps of the titration shown in Figure 8.43. Figure 8.42 gives another example of the ionic strength effect in a two sweep titration of a CMP silica slurry.

Ionic strength is important not just as a factor affecting ς-potential. Increasing ionic strength requires some modifications of the Electroacoustic measurement technique. The problem is related to the Ion Vibration Current (IVI), which becomes comparable to the Colloid Vibration Current at high ionic strength because of the higher ion concentration. The IVI plays the role of a background for the CVI measurement. It must be subtracted from the total measured current (CVI + IVI) to reveal the true CVI. We discuss this background subtraction and the results in detail in the next section. Here we would like to stress that the background IVI is one factors that might affect the position of the IEP. If it is not accounted for at high ionic strength, meaningless titration results can be produced.

Summarizing, we conclude that there are three major factors that must be taken into account for performing a proper titration experiment:
1. agitation;
2. equilibration time;
3. background electroacoustic subtraction.

The method for resolving this last factor is covered in detail in the following section.

8.6. Colloids with high ionic strength - Electroacoustic background.

The measured electroacoustic signal (*Total Vibration Current* (TVI)) contains a signal contribution from ions (*Ion Vibration Current* (IVI)) and a signal contribution from the colloidal particles (*Colloid Vibration Current* (CVI)). In many real world systems the IVI is negligible relative to the CVI, and can be neglected. However, this assumption may not be valid when:
- there is a low particle-liquid density contrast;
- when there is a low volume fraction of particles;
- for large particle size;
- or for high ionic strength, meaning a high concentration of ions.

The first three factors decrease CVI, whereas the last one increases IVI.

When the IVI is of comparable magnitude to CVI, it should be subtracted from the measured electroacoustic signal to allow proper calculation of the ς-potential. This subtraction is not trivial since the measured alternating current is a vector, not a scalar. Figure 8.52 illustrate one possible relationship between CVI and IVI. They are presented as vectors on the complex plane of real and imaginary components. We intentionally chose the case when the magnitude of the IVI and CVI are approximately the same, but are opposite in phase. This allows us to illustrate several potential complications caused by the IVI.

First of all, one can see that the measured TVI is much smaller in magnitude than the actual CVI. This means that we would get a much smaller absolute value for ζ-potential if we use the TVI as is for calculating it. The error in the absolute value might be tolerable in many cases. Unfortunately, the IVI affects not only the magnitude but also the phase. This creates problems with the apparent sign of the ζ-potential.

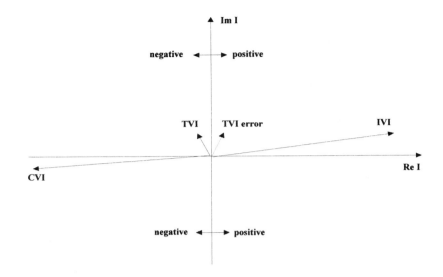

Figure 8.52. Scheme illustrating relationship between CVI and IVI on the complex currents plane.

There is a convention in determining sign of the ς-potential from the TVI phase. This convention is introduced when we calibrate the instrument using 10%wt silica Ludox TM-50 at pH 9.3. We know from independent microelectrophoretic measurement that these silica particles have a ς-potential of −38 mV. The absolute value of this known ς-potential allows us to calibrates the magnitude of the measured electroacoustic signal. The fact that the ς-potential is negative allows us to calibrate the phase. We assign a phase of 180 degrees to this negative silica particles. Positive particles would therefore generate a signal with a phase close to 360 degrees (or 0 degrees).

According to this convention, the ς-potential changes sign when the phase of the measured electroacoustic signal crosses either 90 or 270 degrees, i.e. the vector crosses the boundary between the right and left half planes of Figure 8.52.

This analysis indicates that if the measured TVI phase is in the vicinity of the 270 or 90 degrees we might have problem with determining sign of the ς-potential.

These problems become especially pronounced near the IEP where the CVI becomes smaller due to the decrease in the ς-potential. Failure to subtract any IVI near the IEP might lead to errors in the apparent position of the IEP. These errors would increase with increasing IVI. This is one of the reasons why the case of colloids with the high ionic strength is of general interest for electroacoustic characterization.

Table 8.6. Ion Vibration Current for four different electrolytes at different concentrations.

	Magnitude in absolute units	Magnitude relative to Ludox in %	Magnitude measurement error %	Phase in degrees
silica Ludox, calibration	351.7	100.0	0.7	180
distilled water	3.28	0.9	1.8	325
0.01M KCl	8.98	2.6	0.8	343
0.1M KCl	38.75	11.0	1.2	352
0.5M KCl	77.63	22.1	0.8	358
1 M KCl	85.04	24.2	0.8	1
0.01M NaCl	7.72	2.2	3.4	343
0.1M NaCl	25.96	7.4	1.7	352
0.5M NaCl	48.04	13.7	1.5	1.5
1 M NaCl	52.3	14.9	0.4	2.5
0.01M NaI	2.35	0.7	3.0	250.2
0.1M NaI	21.32	6.1	0.8	204
0.5M NaI	54.13	15.4	0.2	209
1 M NaI	61.8	17.6	0.2	212
0.01M RuCl	16.28	4.6	0.6	352
0.1M RuCl	78.45	22.3	0.1	4.2
0.5M RuCl	128.09	36.4	0.8	13.5
1 M RuCl	146.17	41.6	0.9	15.5

A further analysis will show that there are also substantial effects on the Double Layer structure at high ionic strength. We address this general colloidal-chemical issue later in this section.

The logical starting point for understanding the situation is a measurement of the IVI with no colloidal particles present in the solution. This would allow us to determine the magnitude of the effect, and give us some idea of the ionic strength range where IVI becomes important.

Measurement of IVI requires very highly sensitive equipment because the electroacoustic signals produced by ion vibration alone are very weak. Fortunately, the Zeta Potential Probe of Dispersion Technology DT-1200 is sensitive enough to measure electroacoustic signal even in distilled water. This high sensitivity is related to the pulse technique that is described in the Chapter 7. Briefly, the number of pulses collected is a function of the signal-to-noise ratio. The weaker the signal, the more pulses the sensor sends through the sample in order to reach a defined target signal-to-noise ratio.

Table 8.6 shows results of these measurements for distilled water as well as four different strong electrolytes. For comparison, we also present calibration data for silica Ludox. The second column of this table presents the relative magnitude in arbitrary units. In order to simplify a comparison with colloids, we present the IVI for all electrolytes normalized by the electroacoustic signal generated by silica Ludox – Column 3 in the Table 8.6.

It is seen that the IVI is a very small fraction of the Ludox CVI, if the ionic strength is below 0.1 M. This roughly defines the range where the IVI should be taken into account – ionic strength above 0.1 M.

The data in Table 8.6 contains information about solvated layers of ions. This information can be extracted in principle using the Bugosh IVI theory [42]. This goes back 40 years to the early papers on Ion Vibration Potential by the Yeager-Zana group [43-51]. Modern experimental techniques allows us to collect such electroacoustic data for ions much faster and easier than at the time of their pioneering work.

The ability to measure IVI opens up the possibility to characterize the colloidal Double Layer at high ionic strength. This problem has very interesting history.

The first experimental work known to us was in Russia by Deinega *et al* [52]. They reported electrophoretic mobility for phenol formaldehyde and aniline formaldehyde resins in zinc sulfate solutions at increasing ionic strengths up to 1M. The electrophoretic mobility reached a minimum value at about 0.1 M and then increased again.

The authors of this work related their results to an insolvent structured water layer near the hydrophilic surfaces. The Dukhin-Shilov electrokinetic theory [54] was suggested in 1983 for explaining electrokinetic effects when the traditional diffuse layer has collapsed due to high ionic strength. Electrokinetic phenomena in such systems exists because of the electric charge separations due to the insolvent properties of the surface structured water layer. This theory has been confirmed experimentally using the electroosmotic method by Alekssev *et al* [53].

Neither microelectrophoretic nor electroosmosis techniques are convenient or reliable for work at high ionic strength. Electroacoustics introduces a much easier and more reliable way to perform measurements at high ionic strength.

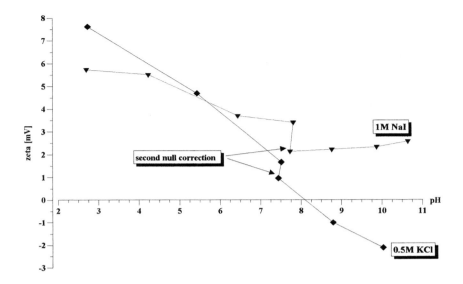

Figure 8.53 Manual titration of 2%vl anatase (200 nm) at high ionic strength.

The first measurements using electroacoustics to study high ionic strength colloidal were done by Rosenholm at Abo University and Kosmulski at Lublin University. They and their co-authors published a series of papers [2-6] presenting their results obtained with the ESA technique using a Matec. They observed measurable electroacoustic effects at high ionic strength, which indicated the presence of a Double Layer beyond the limits of the classical DL theory. They also observed an interesting shift of the IEP to higher pH. They also discovered two problems that are inherent to the ESA technique at high ionic strength. Later developers of the Acoustosizer [55] confirmed their conclusions. Briefly, these problems are that:
- the electric field, the driving force in the ESA mode, causes a very large current at high ionic strength, which overloads the driving amplifier; and
- the high electric current set up magnetic effects.

Neither of these problems exists in the CVI mode of electroacoustics that is used in Dispersion Technology instruments. The total acoustic power introduced into the sample is exceedingly small, and the resultant electric currents are similarly weak. Nevertheless, the sophisticated pulse techniques that are employed in these instruments allow such weak signals to be accurately measured.

There is another feature of the Dispersion Technology instruments which makes them very suitable for work at high ionic strength - background subtraction. The true CVI can be computed by subtracting a previously measured IVI from the measured Total Vibration Current (TVI).

The first step in using background subtraction is to extract an equilibrium solution. This might be achieved either by centrifugation or by simple sedimentation if the particles or flocs are sufficiently large. The resultant supernate is then a particle-free a solvent that has a chemical composition that was in equilibrium with the surface of the particles.

The next step is to measure the electroacoustic signal of this particle-free supernate. Since this vector signal corresponds to the IVI, we can save it and use this as a background reference. This IVI reference is then subtracted from subsequent TVI measurements of the colloidal dispersion to yield the true Colloid Vibration Current. This background subtraction achieves the following two goals:

- The number of collected pulses is automatically set appropriately to achieve the desired signal to noise target for the actual magnitude of the CVI signal, which might be much smaller in magnitude than the magnitude of the CVI+IVI signal, and therefore require a collection of more pulse data. This provides much more precise CVI data near the isoelectric point than would be the case in the absence of such subtraction.
- The uncertainty of the sign of the ς-potential near the isoelectric point is eliminated.

In order to show Dispersion Technology instrument's ability to work at high ionic strength, we performed a manual titration of anatase much as described in Section 8.1. Figure 8.53 shows the results of two pH-titrations for two solutions, at 0.5 M and 1 M ionic strength respectively.

It can be seen that IEP shifts to higher pH. This was observed initially by Rosenholm, Kosmulski *et al* [2-6], and later by Rowlands *et al* [55]. It completely disappears in a 1 M solution. Apparently the DL at this high ionic strength exists completely as a result of structural and insolvent properties of the water surface layer.

The lower pH points have been measured assuming the same IVI background as for the starting pH point. In order to test this assumption, we took a supernate sample at a pH point between 7 and 8, and redefined the IVI reference. One can see that the correction is measurable, but it would not affect the general shape of the titration curves, and can thus be ignored. In principle one could run a manual titration redefining the background IVI reference at each pH point.

We have also made the observation that this measurement requires special care of the electrode surfaces. For some colloidal systems, a deposit of particles tends to build up on the electrode surfaces, which then affects the value of the measured electroacoustic signal. This is the reason why we chose to run manual titrations at this high ionic strength, stopping at each new pH to clean the electrodes. Alternatively, one could run automatic titrations, allowing sufficient equilibration time between measurements to clean the electrodes.

We conclude that electroacoustics using the CVI mode offers an easy, fast and reliable way to characterize:
- the properties of ions in solutions;
- the properties of the Double Layer at high ionic strength;
- the properties of a structured insolvent hydrophilic water layers; and
- the surface properties of materials, such as cement, that exist only at high ionic strength.

This application might also be important in the future for studying various biological systems with relatively high ionic strength.

8.7 Effect of air bubbles

Air contained in sample may affect both the attenuation and sound speed. Air may be present either as bubbles or as air trapped at the interstices between particle agglomerates. Here we consider how such air might effect the measurements and how we can gain further information about this air by interpreting such data.

An acoustic theory describing sound propagation through a bubbly liquid was created by Foldy [134], and confirmed experimentally 135, 136].

The effect of air bubbles on the attenuation and sound speed depends both on the bubble size and sound frequency. For instance, a 100 micron bubble has a resonance frequency of about 60 KHz. This frequency is reciprocally proportional to the bubble diameter. A bubble of 10 micron diameter will have a resonance frequency of about 0.6 MHz.

Acoustic spectroscopy typically measures the attenuation spectra over a range of frequencies from 1 to 100 MHz. The size of the bubbles must be well below 10 micron in order to affect the complete frequency range of acoustic spectrometer.

Bubbles of sizes below 10 micron are very unstable, as is known from general colloid chemistry and the theory of flotation. Colloid-sized gas bubbles have astonishingly short lifetimes, normally between 1 μs and 1 ms [137]. They simply dissolve in the liquid because of high curvature.

Thus, air bubbles can only affect the low frequency part of the acoustic spectra, typically below 10 Mhz. The frequency range from 10Mhz to 100 MHz is available for particle characterization, even in bubbly liquids. For bubble free samples, the entire spectral range can be used, providing more detailed particle sizing capabilities and more robust characterization

The low frequency part of the attenuation spectra, from 1 MHz to 100 MHz, can be used to characterize the amount of air in the system. Air in colloidal systems can have a surprising and dramatic effect on the rheological properties, which is often not understood. Hydrophobic systems, such as carbon black or graphite, may contain air trapped at the interstices of particle agglomerates. For instance, bubbles can be centers of aggregation which might make them an important factor controlling stability. The low frequency portion of the attenuation spectra can be used to quantify the wettability of such dispersions, while at the same time the particle size can be determined using the high frequency portion.

For the vast majority of samples, the presence of air can be ignored. This conclusion was confirmed with thousands of measurements performed with hundreds of different systems

Table of Applications.

Particles	Liquid	Solid content	Particle size [microns] Attenuation	• potential [mV] CVI/P-ESA	Application	Authors, reference year
silver iodide	water		sol, NotM	CVP, own design	first CVP experiment	Rutgers and Rigole [56] 1958
silica Ludox	water	<27%wt	0.03, NotM	CVP, own design	first CVP tests	Yeager and oth. [57] 1953
silica Ludox	water	<30%wt	0.03, NotM	CVP pulse, own design	first pulse measurement	Zana and Yeager [58], 1967
rutile	water	<30%vl	0.3, NotM	CVP PenKem 8000	vfr test, microelectrophoresis	Marlow and oth. [68] 1983
coal	water	<40%vl		CVP PenKem 8000	vfr test, surface properties	Marlow and Rowell [67] 1988
latex	water		Not M	ESA own design	experimental test of inertia correction	Hunter [60] 1988

Table of Applications.						
silver iodide, bitumen	water	<10%wt	NotM	ESA, CVP Matec 8000	theory test	Babchin and oth. [24] 1989
water	crude oil	1-10%vl	NotM	CVP Matec 8000	theory test	Babchin and oth, [24] 1989
water	crude oil	1-10%	NotM	CVP Matec 8000	theory test	Issacs and oth. [25] 1990
cobalt phosphate	water	<5%vl	0.19, NotM	ESA Matec 8000	test O'Brien theory	O'Brien and oth [61] 1990
titanium dioxide	water	<5%vl	0.35, Not M	ESA Matex 8000	test O'Brien theory	O'Brien [61] 1990
latex, alumina, silica	water	<5%vl	0.3, NotM 0.2, NotM 0.03, NotM	ESA, Matec 8000	vfr test	James and oth. [64] 1991
toluene	water	<60%vl	0.1-0.25, NotM	ESA Matec 8000	vfr effects	Goetz and El-Aasser [26] 1992
alumina, silicon nitride	water	1.5%vl	0.2- 0.3, NotM	ESA Matec 8000	polydispersity, aggregation	James and oth [63] 1992
pigment	water		<0.3, NotM	ESA Matec 8000	vfr test, electric surface properties	Texter [65] 1992
rutile	water		0.3 NotM	ESA Matec 8000	vfr test, polydispersity	Scales and oth [66] 1992
carbon black	dodecane, isopar	<15%vl	0.1 NotM	ESA Matec 8000	charging in non-aqueous media	Kornbrekke and oth. [74] 1993
diamond	water methanol	0.5%vl	<0.5 NotM	ESA Matec 8000	interfacial chemistry	Valdes [75] 1993
alumina silica	water	5%vl	4 1.6 NotM	ESA Matec 8000	background signal	Desai and oth. [76] 1993

Table of Applications.

silicon nitride, yttrium oxide	water	<45%vl	NotM	ESA Matec 8000	characterization of ceramic slurries	Pollinger [77] 1993
silicon nitride	water	<6%vl	0.4 NotM	ESA Matec 8000	modification of the surface charge	Hackley and Malghan [78] 1993
PCC	water	<20%wt	0.7 NotM	ESA Matec 8000	rheology controlled by surface charge	Huang and oth. [79] 1993
coal	water	5%vl	12 NotM	ESA Matec 8000	electrokinetics of coal	Polat and oth. [80] 1993
cement	water	24%vl	3-10 Notm	ESA Matec 8000	interparticles effects in cement	Darwin and oth [81] 1993
ink	toluene		<0.65 NotM	ESA Matec 8000	interaction in inks	Krishnan [82] 1993
phosphors	water	<2%vl	NotM	ESA Matec 8000	electrokinetics of phosphors	Dutta [83] 1993
toner	isopar		NotM	ESA Matec 8000	charging in hydrocarbons	Larson [84] 1993
kaolin	water	1.5%vl	<2	ESA Acoustosizer	surface conductivity and shape	Rowlands and O'Brien [69] 1995
fat	water	10% and 20%	0.25 micron	ESA Acoustosizer	size, zeta of emulsion	Carasso and oth [73] 1995
anatase	water	<2%vl	0.2	ESA Acoustosizer	zeta and rheology	Kosmulski and Rosenholm [2] 1996
silica	water	<10%wt	0.5-1, NotM	ESA Acoustosizer	polymer adsorption	Carasso and oth. [59] 1997
aluminum hydroxide	water	15%vl	0.5	ESA Acoustosizer	high ionic strength	Rowlands and oth. [55] 1997

				Table of Applications.			
rutile	water	7%vl	0.3-2	CVI	aggregation	Dukhin and Goetz [1] 1998	
alumina		4%vl	0.1-1	DT1200			
dairy fat,	water	<38%vl	1.5	ESA	emulsions	O'Brien and oth. [27] 1998	
intravenous fat,		<22%vl	0.4	Acoustosizer			
bitument		60%	4-5				
silica,	water	<5%vl	<100 nm	ESA	test CMP	Carasso and oth. [38] 1998	
slumina			Ultrasizer	Acoustosizer			
CMP							
zirconia,	water	10%vl	0.4	ESA	experimental procedure	Bergstrom and oth. [86] 1998	
silicon nitride			0.5 NotM	Acoustosizer			
silica	water	46%wt	0.05-0.1 NotM	ESA Matec 8000	sol-gel transition	Valdes and oth [87] 1998	
silica	water	<45%vl	1.5	PenKem 8000	vfr effects	Hinze and oth. [88] 1998	
rutile		<26%vl	0.3				
anatase	water	<2%vl	0.2	ESA Acoustosizer	zeta and rheology	Kosmulski and oth. [4] 1999	
polyelectrolyte	water	<0.4%wt	NotM	ESA Matec 8000	ionic strength effect	Walldal and Akerman [72] 1999	
anatase	water	10%wt	0.5, NotM	ESA Acoustosizer	correlation zeta and rheology	Gustafsson and oth. [6] 2000	
hematite	water	0.5%vl	0.1, NotM	ESA, Acoustosizer	test Stern model	Gunnarsson and oth [62] 2001	
latex	water	<15%vl	0.165 nm NotM	ESA Matec 8000	vfr effect	Rasmusson [70] 2001	
silica		<25%vl	0.137 NotM	Acoustosizer			
sunflower oil	water	<50%vl	0.1-3	ESA Acoustosizer	vfr effect, particle sizing	Kong and oth. [29] 2001	

				Table of Applications.			
antimony-doped Tin(IV) barium sulfate	water			ESA Acoustosizer	theory for semiconductors	Guffond and oth [71] 2001
alumina titania	water	1%, 5%, 10% vl	0.3micr 0.05 m NotM	CVI, DT1200 ESA, Matec	statistics of IEP	Hackley and oth. [41] 2001
aluimna, titania silica	water	<30%vl	<10 PenKem 8000	NotM	verification PSD by acoustics	Stintz and oth. [85] 1998
lycopodium	water	2%vl	30 Own design	NotM	test Sewell theory	Hartmann and Focke [97] 1940
kaolin quartz	water	<5%vl	0.9 2.2 Own design	NotM	theory test	Urick [95] 1948
metal cylinders	water	single	Own design	NotM	scatter angular dependence	Faran [101] 1951
milk	water		Own design	NotM	attenuation measurement	Hueter and oth. [124] 1953
blood albumin hemoglobin	plasma	<80%vl	Own design	NotM	protein content determines acoustic properties	Carstensen and Schwan [122] 1953
lycopodium quartz	water	<1.3%vl <0.54%v	30 2.15 Own design	NotM	test Epstein theory	Stakutis and oth. [93] 1955
glass silica	water	<70%vl	95-155 <14	NotM	test Sewell theory	Busby and Richardson [96] 1956
benzene water benzene	water benzen glyceri	<40%	Own design	NotM	test theory for emulsions	Allison and Richardson [98] 1958

Table of Applications.

sea water	water		Own design	NotM	acoustic properties	Murphy and oth. [130] 1958
hemoglobin blood	water plasma	<30%	Own design	NotM	relaxation as dielectric	Carstensen and Schwan [121] 1959
brine shrimp	water	<0.3%	<200 Own design	NotM	theory test	Gruber and Meister [127] 1960
sea sediments	water		Own design	NotM	acoustics of sediments	Shumway [129] 1960
polymers	water		Own design	NotM	rheology-acoustics	Mason [94] 1961
hemoglobin	water		Own design	NotM	test Schwan theory	Edmonds [120] 1962
plankton	water	<35%	25-35 115 0.05 own design	NotM	scattering theory	Watson and Meister [125] 1963
mud	water		Own design	NotM	acoustic properties	Wood and Weston [128] 1964
poly-saccharides	water		Own design	NotM	similar to hemoglobulin	Hawley and oth. [126] 1965
sediments	water	<30%	Own design	NotM	acoustic properties	Hampton [132] 1967
sediments	water	<40%vl	Own design	NotM	mechanism of attenuation	Duykers [92] 1967
sea sediments	water		Own design	NotM	acoustic properties	McLeroy and DeLoach [131] 1968
shaped clays	water	<25%vl	<100 own design	NotM	shape effect	Blue and McLeroy [91] 1968
albumin	water	<10%	Own design	NotM	polymer conformation e	Kessler and Dunn [123] 1969

Table of Applications.							
silicon carbide;	gelatin	0.01%	15	NotM	scatter angular dependence	Koltsova and Mikhailov	
graphite	gelatin		760 Own design			[100] 1970	
latex	water	<50%	0.04-0.65	NotM	theory test	Allegra and Hawley	
toluene	water	<20%	0.3-0.4			[102] 1971	
headecane	water	<30%	0.7 Own design				
saturated sand	water		180 own design	NotM	Biot theory test	Hovem and Ingram [133] 1979	
ocean sediment	water		NotM	NotM	acoustic properties of ocean sediment	Mitchell and Focke [107] 1980	
fibers	water	single	<120 own design	NotM	theory test	Haberger [117] 1982	
polymers	water		Own design attenuation	Not M	rheology of polymers	Hauptmann and oth [119] 1982	
latex	water	<78%	NotM	NotM	s.speed-compressibility-inteface	Barret-Gultepe and oth [116] 1983	
surfactant	water		Own design attenuation	NotM	CMC and other properties	Borthakur and Zana [118] 1987	
coal	water	<50%wt	4 own design	NotM	stability	Barret-Gultepe and oth.[113] 1989	
sunflower oil	water	<50%wt	0.14-0.74 own design	NotM	theory test	McClements and Povey [9] 1989	

Table of Applications.

oil	water	5%vl	<0.3 own design	NotM	theory test	Javanaud and oth. [115] 1990
hexadecane	water	20%vl	0.36;3.3 own design	NotM	monitoring crystallization	Dickinson and oth. [23] 1990
silicon carbide	water; ethylene glycol	<20%vl	1.5;3 own design	NotM	theory test	Harker and oth. [114] 1991
kaolin	water	<30%vl	<2 own design	NotM	theory test	Sherman [108] 1991
hexadecane	water	<56%wt	<1.8 own design	NotM	teory test	McClements [8] 1992
latex	water	<45%vl	0.2-0.6 own design	NotM	theory test	Holmes and oth. [103] 1993
kaolin	water	<24%vl	0.85 own design	NotM	determine concentration	Greenwood and oth. [110] 1993
hexadecane	water	<20%wt	0.55 own design	NotM	flocculation test	McClements [12] 1994
emulsifier	water	<1%	NotM	NotM	CMC using sound speed	Durackova and oth. [106] 1995
corn oil	water	<50%wt	<0.4 own design	NotM	temperature effect	Chanamai and oth. [21] 1998
kaolin	water	<8%vl	<2.5 bandwidth & Acousts.	Acoustosizer	effect flocculation confirm coupled phase model	Austin and Challis [109] 1998
corn oil	water	<50%wt	0.12; 0.58 Ultrasizer	NotM	core-shell model test	Chanamai and oth. [22] 1999

			Table of Applications.			
hexadecane	water	<53%vl	Own design	NotM	core-shell model theory test	McClements and oth. [16] 1999
latex	water	<27%vl	0.25 Ultrasizer	NotM	theory test	Hipp and oth. [104] 1999
water	heptane	<20%	0.05-0.015 DT1200	NotM	theory for emulsions	Wines and oth. [30] 1999
silica CMP	water	<25%wt	0.1 and 1 DT1200	NotM	resolution test	Dukhin and Goetz [34] 1999
zirconia, aluimna silica PCC	water	<10%vl	0.3 2 0.5 0.6 DT1200	NotM	mixed dispersions	Dukhin and Goetz [35] 2000
alumina	water	<40%vl	<5 DT1200	NotM	stability	Takeda and Goetz [31] 1998
hematite anatease alumina	water	10%vl	<2 DT1200	NotM	bimodality	Takeda [32] 1998
silicon	water	11%wt	<10 PenKem 8000	NotM	test modified log-normal PSD	Dukhin and oth. [89] 1998
alumina	water	<40%	<1 DT1200	NotM	bimodality, stability	Takeda and oth. [33] 1999
asphaltines	oil		Own design	NotM	stability	Jones and oth. [20] 1999
silica latex	water	1%wt	0.1;0.3 0.25 Ultrasizer	NotM	role of the material properties, sensetivity	Hipp and oth. [112] 1999
alumina	water	<40%vl	<2 DT1200	NotM	verification psd at high loads	Hayashi and oth. [99] 2000

Table of Applications.						
oil	water	<70%vl	<3 DT1200	NotM	material properties	Babick and oth. [111] 2000
latex	water	<37%vl	0.065-0.093 own design	NotM	Wide frequency spectra	Stieler and oth. [105] 2001
water oil	oil water		<100 Ultrasizer	NotM	emulsion stability	Kippax and Higgs [90] 2001
alumina	ethanol	1%vl	0.27 DT1200	DT1200	adsorption	Van Tassel and Randall [134] 2001

REFERENCES.

1. Dukhin, A.S., and Goetz, P.J. "Characterization of aggregation phenomena by means of acoustic and electroacoustic spectroscopy", Colloids and Surfaces, 144, 49-58 (1998)

2. Kosmulski, M. and Rosenholm, J.B. "Electroacoustic study of adsorption of ions on anatase and zirconia from very concentrated electrolytes", J. Phys. Chem., 100, 28, 11681-11687 (1996)

3. Kosmulski, M. "Positive electrokinetic charge on silica in the presence of chlorides", JCIS, 208, 543-545 (1998)

4. Kosmulski, M., Gustafsson, J. and Rosenholm, J.B. "Correlation between the Zeta potential and Rheological properties of Anatase dispersions", JCIS, 209, 200-206 (1999)

5. Kosmulski, M., Durand-Vidal, S., Gustafsson, J. and Rosenholm, J.B. "Charge interactions in semi-concentrated titania suspensions at very high ionic strength", Colloids and Surfaces A, 157, 245-259 (1999)

6. Gustafsson, J., Mikkola, P., Jokinen, M., Rosenholm, J.B. "The influence of pH and NaCl on the zeta potential and rheology of anatase dispersions", Colloids and Surfaces A, 175, 349-359 (2000)

7. McClements, J.D. and Povey, M.J. "Ultrasonic velocity as a probe of emulsions and suspensions", Adv. Colloid Interface Sci., 27, 285-316 (1987)

8. McClements, D.J. "Comparison of Multiple Scattering Theories with Experimental Measurements in Emulsions", The Journal of the Acoustical Society of America, vol.91, 2, pp. 849-854, (1992)

9. McClements, D.J. and Povey, M.J.W. "Scattering of Ultrasound by Emulsions", J. Phys. D: Appl. Phys., 22, 38-47 (1989)

10. McClements, D.J. "Ultrasonic Characterization of Emulsions and Suspensions", Adv. Colloid Int. .Sci., 37, 33-72 (1991)

11. McClements, J.D. "Characterization of emulsions using a frequency scanning ultrasonic pulse echo reflectometer", Proc. Inst. Acoust., 13, 71-78 (1991)

12. McClements, J.D. "Ultrasonic Determination of Depletion Flocculation in Oil-in-Water Emulsions Containing a Non-Ionic Surfactant", Colloids and Surfaces, 90, 25-35 (1994)

13. McClements, J.D. and Powey, M.J.W., "Scattering of ultrasound by emulsions", J. Phys. D: Appl. Phys., 22, 38-47 (1989)

14. McClements, J.D. "Principles of ultrasonic droplet size determination in emulsions:, Langmuir, 12, 3454-3461 (1996)

15. McClements, J.D. and Coupland, J.N. "Theory of droplet size distribution measurement in emulsions using ultrasonic spectroscopy", Colloids and Surfaces A, 117, 161-170 (1996)

16. McClements, J.D., Hermar, Y. and Herrmann, N. "Incorporation of thermal overlap effects into multiple scattering theory", J. Acoust. Soc. Am., 105, 2, 915-918 (1999)

17. McClements, D.J. "Ultrasonic characterization of food emulsions", in Ultrasonic and Dielectric Characterization Techniques for Suspended Particulates, Ed. V.A. Hackley and J. Texter, Amer. Ceramic Soc., 305-317, (1998)

18. Povey, M., "The Application of Acoustics to the Characterization of Particulate Suspensions", in Ultrasonic and Dielectric Characterization Techniques for Suspended Particulates, ed. V. Hackley and J. Texter, Am. Ceramic Soc., Ohio, (1998)

19. Povey, M.J.W. "Ultrasonic Techniques for Fluid Characterization", Academic Press, San Diego, 1997

20. Jones, G.M., Povey, M.J.W. and Campbell, J. "Measurement and control of asphaltene agglomeration in hydrocarbon liquids", US Patent 5,969,237 (1999)

21. Chanamai, R., Coupland, J.N. and McClements, D.J. "Effect of Temperature on the Ultrasonic Properties of Oil-in-Water Emulsions", Colloids and Surfaces, 139, 241-250 (1998)

22. Chanamai, R., Hermann, N. and McClements, D.J. "Influence of thermal overlap effects on the ultrasonic attenuation spectra of polydisperse oil-in-water emulsions", Langmuir, 15, 3418-3423 (1999)

23. Dickinson, E., McClements, D.J. and Povey, M.J.W. "Ultrasonic investigation of the particle size dependence of crystallization in n-hexadecane-in-water emulsions", J. Colloid and Interface Sci., 142,1, 103-110 (1991)

24. Babchin, A.J., Chow, R.S. and Sawatzky, R.P. "Electrokinetic measurements by electroacoustic methods", Adv. Colloid Interface Sci., 30, 111 (1989)

25. Issacs, E.E., Huang, H., Babchin, A.J. and Chow, R.S. "Electroacoustic method for monitoring the coalescence of water-in-oil emulsions", Colloids and Surfaces, 46, 177-192 (1990)

26. Goetz, R.J. and El-Aasser, M.S. "Effects of dispersion concentration on the electroacoustic potentials of o/w miniemulsions", JCIS, 150, 2, 436-451 (1992)

27. O'Brien, R.W., Wade, T.A., Carasso, M.L., Hunter, R.J., Rowlands, W.N. and Beattie, J.K. "Electroacoustic determination of droplet size and zeta potential in concentrated emulsions", JCIS, (1996)

28. O'Brien, R.W., Cannon, D.W. and Rowlands, W.N. "Electroacoustic determination of particle size and zeta potential", JCIS, 173, 406-418 (1995)

29. Kong, L., Beattie, J.K. and Hunter, R.J. "Electroacoustic study of concentrated oil-in-water emulsions", JCIS, 238, 70-79 (2001)

30. Wines, T.H., Dukhin A.S. and Somasundaran, P. "Acoustic spectroscopy for characterizing heptane/water/AOT reverse microemulsion", JCIS, 216, 303-308 (1999)

31. Takeda, Shin-ichi and Goetz. P.J. "Dispersion/Flocculated Size Characterization of Alumina Particles in Highly Concentrated Slurries by Ultrasound Attenuation Spectroscopy", Colloids and Surfaces, 143, 35-39 (1998)

32. Takeda, Shin-ichi, "Characterization of Ceramic Slurries by Ultrasonic Attenuation Spectroscopy", Ultrasonic and Dielectric Characterization Techniques for Suspended Particulates, Edited by V.A. Hackley and J. Texter, American Ceramic Society, Westerville, OH, (1998)

33. Takeda, Shin-ichi, Chen, T. and Somasundaran P. "Evaluation of Particle Size Distribution for Nanosized Particles in Highly Concentrated Suspensions by Ultrasound Attenuation Spectroscopy", Colloids and Surfaces, (1999)

34. Dukhin, A.S. and Goetz, P.J "Characterization of Chemical Polishing Materials (monomodal and bimodal) by means of Acoustic Spectroscopy ", Colloids and Surfaces, 158, 343-354, (1999).

35. Dukhin, A.S. and Goetz, P.J. "Characterization of concentrated dispersions with several dispersed phases by means of acoustic spectroscopy", Langmuir (2000)

36. Bleier, A., "Secondary minimum interactions and heterocoagulation encountered in the aqueous processing of alumina-zirconia ceramic composites", Colloids and Surfaces, 66, 157-179 (1992)

37. Antony, L., Miner, J., Baker, M., Lai, W., Sowell, J., Maury, A., Obeng, Y. "The how's and why's of characterizing particle size distribution in CMP slurries", Electrochemical Soc., Proc., 7, 181-195 (1998)

38. Carasso, M., Valdes, J., Psota-Kelty, L.A. and Anthony, L. "Characterization of concentrated CMP slurries by acoustic techniques", Electrochemical Society Proc., vol.98-7, 145-197 (1998)

39. Osseo-Asare, K. and Khan, A. "Chemical-mechanical polishing of tungsten: An electrophoretic mobility investigation of alumina-tungstate interactions", Electrochemical Soc. Proc., 7, 139-144 (1998)

40. Webster's Unabridged Dictionary, Second Edition, Simon & Schuster, (1983)

41. Hackley, V.A., Patton, J., Lum, L.S.H., Wasche, R.J., Naito, M., Abe, H., Hotta, Y., and Pendse, H. "Analysis of the isoelectric point in moderately concentrated alumina suspensions using electroacoustic and streaming potential methods", NIST, Fed.Inst. for Material Research-Germany, JFCC-Japan, University of Main., Report (2001)

42. Bugosh, J., Yeager, E., and Hovorka, F., "The application of ultrasonic waves to the study of electrolytes. I.A modification of Debye's equation for the determination of the masses of electrolyte ions by means of ultrasonic waves", J. Chem. Phys. 15, 8, 592-597 (1947)

43. Yeager, E., Bugosh, J., Hovorka, F. and McCarthy, J., "The application of ultrasonics to the study of electrolyte solutions. II. The detection of Debye effect", J. Chem. Phys., 17, 4, 411-415 (1949)

44. Yeager, E., and Hovorka, F., "The application of ultrasonics to the study of electrolyte solutions. III. The effect of acoustical waves on the hydrogen electrode", J. Chem. Phys., 17, 4, 416-417 (1949)

45. Yeager, E., and Hovorka, F., "Ultrasonic waves and electrochemistry. I. A survey of the electrochemical applications of ultrasonic waves", J. Acoust. Soc. Am., 25, 3, 443-455 (1953)

46. Yeager, E., Dietrick, H., and Hovorka, F., J. "Ultrasonic waves and electrochemistry. II. Colloidal and ionic vibration potentials", J. Acoust. Soc. Am., 25, 456 (1953)

47. Zana, R., and Yeager, E., "Determination of ionic partial molal volumes from Ionic Vibration Potentials", J. Phys. Chem, 70, 3, 954-956 (1966)

48. Zana, R., and Yeager, E., "Quantative studies of Ultrasonic vibration potentials in Polyelectrolyte Solutions", J. Phys. Chem, 71, 11, 3502-3520 (1967)

49. Zana, R., and Yeager, E., "Ultrasonic vibration potentials in Tetraalkylammonium Halide Solutions", J. Phys. Chem, 71, 13, 4241-4244 (1967)

50. Zana, R., and Yeager, E., "Ultrasonic vibration potentials and their use in the determination of ionic partial molal volumes", J. Phys. Chem., 71, 13, 521-535 (1967)

51. Zana, R., and Yeager, E., "Ultrasonic vibration potentials", Mod. Aspects of Electrochemistry, 14, 3-60 (1982)

52. Deinega, Yu.F., Polyakova, V.M., Alexandrova, L.N. "Electrophoretic mobility of particles in concentrated electrolyte solutions", Colloid. Zh., 48, 3, 546-548 (1986)

53. Alekseev, O.L., Boiko, Yu.P., Ovcharenko, F.D., Shilov, V.N., Chubirka, L.A. "Electroosmosis and some properties of the boundary layers of bounded water", Kolloid. Zh., 50, 2, 211-216 (1988)

54. Dukhin, S.S. and Shilov, V.N. In the book "Uspekhi Kolloidnoy Himii", Kiev, (1983)

55. Rowlands, W.N., O'Brien, R.W., Hunter, R.J., Patrick, V. "Surface properties of aluminum hydroxide at high salt concentration", JCIS, 188, 325-335 (1997)

56. Rutgers, A.J. and Rigole, W. "Ultrasonic vibration potentials in colloid solutions, in solutions of electrolytes and pure liquids", Trans. Faraday Soc., 54, 139-143 (1958)

57. Yeager, E., Dietrick, H., and Hovorka, F., J. "Ultrasonic waves and electrochemistry. II. Colloidal and ionic vibration potentials", J. Acoust. Soc. Am., 25, 456 (1953)

58. Zana, R., and Yeager, E., "Ultrasonic vibration potentials", Mod. Aspects of Electrochemistry, 14, 3-60 (1982)

59. Carasso, M.L., Rowlands, W.N. and O'Brien, R.W. "The effect of neutral polymer and nonionic surfactant adsorption on the electroacoustic signals of colloidal silica", JCIS, 193, 200-214 (1997)

60. Hunter, R.J., Liversidge lecture, J. Proc. R. Soc. New South Wales, 121, 165-178 (1988)

61. O'Brien, R.W., Midmore, B.R., Lamb, A. and Hunter, R.J. "Electroacoustic studies of Moderately concentrated colloidal suspensions", Faraday Discuss. Chem. Soc., 90, 1-11 (1990)

62. Gunnarsson, M., Rasmusson, M., Wall, S., Ahlberg, E., Ennis, J. "Electroacoustic and potentiometric studies of the hematite/water interface", JCIS, 240, 448-458 (2001)

63. James, M., Hunter, R.J. and O'Brien, R.W. "Effect of particle size distribution and aggregation on electroacoustic measurements of Zeta potential", Langmuir, 8, 420-423 (1992)

64. James, R.O., Texter, J. and Scales, P.J. "Frequency dependence of Electroacoustic (Electrophoretic) Mobilities", Langmuir, 7, 1993-1997 (1991)

65. Texter, J., "Electroacoustic characterization of electrokinetics in concentrated pigment dispersions", Langmuir, 8, 291-305 (1992)

66. Scales, P.J. and Jones, E. "The effect of particle size distribution on the accuracy of electroacoustic mobilities", Langmuir, 8, 385-389 (1992)

67. Marlow, B.J. and Rowell, R.L. "Acoustic and Electrophoreric mobilities of Coal Dispersions", J. of Energy and Fuel, 2, 125-131 (1988)

68. Marlow, B.J., Fairhurst, D. and Pendse, H.P., "Colloid Vibration Potential and the Electrokinetic Characterization of Concentrated Colloids", Langmuir, 4,3, 611-626 (1983)

69. Rowlands, W.N., O'Brien, R.W. "The dynamic mobility and dielectric response of kaolinite particles", JCIS, 175, 190-200 (1995)

70. Rasmusson, M. "Volume fraction effects in Electroacoustic Measurements", JCIS, 240, 432-447 (2001)

71. Guffond, M.C., Hunter, R.J., O'Brien, R.W. and Beattie, J.K. "The dynamic mobility of colloidal semiconducting antimony-doped Tin(IV) oxide particles", JCIS, 235, 371-379 (2001)

72. Walldal, C., Akerman, B. "Effect of ionic strength on the dynamic mobility of polyelectrolytes", Langmuir, 15, 5237-5243 (1999)

73. Carasso, M.L., Rowlands, W.N., Kennedy, R.A. "Electroacoustic determination of droplet size and zeta potential in concentrated intravenous fat emulsions", JCIS, 174, 405-413 (1995)

74. Kornbrekke, R.E., Morrison, I.D. and Oja, T. "Electrokinetic measurements of concentrated suspensions in a polar liquids", in Electroacoustics for Characterization of Particulates and Suspensions, Ed. S.G. Malghan, NIST, 92-111, (1993)

75. Valdes, J.L. "Acoustophoretic characterization of colloidal diamond particles", in Electroacoustics for Characterization of Particulates and Suspensions, Ed. S.G. Malghan, NIST, 111-129, (1993)

76. Desai, F.N., Hammad, H.R., Hayes, K.F. "ESA measurements for silica and alumina; background electrolyte corrections", in S.B. Malghan (Ed.) Electroacoustics for Characterization of Particulates and Suspensions, NIST, 129-143 (1993)

77. Pollinger, J.P. "Oxide and non-oxide powder processing applications using electroacoustic characterization", in S.B. Malghan (Ed.) Electroacoustics for Characterization of Particulates and Suspensions, NIST, 143-161 (1993)

78. Hackley, V.A. and Malghan, S.G "Investigation of parameters and secondary components affecting the electroacoustic analysis of silicon nitride powders", in Electroacoustics for Characterization of Particulates and Suspensions, Ed. S.G. Malghan, NIST, 161-180, (1993)

79. Huang, Y.C., Sanders, N.D., Fowkes, F.M. and Lloyd, T.B. "The impact of surface chemistry on particle electrostatic charging and viscoelasticity of precipitated calcium carbonate slurries", in Electroacoustics for Characterization of Particulates and Suspensions, Ed. S.G. Malghan, NIST, 180-200, (1993)

80. Polat, H., Polat, M. and Chandler, S. "Characterization of coal slurries by ESA technique", in Electroacoustics for Characterization of Particulates and Suspensions, Ed. S.G. Malghan, NIST, 200-219, (1993)

81. Darwin, D.C., Leung, R.Y. and Taylor, T. "Surface charge characterization of Portland cement in the presence of superplasticizers", in Electroacoustics for Characterization of Particulates and Suspensions, Ed. S.G. Malghan, NIST, 238-263, (1993)

82. Krishnan, R. "Role of particle size and ESA in performance of solvent based gravure inks", in Electroacoustics for Characterization of Particulates and Suspensions, Ed. S.G. Malghan, NIST, 263-274, (1993)

83. Dutta, A. "Electrokinetics of Phosphors", in Electroacoustics for Characterization of Particulates and Suspensions, Ed. S.G. Malghan, NIST, 274-301, (1993)

84. Larson, J.R. "Advances in liquid toners charging mechanism: Electroacoustic measurement of liquid toner charge", in Electroacoustics for Characterization of Particulates and Suspensions, Ed. S.G. Malghan, NIST, 301-315, (1993)

85. Stintz, M., Hinze, F. and Ripperger, S. "Particle size characterization of inorganic colloids by ultrasonic attenuation spectroscopy", in Ultrasonic and Dielectric Characterization Techniques for Suspended Particulates, Ed. V.A. Hackley and J. Texter, Amer. Ceramic Soc., 219-229, (1998)

86. Bergstrom, L., Laarz E., Greenwood, R. "Electroacoustic studies of aqueous Ce-ZrO2 and Si3N4 Suspensions", in Ultrasonic and Dielectric Characterization Techniques for Suspended Particulates, Ed. V.A. Hackley and J. Texter, Amer. Ceramic Soc., 229-243, (1998)

87. Valdes, J.L., Patel, S.S., Chen Y.L. "Reaction chemistries in dynamic colloidal systems", in Ultrasonic and Dielectric Characterization Techniques for Suspended Particulates, Ed. V.A. Hackley and J. Texter, Amer. Ceramic Soc., 275-290, (1998)

88. Hinze, F., Stintz, M. and Ripperger, S. "Surface characterization of organic colloids by the Colloid Vibration Potential", in Ultrasonic and Dielectric Characterization Techniques for Suspended Particulates, Ed. V.A. Hackley and J. Texter, Amer. Ceramic Soc., 291-305, (1998)

89. Dukhin, A.S. , Goetz, P.J and Hackley, V., "Modified Log-Normal Particle Size Distribution in Acoustic Spectroscopy", Colloids and Surfaces, 138, 1-9 (1998).

90. Kippax, P. and Higgs, D. "Application of ultrasonic spectroscopy for particle size analysis of concentrated systems without dilution", American Laboratory, 33, 23, 8-10, (2001)

91. Blue, J.E. and McLeroy, E.G. "Attenuation of Sound in Suspensions and Gels" , J. of Acoust. Soc. Amer., 44, 4, 1145-1149 (1968)

92. Duykers, L.R.B. "Sound attenuation in liquid-solid mixtures", J. Acoust. Soc. Amer., 41, 5, 1330-1335 (1967)

93. Stakutis, V.J., Morse, R.W., Dill, M. and Beyer, R.T. "Attenuation of Ultrasound in Aqueous Suspensions", J. Acoust. Soc. Amer., 27, 3 (1955)

94. Mason, W.P. "Dispersion and Absorption of Sound in High Polymers", in Handbuch der Physik., vol.2, Acoustica part1, 1961.

95. Urick R.J. "Absorption of Sound in Suspensions of Irregular Particles", J. Acoust. Soc. Amer., 20, 283 (1948)

96. Busby, J. and Richrdson, E.G. "The Propagation of Utrasonics in Suspensions of Particles in Liquid", Phys. Soc. Proc., v.67B, 193-202 (1956)

97. Hartmann G.K and Focke, A.B. "Absorption of Supersonic Waves in Water and in Aqueous Suspensions", Physical Review, 57, 1, 221-225 (1940)

98. Allison, P.A. and Richardson, E.G. "the Propagation of Ultrasonics in Suspensions of Liquid Globules in Another Liquid", Proc. Phys. Soc., 72, 833-840 (1958)

99. Hayashi, T., Ohya, H., Suzuki, S. and Endoh, S. "errors in Size Distribution Measurement of Concentrated Alumina Slurry by Ultrasonic Attenuation Spectroscopy", J. Soc. Powder Technology Japan, 37, 498-504 (2000)

100. Kol'tsova, I.S. and Mikhailov, I.G. "Scattering of Ultrasound Waves in Heterogeneous Systems", Soviet Physics-Acoustics, 15, 3, 390-393 (1970)

101. Faran, J.J. "Sound Scattering by Solid Cylinders and Spheres", J. Acoust. Soc. Amer., 23, 4, 405-418 (1951)

102. Allegra, J.R. and Hawley, S.A. "Attenuation of Sound in Suspensions and Emulsions: Theory and Experiments", J. Acoust. Soc. Amer., 51, 1545-1564 (1972)

103. Holmes, A.K., Challis, R.E. and Wedlock, D.J. "A Wide-Bandwidth Study of Ultrasound Velocity and Attenuation in Suspensions: Comparison of Theory with Experimental Measurements", J. Colloid and Interface Sci., 156, 261-269 (1993)

104. Hipp, A.K., Storti, G. and Morbidelli, M. "On multiple-particle effects in the acoustic characterization of colloidal dispersions", J. Phys., D: Appl.Phys. 32, 568-576 (1999)

105. Stieler. T., Scholle. F-D., Weiss. A., Ballauff. M., and Kaatze, U. "Ultrasonic spectrometry of polysterene latex suspensions. Scattering and configurational elasticity of polymer chains", Langmuir, 17, 1743-1751 (2001)

106. Durackova, S., Apostolo, M., Ganegallo, S. and Morbidelli, M. "Estimation of critical micellar concentration through ultrasonic velocity measurements", J. of Applied Polymer Science, 57, 639-644 (1995)

107. Mitchell, S.K., and Focke, K.C. "New measurements of compressional wave attenuation in deep ocean sediments", J. Acoust. Soc. Am, 67, 5, 1582-1589 (1980)

108. Sherman, N.E. "Ultrasonic velocity and attenuation in aqueous kaolin dispersions", J. Colloid Interface Sci., 146, 2, 405-414 (1991)

109. Austin, J. and Challis, R.E. "The effect of flocculation on the propagation of ultrasound in dilute kaolin slurries", J. Colloid and Interface Sci., 206, 146-157 (1998)

110. Greenwood, M.S., Mai, J.L. and Good, M.S. "Attenuation measurements of ultrasound in a kaolin-water slurry. A linear dependence upon frequency ", J. Acoust. Soc. Am., 94, 2, 908-916 (1993)

111. Babick, F., Hinze, F., and Ripperger, S. "Dependence of Ultrasonic Attenuation on the Material Properties", Colloids and Surfaces, 172, 33-46 (2000)

112. Hipp, A.K., Storti, G. and Morbidelli, M. "Particle sizing in Colloidal Dispersions by Ultrasound. Model calibration and sensitivity analysis", Langmuir, 15, 2338-2345 (1999)

113. Barrett-Gultepe, M.A., Gultepe, M.E, McCarthy, J.L. and Yeager, E.B. "A study of steric stability of coal-water dispersions by ultrasonic absorption and velocity measurements", JCIS, 132, 1, 145-161 (1989)

114. Harker, A.H., Schofield, P., Stimpson, B.P., Taylor, R.G., and Temple, J.A.G. "Ultrasonic Propagation in Slurries", Ultrasonics, 29, 427-439 (1991)

115. Javanaud, C., Gladwell, N.R., Gouldby, S.J., Hibberd, D.J., Thomas, A. and Robins, M.M. "Experimental and theoretical values of the ultrasonic properties of dispersions: effect of particle state and size distribution", Ultrasonics, 29, 330-337, (1991)

116. Barrett-Gultepe, M.A., Gultepe, M.E. and Yeager, E.B. "Compressibility of colloids. 1. Compressibility studies of aqueous solutions of amphiphilic polymers and their absorbed state on polysterene latex dispersions by ultrasonic velocity measurements", Langmuir, 34, 313-318 (1983)

117. Haberger, C.C. "The attenuation of ultrasound in dilute polymeric fiber suspensions", J. Acoust. Soc. Am., 72, 3, (1982)

118. Borthakur, A., Zana, R. "Ultrasonic absorption studies of aqueous solutions of nonionic surfactants in relation with critical phenomena and micellar dynamics", J. Phys. Chem., 91, 5957-5960, (1987)

119. Hauptmann, P., Sauberlich, R. and Wartewig, S. "Ultrasonic attenuation and mobility in polymer solutions and dispersions", Polymer Bulletin, 8, 269-274 (1982)

120. Edmonds, P.D. "Ultrasonic absorption of haemoglobin solutions", J. Acoust. Soc. Am., 63, 216-219 (1962)

121. Carstensen, E.L. and Schwan, H.P. "Acoustic properties of hemoglobin solutions", J. Acoust. Soc. Amer., 31, 3, 305-311 (1959)

122. Carstensen, E.L., Kam Li and Schwan, H.P. "Determination of acoustic properties of blood and its components", J. Acoust. Soc. Amer., 25, 2, 286-289 (1953)

123. Kessler, L.W. and Dunn, F. "Ultrasonic investigation of the conformal changes of bovine serum albumin in aqueous solution", J. Phys. Chem., 73, 12, 4256-4262 (1969)

124. Hueter, T.F., Morgan, H. and Cohen, M.S. "Ultrasonic attenuation in biological suspensions", J. Acoust. Soc. Amer, 1200-1201 (1953)

125. Watson, J. and Meister, R. "Ultrasonic absorption in water containing plankton in suspension", J. Acoust. Soc. Amer., 35, 10, 1584-1589 (1963)

126. Hawley, S.A. , Kessler, L.W. and Dunn, F. "Ultrasonic absorption in aqueous solutions of high molecular weight polysaccharides", J. Acoust. Soc. Amer., 521-523 (1965)

127. Gruber, G.J. and Meister, R. "Ultrasonic attenuation in water containing brine shrimp in suspension", J. Acoust. Soc. Am., 33, 6, 733-740 (1961)

128. Wood, A.B. and Weston, D.E. "The Propagation of Sound in Mud", Acustica, vol.14, pp.156-162 (1964)

129. Shumway, "Sound Speed and Absorption Studies of Marine Sediments by a resonance Method", Geophysics, vol. 25, no. 3, pp. 659-682 (1960)

130. Murpy, S.R., Garrison, G.R. and Potter, D.S. "Sound Absorption at 50 to 500 kHz from Transmission Measurements in the Sea", J. Acoust. Soc. Amer., vol.30, 9, pp. 871-875, (1958).

131. McLeroy, E.G. and DeLoach, A. "Sound Speed and Attenuation from 15 to 1500 kHz Measured in Natural Sea-Floor Sediments", J. Acoust. Soc. Amer., vol.44, 4, pp. 1148-1150 (1968).

132. Hampton, L. "Acoustic Properties of Sediments", J. Acoust. Soc. Amer., vol.42, 4, 882-890, (1967)

133. Hoverm, J. and Ingram, G., "Viscous Attenuation of Sound in Saturated Sand", J. Acoust. Soc. Amer, vol.66, 6, pp. 1807-1812 (1979)

134. Foldy, L.L "Propagation of sound through a liquid containing bubbles", OSRD Report No.6.1-sr1130-1378, (1944)

135. Carnstein, E.L. and Foldy, L.L "Propagation of sound through a liquid containing bubbles", J. of Acoustic Society of America, 19, 3, 481- 499 (1947)

136. Fox, F.E., Curley S.R. and Larson, G.S. "Phase velocity and absorption measurement in water containing air bubbles" J. of Acoustic Society of America, 27, 3, 534-539 (1957)

137. Ljunggren, S. and Eriksson, J.C. "The lifetime of a colloid sized gas bubble in water and the cause of the hydrophobic attraction", Colloids and Surfaces, 129-130, 151-155 (1997)

138. Kikuta, M. "Method and apparatus for analyzing the composition of an electro-deposition coating material and method and apparatus for controlling said composition", US Patent 5,368,716 (1994)

List of Symbols

a	particle radius
b	cell radius
b_{sh}	radius of the shell in the core-shell model
c	sound speed
c_i	concentration of the ith ion species
C_p	heat capacity at constant pressure
C_{dl}	DL capacitance
C_s	electrolyte concentration
C_{ext}	extinction cross-section
C_{abs}	absorption cross-section
C_{sca}	scattering cross-section
D_i	diffusion coefficient of the ith ion species
Du	Dukhin number
d	particle diameter
e	elementary electric charge
E	electric field strength
$<E>$	macroscopic electric field strength
f	frequency of ultrasound
F	Faraday constant
F^h	hydrodynamic friction force

F^{hook}		Hookean force
k		Bolzmann constant
k		compression wavenumber
k_T		thermal wavenumber
k_s		shear wavenumber
$K_{\pm}^{0}$		limiting conductances of cations and anions
K		conductivity attributed with index
I_r		local current in the cell
$<I>$		macroscopic current
I		intensity of the sound or light
I_t		intensity of the transmitted sound or light
I_i		intensity of the incident sound or light
I_s		intensity of the scattered wave
j		complex unit
$j_m\ (ka)$		Bessel function
h		special function (see Special Functions)
H		special function (see Special Functions)
l_i		cell layer thickness
L		gap in the electroacoustic chamber
L_w		mass load
m_i		mass of the solvated ions
M^*		stress modulus
N_A		Avogadro number

N_p		number of particles
N		number of the volume fractions
$n_m(ka)$		Neumann function
P		pressure
p_{ind}		induced dipole moment
R		gas constant
R_{obs}		observation distance
R_0		Rayleigh distance
R_1		radius of the first Fresnel zone
r		spherical radial coordinate
$t^{\pm}$		transport numbers of cations and anions
t		time
T		absolute temperature
S_{exp}		measured electroacoustic signal
$v^{\pm}$		volume of the solvated ions
V_{dd}^r		radial velocity corresponding to the dipole-dipole interaction
V_h^r		radial velocity corresponding to the electro-hydrodynamic interaction
V_c^r		radial velocity corresponding to the concentration interaction
W_{ext}		extinction energy
W_{abs}		absorption energy
W_{sca}		scattered energy
w		width of the sound pulse
u		speed of the motion attributed according to the index

U_m	amplitude of the oscillation velocity
U_{dd}	energy of the dipole-dipole interaction
X	particle size
x	distance
z_i	valency of the ith ion species
$z^{\pm}$	valencies of the cations and anions
Z	acoustic impedance
y_i	activity coefficient of the ith ion species
α	attenuation specified with index
β	thermal expansion attributed with index
β^p	compressibility
δ_η	viscous depth
δ_T	thermal depth
δ_m	scattering phase angles
ε	dielectric permittivity
ε_0	dielectric permittivity of the vacuum
ϕ	electric potential
ϕ_c	potential of the compression wave
ϕ_T	potential of the thermal wave
ϕ_η	potential of the shear wave
Φ_{IVI}	phase of the Ion Vibration Current
Φ_e	flow resistance

γ		hydrodynamic friction coefficient
γ^h		specific heat ratio
γ_{ind}		Particle electric polarizibility
η		dynamic viscosity
φ		volume fraction
φ_w		weight fraction
κ		reciprocal Debye length
κ^σ		surface conductivity
λ		wave length
μ		electrophoretic mobility
μ_d		dynamic electrophoretic mobility
μ_i		electrochemical potential
μ_i^0		standard chemical potential
ν		kinematic viscosity
$\nu_\pm$		dissociation numbers for cations and anions
θ		spherical angular coordinate
ρ		density attributed according to the index
σ^d		electric charge of the diffuse layer
σ		electric surface charge
σ_l		standard deviation of the log-normal PSD
τ		heat conductance attributed according to the index
$\omega = 2\pi f$		frequency

ω_{MW}		Maxwell-Wagner relaxation frequency
ζ		electrokinetic potential
ψ		angle in the polar coordinates
$\psi_{dl}(x)$		Electric potential in the double layer
ψ^d		Stern potential
Π_s		total scattered power by rigid sphere
Ω_e		porosity
Ω		drag coefficient
Σ		entropy production
Indexes		
i		index of the particle fraction or ion species
p		particles
m		medium
s		dispersion
r		radial component
θ		tangential component
vis		viscous
th		thermal
sc		scattering
int		intrinsic
ef		effective medium

in	acoustic input
out	acoustic output
rod	delay rod properties
sur	surface layer

BIBLIOGRAPHY ALPHABETICAL.

1. Adler, F.T., Sawyer, W.M., and ferry, J.D. "Propagation of transverse waves in viscoelastic media", J. Applied Phys., 20, 1036-1041 (1949)

2. Alba, F. "Method and Apparatus for Determining Particle Size Distribution and Concentration in a Suspension Using Ultrasonics", US Patent No. 5121629 (1992)

3. Alba, F., Higgs, D., Jack, R., and Kippax, P. "Ultrasound Spectroscopy. A Sound Approach to Sizing of Concentrated Particulates", Characterization Techniques for Suspended Particulates, ed. V. Hackley and J. Texter, Am. Ceramic Soc., Ohio, 111-129, (1998)

4. Alekseev, O.L., Boiko, Yu.P., Ovcharenko, F.D., Shilov, V.N., Chubirka, L.A. "Electroosmosis and some properties of the boundary layers of bounded water", Colloid. Zh., 50, 2, 211-216 (1988)

5. Alig, I.; Stieber, F.; Wartewig, S.; Bakhramov, A. D.; Manucarov, Yu. S. "Ultrasonic Absorption and Shear Viscosity Measurements for Solutions of Polybutadiene and Polybutadiene-Block-Polystyrene Copolymers." Polymer, 28, 9, 1543-1546 (1987)

6. All, A.; Hyder, S.; Nain, A. K., "Intermolecular interactions in ternary liquid mixtures by ultrasonic velocity measurement", Indian Journal of Physics, Part B, VOL. 74B, NO. 1, 63-7 (2000)

7. Allegra, J.R. and Hawley, S.A. "Attenuation of Sound in Suspensions and Emulsions: Theory and Experiments", J. Acoust. Soc. Amer., 51, 1545-1564 (1972)

8. Allen, T. "Sedimentation methods of particle size measurement", Plenary lecture presented at PSA'85 (ed. P. J. Lloyd), Wiley, NY (1985)

9. Allen, T., "Particle size measurement", 4th edition, Chapman and Hall, NY, 1990

10. Allison, P.A. and Richardson, E.G. "the Propagation of Ultrasonics in Suspensions of Liquid Globules in Another Liquid", Proc. Phys. Soc., 72, 833-840 (1958)

11. Amari, T., Fujioka, S. and Watanabe, K., "Acoustic properties of aqueous suspensions of clay and calcium carbonate", JCIS., 134,2, 366-350 (1990)

12. Anderson, V.C. "Sound Attenuation from a Fluid Sphere", J. Acoust. Soc. Amer., 22, 4, 426-431 (1950)

13. Andreae, J., Bass, R., Heasell, E. and Lamb, J. "Pulse Technique for Measuring Ultrasonic Absorption in Liquids", Acustica, v.8, p.131-142 (1958)

14. Andreae., J and Joyce, P. "30 to 230 Megacycle Pulse Technique for Ultrasonic Absorption Measurements in Liquids", Brit .J. Appl. Phys., v.13, p.462-467 (1962)

15. Andle, J.C., Vetelino, J.F., Lec, R., McAllister, D.J. "An acoustic plate mode immunosensor", Ultrasonic Symposium, IEEE, (1989)

16. Anson, L.W. and Chivers, R.C. "Thermal effects in the attenuation of ultrasound in dilute suspensions for low values of acoustic radius", Ultrasonic, 28, 16-25 (1990)

17. Antony, L., Miner, J., Baker, M., Lai, W., Sowell, J., Maury, A., Obeng, Y. "The how's and why's of characterizing particle size distribution in CMP slurries", Electrochemical Soc., Proc., 7, 181-195 (1998)

18. Apfel, R.E., "Technique for measuring the adiabatic compressibility, density and sound speed of submicroliter liquid samples", J. Acoust. Soc. Amer., 59, 2, 339-343 (1976)

19. Apostolo, M., Canegallo, S., Siani, A. and Morbidelli, M. "Characterization of emulsion polymerization through ultrasound propagation velocity measurement", Macromol. Symp., 92, 205-222 (1995)

20. Atkinson, C.M and Kytomaa, H.K., "Acoustic Properties of Solid-Liquid Mixtures and the Limits of Ultrasound Diagnostics-1.Experiments", J. Fluids Engineering, 115, 665-675 (1993)

21. Atkinson, C.M and Kytomaa, H.K., "Acoustic Wave Speed and Attenuation in Suspensions", Int. J. Multiphase Flow, 18, 4, 577-592 (1992)

22. Attenborough, K. "Acoustical characterization of rigid fibrous absorbents and granular materials", J. Acoust. Soc. Amer., 73 (3) 1983, pp.785-799

23. Austin, J. and Challis, R.E. "The effect of flocculation on the propagation of ultrasound in dilute kaolin slurries", J. Colloid and Interface Sci., 206, 146-157 (1998)

24. Babchin, A.J., Chow, R.S. and Sawatzky, R.P. "Electrokinetic measurements by electroacoustic methods", Adv. Colloid Interface Sci., 30, 111 (1989)

25. Babchin, A.J., Huang, H., Rispler, K. and Gunter, D. "Method for determining the wetting preference of particulate solids in a multiphase liquid system", US Patent 5,293,773 (1994)

26. Babick, F., Hinze, F., and Ripperger, S. "Dependence of Ultrasonic Attenuation on the Materal Properties", Colloids and Surfaces, 172, 33-46 (2000)

27. Bacri, J. C. , "Measurement of some viscosity coefficients in the nematic phase of a liquid crystal", Journal de Physique Letters, VOL. 35, NO. 9, . L141-2 (1974)

28. Barrett-Gultepe, M.A., Gultepe, M.E, McCarthy, J.L. and Yeager, E.B. "A study of steric stability of coal-water dispersions by ultrasonic absorption and velocity measurements", JCIS, 132, 1, 145-161 (1989)

29. Barrett-Gultepe, M.A., Gultepe, M.E. and Yeager, E.B. "Compressibility of colloids. 1. Compressibility studies of aqueous solutions of amphiphilic polymers and their absorbed state on polysterene latex dispersions by ultrasonic velocity measurements", Langmuir, 34, 313-318 (1983)

30. Beck et al., "Measuring Zeta Potential by Ultrasonic Waves", Tappi, vol.61, 63-65, 1978

31. Becker, H.L. "Computerized sonic portable testing laboratory", US Patent 5,546,792

32. Bedwell B., and Gulari, E., *in* "Solution Behavior of Surfactants" (K.L. Mittal, Ed.), Vol. 2., Plenum Press, New York, 1982.

33. Belyaev, V. V. , "Physical methods for measuring the viscosity coefficients of nematic liquid crystals", Physics-Uspekhi, VOL. 44, NO. 3, 255-6, 281-4 (2001)

34. Bergstrom, L., Laarz E., Greenwood, R. "Electroacoustic studies of aqueous Ce-ZrO2 and Si3N4 Suspensions", in Ultrasonic and Dielectric Characterization Techniques for Suspended Particulates, Ed. V.A. Hackley and J. Texter, Amer. Ceramic Soc., 229-243, (1998)

35. Bhardwaj, M.C. "Advances in ultrasound for materials characterization", Advanced Ceramic Materials, 2, 3A, 198-203 (1987)

36. Biot, M.A "Theory of Propagation of Elastic Waves in a Fluid Saturated Porous Solid. I. Low Frequency range. ", J. Appl. Phys., 26, 182, pp. 168-179 (1955)

37. Biot, M.A "Theory of Propagation of Elastic Waves in a Fluid Saturated Porous Solid. I. High Frequency range. ", J. Appl. Phys., 26, 182, pp. 179-191 (1955)

38. Biot, M.A. "Theory of propagation of elastic waves in a fluid saturated porous solid." J. Acoustic Soc. Am., 28, p.171-191, (1956)

39. Blackstock, D. "Fundamentals of Physical Acoustics", J. Wiley & Sons, NY, 541p, (2000)

40. Bleier, A., "Secondary minimum interactions and heterocoagulation encountered in the aqueous processing of alumina-zirconia ceramic composites", Colloids and Surfaces, 66, 157-179 (1992)

41. Blue, J.E. and McLeroy, E.G. "Attenuation of Sound in Suspensions and Gels" , J. Acoust. Soc. Amer., 44, 4, 1145-1149 (1968)

42. Bockris, J.O'M, and Saluja, P.P.S. "Reply to the comments of Yager and Zana on the paper "Ionic Solvation Numbers from Compressibility and Ionic Vibration Potential Measurements", J. Phys. Chem., 79,12, 230-233 (1975)

43. Bohren, C., and Huffman, D. "Absorption and Scattering of Light by Small Particles", J. Wiley & Sons, (1983)

44. Bondarenko, M.P. and Shilov, V.N. "About some relations between kinetic coefficients and on the modeling of interactions in charged dispersed systems and membranes", JCIS, 181, 370-377 (1996)

45. Booth, F. and Enderby, J. "On Electrical Effects due to Sound Waves in Colloidal Suspensions", Proc. of Amer. Phys. Soc., 208A, 32 (1952)

46. Borsay, F. and Yeager, E. "Generation of ultrasound at metal-electrolyte interfaces", J. Acoust. Soc. Am., 64, 1, 240-242 (1978)

47. Borthakur, A., Zana, R. "Ultrasonic absorption studies of aqueous solutions of nonionic surfactants in relation with critical phenomena and micellar dynamics", J. Phys. Chem., 91, 5957-5960, (1987)

48. Brinkman, H.C. "A calculation of viscous force exerting by a flowing fluid on a dense swarm of particles", Appl. Sci. Res., A1, 27 (1947)

49. Buckin, V. and Kudryashov, E. "Ultrasonic shear wave rheology of weak particle gels", Adv. Of Colloid and Interface Sci., 89-90, 401-423 (2001)

50. Bugosh, J., Yeager, E., and Hovorka, F., "The application of ultrasonic waves to the study of electrolytes. I.A modification of Debye's equation for the determination of the masses of electrolyte ions by means of ultrasonic waves", J. Chem. Phys. 15, 8, 592-597 (1947)

51. Buiochi, F.; Higuti, T.; Furukawa, C. M.; Adamowski, J. C. "Ultrasonic measurement of viscosity of liquids", 2000 IEEE Ultrasonics Symposium. Proceedings. An International Symposium (Cat. No.00CH37121) , Ed.: Schneider, S. C.; Levy, M.; McAvoy, B. R., vol.1, 525-8 (2000)

52. Bujard, Martial R. " Method of measuring the dynamic viscosity of a viscous fluid utilizing acoustic transducer", US PATENT NUMBER- 04862384, (1989)

53. Butler, J.P., Reeds, J.A. and Dawson, S.V. "Estimating solutions of First kind integral equations with nonnegative constraints and optimal smoothing", SIAM J. Numer. Anal., 18, 3, 381-397 (1981)

54. Busby, J. and Richrdson, E.G. "The Propagation of Utrasonics in Suspensions of Particles in Liquid", Phys. Soc. Proc., v.67B, 193-202 (1956)

55. Cabos, P.C., and Delord, P., J. App. Cryst. 12, 502 (1979).

56. Cady, W.G. "A theory of the crystal transducer for plane waves", J. Acoust. Soc. Am., 21, 2, 65-73 (1949)

57. Canegallo. S., Apostolo. M, Storti, G. and Morbidelli, M. "On-line conversion monitoring through ultrasound propagation velocity measurements in emulsion polymerization", J. of Applied Polymer Science, 57, 1333-1346 (1995)

58. Cannon D.W. "New developments in electroacoustic method and instrumentation", in S.B. Malghan (Ed.) Electroacoustics for Characterization of Particulates and Suspensions, NIST, 40-66 (1993)

59. Cannon, D.W. and O'Brien, R.W. "Device for determining the size and charge of colloidal particles by measuring electroacoustic effect", US Patent 5,245,290 (1993)

60. Carnstein, E.L. and Foldy, L.L "Propagation of sound through a liquid containing bubbles", J. of Acoustic Society of America, 19, 3, 481- 499 (1947)

61. Carstensen, E.L. and Schwan, H.P. "Acoustic properties of hemoglobin solutions", J. Acoust. Soc. Am., 31, 3, 305-311 (1959)

62. Carstensen, E.L. and Schwan, H.P. "Absorption of sound arising from the presence of intact cells in blood", J. Acous. Soc. Am., 31, 2, 185-189 (1959)

63. Carstensen, E.L., Kam Li and Schwan, H.P. "Determination of acoustic properties of blood and its components", J. Acoust. Soc. Am., 25, 2, 286-289 (1953)

64. Carasso, M.L., Rowlands, W.N. and O'Brien, R.W. "The effect of neutral polymer and nonionic surfactant adsorption on the electroacoustic signals of colloidal silica", JCIS, 193, 200-214 (1997)

65. Carasso, M.L., Rowlands, W.N., Kenedy, R.A. "Electroacoustic determination of droplet size and zeta potential in concentrated intravenous fat emulsions", JCIS, 174, 405-413 (1995)

66. Carasso, M., Valdes, J., Psota-Kelty, L.A. and Anthony, L. "Characterization of concentrated CMP slurries by acoustic techniques", Electrochemical Society Proc., vol.98-7, 145-197 (1998)

67. Carasso, M., Patel, S., Valdes, J., White, C.A. "Process for determining characteristics of suspended particles", US Patent 6,119,510 (2000)

68. Chanamai, R., Coupland, J.N. and McClements, D.J. "Effect of Temperature on the Ultrasonic Properties of Oil-in-Water Emulsions", Colloids and Surfaces, 139, 241-250 (1998)

69. Chanamai, R., Hermann, N. and McClements, D.J. "Influence of thermal overlap effects on the ultrasonic attenuation spectra of polydisperse oil-in-water emulsions", Langmuir, 15, 3418-3423 (1999)

70. Chang, C. and Powell, R. "Effect of particle size distribution on the rheology of concentrated bimodal suspensions", J. Rheology, 38, 1, 85-98 (1994)

71. Chow, J.C.F., "Attenuation of Acoustic Waves in Dilute Emulsions and Suspensions", J. Acoust. Soc. Amer., 36, 12, 2395-2401 (1964)

72. Christoforou, C.C., Westermann-Clark, G.B. and J.L. Anderson, "The Streaming Potential and Inadequacies of the Helmholtz Equation", J. Colloid and Interface Science, 106, 1, 1-11 (1985)

73. Chu, B., and Liu, T. "Characterization of nanoparticles by scattering techniques", J. of Nanoparticle Research, 2, 29-41, (2000)

74. Chynoweth, A.G. and Scheider, W.G., "Ultrasonic Propagation in Binary Liquid Systems near Their Critical Solution Temperature", J. of Chemical Physics, vol.19, 12, 1566-1569 (1951)

75. Coca, J., Bueno, J.L. and Sastre, H. "Electrokinetic behavior of coal particles suspensions", J. Chem. Tech. Biotechnol., 2, 637-642 (1982)

76. Cohen-Bacrie, C. "Estimation of viscosity from ultrasound measurement of velocity", 1999 IEEE Ultrasonics Symposium. Proceedings. International Symposium (Cat. No.99CH37027), Ed.: Schneider, S. C.; Levy, M.; McAvoy, B. R., vol.2, 1489-92 (1999)

77. Colic, M. and Morse, D. "The elusive mechanism of the magnetic memory of water", Colloids and Surfaces, 154, 167-174 (1999)

78. Costley, R. D.; Ingham, W. M.; Simpson, J. A., "Ultrasonic velocity measurement of molten liquids." Ceramic Transactions, 89, 241-251 (1998)

79. Crupi, V., Maisano, G., Majolino, D., Ponterio, R., Villari, V., and Caponetti, E., *J. Mol. Struct.* 383, 171 (1996).

80. Cushman, et all. US Patent 3779070, (1973)

81. Dann, M.S., Hulse, N.D. "Method and apparatus for measuring or monitoring density or rheological properties of liquids or slurries", UK Patent GB, 2 181 243 A (1987)

82. D'Arrigo, G., Mistura, L. and Tartaglia, P. " Sound Propagation in the Binary System Aniline-Cyclohaxane in the Critical Region", Physical review A, vol.1, 2, 286-295 (1974)

83. Darwin, D.C., Leung, R.Y. and Taylor, T. "Surface charge characterization of Portland cement in the presence of superplasticizers", in Electroacoustics for Characterization of Particulates and Suspensions, Ed. S.G. Malghan, NIST, 238-263, (1993)

84. De Maeyer, L. Eigen, M., and Suarez, J., "Dielectric Dispersion and Chemical Relaxation", J. Of the Amer.Chem.Soc., 90, 12, 3157-3161 (1968)

85. Debye, P. "A method for the determination of the mass of electrolyte ions", J. Chem. Phys., 1,13-16, (1933)

86. Deinega, Yu.F., Polyakova, V.M., Alexandrove, L.N. "Electrophoretic mobility of particles in concentrated electrolyte solutions", Colloid. Zh., 48, 3, 546-548 (1986)

87. DeLacey, E.H.B. and White, L.R., "Dielectric Response and Conductivity of Dilute Suspensions of Colloidal Particles", J. Chem. Soc. Faraday Trans., 77, 2, 2007 (1982)

88. Derjaguin , B.V. and Landau, L., Acta Phys. Chim, USSR, 14, 633 (1941)

89. Derouet, B., and Denizot, F., C.R. Acad. Sci., Paris, 233,368 (1951)

90. Desai, F.N., Hammad, H.R., Hayes, K.F. "ESA measurements for silica and alumina; background electrolyte corrections", in S.B. Malghan (Ed.) Electroacoustics for Characterization of Particulates and Suspensions, NIST, 129-143 (1993)

91. Dickinson, E., McClements, D.J. and Povey, M.J.W. "Ultrasonic investigation of the particle size dependence of crystallization in n-hexadecane-in-water emulsions", J. Colloid and Interface Sci., 142,1, 103-110 (1991)

92. Dietrick, H., Yeager, E., Bugosh, J., Hovorka, F. "Ultrasonic waves and electrochemistry. III. An elecrokinetic effect produced by ultrasonic waves", J. Acoust. Soc. Am., 25, 3, 461-465 (1953)

93. Doppler, "Theorie des farbigen Lichtes der Doppelsterne", Prag, (1842)

94. Dukhin, A. S. and Murtsovkin, V. A. "Pairwise Interaction of Particles in an Electric Field. 2. Influence of Polarization of the Double Layer of Dielectric Particles on their Hydrodynamic Interaction in Stationary Electric Field", Kolloidn. Zh., 48, 240-247 (1986)

95. Dukhin, A. S. "Biospecifical Mechanism of Double Layer Formation in Living Biological Cells and Peculiarities of Cell Electrophoresis". Colloids and Surfaces, 73, 29-48 (1993).

96. Dukhin, A. S., Shilov, V.N. and Borkovskaya Yu. "Dynamic Electrophoretic Mobility in Concentrated Dispersed Systems. Cell Model.", Langmuir, 15, 10, 3452-3457 (1999)

97. Dukhin, A.S. and Goetz, P.J. "Acoustic Spectroscopy for Concentrated Polydisperse Colloids with High Density Contrast", Langmuir, 12 [21] 4987-4997 (1996)

98. Dukhin, A.S. and Goetz, P.J. "Method and device for characterizing particle size distribution and zeta potential in concentrated system by means of Acoustic and Electroacoustic Spectroscopy", patent USA, 09/108,072, (2000)

99. Dukhin, A.S. and Goetz, P.J. "Method and device for Determining Particle Size Distribution and Zeta Potential in Concentrated Dispersions", patent USA, pending

100. Dukhin, A.S. and Goetz, P.J. "Method for Determining Particle Size Distribution and Structural Properties of Concentrated Dispersions", patent USA, pending

101. Dukhin, A.S. and Goetz, P.J. "Method for Determining Particle Size Distribution and Mechanical Properties of Soft Particles in Liquids", patent USA, pending

102. Dukhin, A.S. and Goetz, P.J. "New Developments in Acoustic and Electroacoustic Spectroscopy for Characterizing Concentrated Dispersions", Colloids and Surfaces, 192, 267-306 (2001)

103. Dukhin, A.S. and Goetz, P.J "Characterization of Chemical Polishing Materials (monomodal and bimodal) by means of Acoustic Spectroscopy ", Colloids and Surfaces, 158, 343-354, (1999).

104. Dukhin, A.S., and Goetz, P.J. "Acoustic and Electroacoustic Spectroscopy", Langmuir, 12, 19, 4336-4344 (1996)

105. Dukhin, A.S. and Goetz, P.J. "Acoustic and Electroacoustic Spectroscopy", Characterization Techniques for Suspended Particulates, ed. V. Hackley and J. Texter, Am. Ceramic Soc., Ohio, 77-97, (1998)

106. Dukhin, A.S. , Goetz, P.J and Hackley, V., "Modified Log-Normal Particle Size Distribution in Acoustic Spectroscopy", Colloids and Surfaces, 138, 1-9 (1998).

107. Dukhin, A.S., Goetz., P.J. and Hamlet, C.W. "Acoustic Spectroscopy for Concentrated Polydisperse Colloids with Low Density Contrast", Langmuir, 12, 21, 4998-5004, (1996).

108. Dukhin, A.S., and Goetz, P.J. "Acoustic and Electroacoustic Spectroscopy for Characterizing Concentrated Dispersions and Emulsions", Adv. In Colloid and Interface Sci., 92, 73-132 (2001)

109. Dukhin, A.S., and Goetz, P.J. "Characterization of aggregation phenomena by means of acoustic and electroacoustic spectroscopy", Colloids and Surfaces, 144, 49-58 (1998)

110. Dukhin, A.S., Goetz, J.P., Wines, T.H. and Somasundaran, P. "Acoustic and electroacoustic Spectroscopy", Colloids and Surfaces, A, 173, 1-3, pp.127-159, (2000)

111. Dukhin, A.S., Ohshima, H., Shilov, V.N. and Goetz, P.J. "Electroacoustics for Concentrated Dispersions", Langmuir, 15,10, 3445-3451, (1999)

112. Dukhin, A.S., Shilov, V.N, Ohshima, H., Goetz, P.J "Electroacoustics Phenomena in Concentrated Dispersions. New Theory and CVI Experiment", Langmuir, 15, 20, 6692-6706, (1999)

113. Dukhin, A.S., Shilov, V.N, Ohshima, H., Goetz, P.J "Electroacoustics Phenomena in Concentrated Dispersions. Effect of the Surface Conductivity", Langmuir, 16, 2615-2620 (2000)

114. Dukhin, A.S. and Goetz, P.J. "Characterization of concentrated dispersions with several dispersed phases by means of acoustic spectroscopy", Langmuir, *16*(20); 7597-7604 (2000)

115. Dukhin, A.S., Goetz, P.J. and Truesdail, S.T. "Surfactant titration of kaolin slurries using ς-potential probe", Langmuir, 17, 964-968 (2001)

116. Dukhin, S.S. "Electrochemical characterization of the surface of a small particle and nonequilibrium electric surface phenomena", Adv. Colloid Interface Sci., 61, 17-49 (1995)

117. Dukhin, S.S. and Semenikhin, N.M. "Theory of double layer polarization and its effect on the electrokinetic and electrooptical phenomena and the dielectric constant of dispersed systems", Kolloid. Zh., 32, 360-368 (1970)

118. Dukhin, S.S. and Derjaguin, B.V. "Electrokinetic Phenomena" in "Surface and Colloid Science", E.Matijevic (Ed.), John Wiley & Sons, NY, v.7 (1974)

119. Dukhin, S.S. and Shilov V.N. "Dielectric phenomena and the double layer in dispersed systems and polyelectrolytes", John Wiley and Sons, NY, (1974)

120. Dukhin, S.S., Semenikhin, N.M., Shapinskaya, L.M. Transl. Dokl.Phys.Chem., 193, 540 (1970)

121. Dukhin, S.S. and Shilov, V.N. In the book "Uspekhi Kolloidnoy Himii", Kiev, (1983)

122. Dukhin, S.S., Churaev, N.V., Shilov, V.N. and Starov, V.M. "Problems of the reverse osmosis modeling", Uspehi Himii, (Russian) 43, 6 , 1010-1023 (1988), English, 43, 6 (1988).

123. Durackova, S., Apostolo, M., Ganegallo, S. and Morbidelli, M. "Estimation of critical micellar concentration through ultrasonic velocity measurements", J. of Applied Polymer Science, 57, 639-644 (1995)

124. Dutta, A. and Dullea, L.V "Effect of polyacrylate absorption on ESA of aqueous dispersions of phosphors", in Ultrasonic and Dielectric Characterization Techniques for Suspended Particulates, Ed. V.A.Hackley and J.Texter, Amer. Ceramic Soc., 257-275, (1998)

125. Dutta, A. "Electrokinetics of Phosphors", in Electroacoustics for Characterization of Particulates and Suspensions, Ed. S.G. Malghan, NIST, 274-301, (1993)

126. Duykers, L.R.B. "Sound attenuation in liquid-solid mixtures", J. Acous. Soc. Amer., 41, 5, 1330-1335 (1967)

127. Ebel, J.P., Anderson, J.L. and Prieve, D.C. "Diffusiophoresis of latex particles in electrolyte gradients", Langmuir, 4, 396-406 (1988)

128. Edmonds, P.D., Pearce, V.F. and Andreae, J.H. "1.5 to 28.5 Mc/s Pulse Apparatus for Automatic Measurement of Sound Absorption in Liquids and Some Results for Aqueous and other Solutions", Brit. J. Appl. Phys., vol.13, pp. 551-560 (1962)

129. Edmonds, P.D. "Ultrasonic absorption of haemoglobin solutions", J. Acoust. Soc. Am., 63, 216-219 (1962)

130. Edmonds, P.D., "Ultrasonic absorption cell for normal liquids", The Review of Scientific Instruments, 37, 3, 367-370 (1966)

131. Eicke, H.-F. *in* "Microemulsions" (I.D. Rob,Ed.), p.10, Plenum Press, New York, (1982)

132. Eicke, H.-F., and Rehak, J., *Helv. Chim. Acta,* 59, 8, 2883 (1976)

133. Eigen and deMaeyer, in "Techniques of Organic Chemistry", (ed. Weissberger) Vol. VIII Part 2, Wiley, p.895 (1963)

134. Eigen., "Determination of general and specific ionic interactions in solution", Faraday Soc. Discussions, No.24, p.25 (1957)

135. Enderby, J.A. "On Electrical Effects Due to Sound Waves in Colloidal Suspensions", Proc. Roy. Soc., London, A207, 329-342 (1951)

136. Endo, S. "Characterization for size and morphology of particle in dense suspension system." Ceramics Japan, 34, 10,.834-837 (1999)

137. Ennis, JP, Shugai, AA and Carnie, SL, "Dynamic mobility of two spherical particles with thick double layers"; Journal of Colloid and Interface Science, 223,21-36, (2000)

138. Ennis, JP, Shugai, AA and Carnie, SL, "Dynamic mobility of particles with thick double layers in a non-dilute suspension"; Journal of Colloid and Interface Science, 223, 37-53, (2000).

139. Epstein, P.S. and Carhart R.R., "The Absorption of Sound in Suspensions and Emulsions", J. Acoust. Soc. Amer., 25, 3, 553-565 (1953)

140. Erwin, L. and Dohner, J.L. "On-line measurement of fluid mixtures", US Patent 4,509,360 (1985)

141. Estrela-L'opis, V.R. and Razilov, I.A. "Polarization of a double layer of colloidal particles in a mixture of electrolytes in an alternating electric field", Kolloidn Zh., 49, 6, 1155-1165 (1987)

142. Evans, J.M. and Attenborough, K. "Coupled Phase Theory for Sound Propagation in Emulsions", J.Acoust.Soc.Amer., 102, 1, 278-282 (1997)

143. Faran, J.J. "Sound Scattering by Solid Cylinders and Spheres", J. Acoust. Soc. Amer., 23, 4, 405-418 (1951)

144. Ferry, J.D., Sawyer, M.W. and Ashworth, J.N. "Behavior of Concentrated polymer Solutions under Periodic Stresses", J. of Polymer Sci., 2, 6 (1947) 593-611

145. Fischer, H. J.; Tenor, P., " Mean molecular weight of synthetics in solution from ultrasonic measurements", Rheologica Acta, VOL. 15, NO. 7-8, 434-5 (1976)

146. Fisker, R., Carstensen, J.M., Hansen, M.F., Bodker, F. and Morup, S., "Estimation of nanoparticle size distribution by image analysis", J. of Nanoparticle Research, 2, 267-277 (2000)

147. Fletcher, P.D.I., Robinson, B.H., Bermejo-Barrera, F., Oakenfull, D.G., Dore, J.C., and Steytler, D.C. *in* "Microemulsions" (I.D. Rob, Ed.), p. 221, Plenum Press, New York, (1982)

148. Foldy, L.L "Propagation of sound through a liquid containing bubbles", OSRD Report No.6.1-sr1130-1378, (1944)

149. Fox, F.E., Curley S.R. and Larson, G.S. "Phase velocity and absorption measurement in water containing air bubbles" J. of Acoustic Society of America, 27, 3, 534-539 (1957)

150. Fritz, G.; Scherf, G.; Glatter, O. , "Applications of densiometry, ultrasonic speed measurements, and ultralow shear viscosimetry to aqueous fluids", Journal of Physical Chemistry B, VOL. 104, NO. 15, 3463-70 (2000)

151. Fry, W.J. and Fry, R.B. "Determination of Absolute Sound Levels and Acoustic Absorption Coefficients by Thermocouple Probes-Theory", J. Acoust. Soc. Amer., 26, 3, 294-311 (1954)

152. Fry, W.J. and Fry, R.B. "Determination of Absolute Sound Levels and Acoustic Absorption Coefficients by Thermocouple Probes-Experiment", J. Acoust. Soc. Amer., 26, 3, 311-317 (1954)

153. Fuchs, R., "Theory of the optical properties of ionic crystal cubes", Phys. Rev., B11, 1732-1740 (1975)

154. Gaunaurd, G.; Scharnhorst, K. P.; Uberall, H. "New method to determine shear absorption using the viscoelastodynamic resonance-scattering formalism", Journal of the Acoustical Society of America, VOL. 64, NO. 4, 1211-12 (1978)

155. Gibson, R.L. and Toksoz, M.N., "Viscous Attenuation of Acoustic Waves in Suspensions", J. Acoust. Soc. Amer., 85, 1925-1934 (1989)

156. Goetz, R.J. and El-Aasser, M.S. "Effects of dispersion concentration on the electroacoustic potentials of o/w miniemulsions", JCIS, 150, 2, 436-451 (1992)

157. Gonzalez-Fernandez, C.F., Espinosa-Jimenez, M., Gonzalez-Caballero, F. "The effect of packing density cellulose plugs on streaming potential phenomena", Colloid and Polymer Sci., 261, 688-693 (1983)

158. Goodwin, J.W., Gregory, T., and Stile, J.A., "A Study of Some of the Rheological Properties of Concentrated Polystyrene Latices", Adv. In Colloid and Interface Sci., 17, 185-195 (1982)

159. Greenwood, M.S., Mai, J.L. and Good, M.S. "Attenuation measurements of ultrasound in a kaolin-water slurry. A linear dependence upon frequency ", J. Acoust. Soc. Am., 94, 2, 908-916 (1993)

160. Groves, J. and Sears, A. "Alternating Streaming Current Measurements", J. Colloid and Interface Sci., 53, 1, 83-89 (1975)

161. Gruber, G.J. and Meister, R. "Ultrasonic attenuation in water containing brine shrimp in suspension", J.Acoust.Soc.Am., 33, 6, 733-740 (1961)

162. Gulari, E., Bedwell, B. and Alkhafaji, S., *J. Colloid Interface Sci.*, 77,1, 202 (1980).

163. Guo Min, "A novel method to measure the viscosity factor of liquids", Wuli, VOL. 30, NO. 4, 220-2 (2001)

164. Gunnarsson, M., Rasmusson, M., Wall, S., Ahlberg, E., Ennis, J. "Electroacoustic and potentiometric studies of the hematite/water interface", JCIS, 240, 448-458 (2001)

165. Gustafsson, J., Mikkola, P., Jokinen, M., Rosenholm, J.B. "The influence of pH and NaCl on the zeta potential and rheology of anatase dispersions", Colloids and Surfaces A, 175, 349-359 (2000)

166. Guffond, M.C., Hunter, R.J., O'Brien, R.W. and Beattie, J.K. "The dynamic mobility of colloidal semiconducting antimony-doped Tin(IV) oxide particles", JCIS, 235, 371-379 (2001)

167. Guffond, M.C., Hunter, R.J., and Beattie, J.K. "Electroacoustic properties of barium sulfate particles coated with conductive layer of antimony-doped Tin(IV) oxides", ChemMater. 13, 2619-2625 (2001)

168. Haberger, C.C. "The attenuation of ultrasound in dilute polymeric fiber suspensions", J. Acous. Soc. Am., 72, 3, (1982)

169. Hackley, V.A. and Malghan, S.G. "The surface chemistry of silicon nitride powder in the presence of dissolved ions", J. Mat. Sci., 29, 4420-4430 (1994)

170. Hackley, V.A. and Ferraris, C.F. "The Use of Nomenclature in Dispersion Science and Technology", NIST, special publication 960-3 (2001)

171. Hackley, V.A. and Paik, U. "Electroacoustic analysis in processing of advanced ceramics", in Ultrasonic and Dielectric Characterization Techniques for Suspended Particulates, Ed. V.A. Hackley and J. Texter, Amer. Ceramic Soc., 191-205, (1998)

172. Hackley, V.A. and Malghan, S.G "Investigation of parameters and secondary components affecting the electroacoustic analysis of silicon nitride powders", in Electroacoustics for Characterization of Particulates and Suspensions, Ed.S.G.Malghan, NIST, 161-180, (1993)

173. Hackley, V.A., Patton, J., Lum, L.S.H., Wasche, R.J., Naito, M., Abe, H., Hotta, Y., and Pendse, H. "Analysis of the isoelectric point in moderately concentrated alumina suspensions using electroacoustic and streaming potential methods", NIST, Fed.Inst. for Material Research-Germany, JFCC-Japan, University of Main., Report (2001)

174. Hagin, F. "A stable approach to solving one-dimensional inverse problem", SIAM J. Appl. Math., 40, 3, 439-453 (1981)

175. Hampton, L. "Acoustic Properties of Sediments", J. Acoust. Soc. Amer., vol.42, 4, 882-890, (1967)

176. Hamstead, P.J. "Apparatus for determining the time taken for sound energy to cross a body of fluid in a pipe", US Patent 5,359,897 (1994)

177. Han, W., and Pendse, H.P. "Unified Coupled Phase Continuum Model for Acoustic Attenuation in Concentrated Dispersions", Characterization Techniques for Suspended Particulates, ed. V.Hackley and J. Texter, Am. Ceramic Soc., Ohio, 129-155, (1998)

178. Handbook of Chemistry and Physics, Ed. R. Weast, 70th addition CRC Press, Florida, (1989)

179. Happel J. and Brenner, H, "Low Reynolds Number Hydrodynamics", Martinus Nijhoff Publishers, Dordrecht, The Netherlands, (1973)

180. Happel J., "Viscous flow in multiparticle systems: Slow motion of fluids relative to beds of spherical particles", AICHE J., 4, 197-201 (1958)

181. Harker, A.H. and Temple, J.A.G., "Velocity and Attenuation of Ultrasound in Suspensions of Particles in Fluids", J. Phys. D.: Appl. Phys., 21, 1576-1588 (1988)

182. Harker, A.H., Schofield, P., Stimpson, B.P., Taylor, R.G., and Temple, J.A.G. "Ultrasonic Propagation in Slurries", Ultrasonics, 29, 427-439 (1991)

183. Hartmann G.K and Focke, A.B. "Absorption of Supersonic Waves in Water and in Aqueous Suspensions", Physical Review, 57, 1, 221-225 (1940)

184. Hawley, S.A. , Kessler, L.W. and Dunn, F. "Ultrasonic absorption in aqueous solutions of high molecular weight polysaccharides", J. Acoust. Soc. Am., 521-523 (1965)

185. Hauptmann, P., Sauberlich, R. and Wartewig, S. "Ultrasonic attenuation and mobility in polymer solutions and dispersions", Polymer Bulletin, 8, 269-274 (1982)

186. Hay, A.E. and Mercer, D.G. "On the Theory of Sound Scattering and Viscous Absorption in Aqueous Suspensions at Medium and Short wavelength", J. Acoust. Soc. Amer., 78, 5 1761-1771 (1985)

187. Hayashi, T., Ohya, H., Suzuki, S. and Endoh, S. "Errors in Size Distribution Measurement of Concentrated Alumina Slurry by Ultrasonic Attenuation Spectroscopy", J. Soc. Powder Technology Japan, 37, 498-504 (2000)

188. Hellwege, K.H. Ed. Numerical data and functional relationships in Science and Technology, Vol. 5, Molecular Acoustics, Ed. W. Schaaffs, Berlin, NY, 1967

189. Hemar, Y, Herrmann, N., Lemarechal, P., Hocquart, R. and Lequeux, F. "Effect medium model for ultrasonic attenuation due to the thermo-elastic effect in concentrated emulsions", J. Phys. II 7, 637-647 (1997)

190. Henry, Report of the Lighthouse Board of the United States for the year 1874.

191. Hermans, J., Philos. Mag., 25, 426 (1938)

192. Hertz, T. G.; "Viscosity measurement of an enclosed liquid using ultrasound", Rev. Scientific Instruments , VOL. 62, 2 (1991)

193. Heywood, H, "The origins and development of Particle Size Analysis", in "Particle size analysis", ed. Groves, M.J and oth., The Society for analytical chemistry, London, (1970)

194. Hidalgo-Alvarez, R., de las Nieves, F.J., Pardo, G., "Comparative Sedimentation and Streaming Potential Studies for ζ-potential Determination", J.Colloid and Interface Sci., 107, 2, 295-300 (1985)

195. Hidalgo-Alvarez, R., Moleon, J.A., de las Nieves, F.J. and Bijsterbosch, B.H., "Effect of Anomalous Surface Conductance on ζ-potential Determination of Positively Charged Polystyrene Microspheres", J. Colloid and Interface Sci., 149, 1, 23-27 (1991)

196. Hilderbrand, B.P., Davis, T.J., Boland, A.J. and Silta, R.L. "A portable digital ultrasonic holography system for imaging flaws in heavy section materials", IEEE Transactions on Sonic and Ultrasonics; SU-31, 4, 287-293 (1984)

197. Hinze, F., Stintz, M. and Ripperger, S. "Surface characterization of organic colloids by the Colloid Vibration Potential", in Ultrasonic and Dielectric Characterization Techniques for Suspended Particulates, Ed. V.A. Hackley and J. Texter, Amer. Ceramic Soc., 291-305, (1998)

198. Hipp, A.K., Storti, G. and Morbidelli, M. "On multiple-particle effects in the acoustic characterization of colloidal dispersions", J. Phys., D: Appl. Phys. 32, 568-576 (1999)

199. Hipp, A.K., Storti, G. and Morbidelli, M. "Particle sizing in Colloidal Dispersions by Ultrasound.Model calibration and sensitivity analysis", Langmuir, 15, 2338-2345 (1999)

200. Hodne, H. and Beattie, J.K. "Verification of the electroacoustic standard: Comparison of the dynamic mobility of silicododecamolybdate and silicododecatungstate acids and salts", Langmuir, 17, 3044-3046 (2001)

201. Holmes, A.K., Challis, R.E. and Wedlock, D.J. "A Wide-Bandwidth Study of Ultrasound Velocity and Attenuation in Suspensions: Comparison of Theory with Experimental Measurements", J. Colloid and Interface Sci., 156, 261-269 (1993)

202. Holmes, A.K., Challis, R.E. and Wedlock, D.J. "A Wide-Bandwidth Ultrasonic Study of Suspensions: The Variation of Velocity and Attenuation with Particle Size", J. Colloid and Interface Sci., 168, 339-348 (1994)

203. Holmes, A.K., and Challis, R.E "Ultrasonic scattering in concentrated colloidal suspensions", Colloids and Surfaces A, 77, 65-74 (1993)

204. Hoverm, J. and Ingram, G., "Viscous Attenuation of Sound in Saturated Sand", J. Acoust. Soc. Amer, vol.66, 6, pp. 1807-1812 (1979)

205. Hoverm, J., "Viscous attenuation of sound in suspensions and high porosity marine sediments", J. Acoust. Soc. Amer., vol. 67, 5, pp. 1559-1563 (1980)

206. Hoy, C. L. C.; Leung, W. P.; Yee, A. F. "Ultrasonic measurements of the Elastic Moduli of Liquid Crystalline Polymers."- Polymer, 33, 8, 1788-1791 (1992)

207. Hueter, T.F., Morgan, H. and Cohen, M.S. "Ultrasonic attenuation in biological suspensions", J. Acoust. Soc. Amer, 1200-1201 (1953)

208. Hulusi, T. H. , "Effect of pulse-width on the measurement of the ultrasoinic shear absorption in viscous liquids", Acustica, VOL. 40, NO. 4, 269-71 (1978)

209. Huang, Y.C., Sanders, N.D., Fowkes, F.M. and Lloyd, T.B. "The impact of surface chemistry on particle electrostatic charging and viscoelasticity of precipitated calcium carbonate slurries", in Electroacoustics for Characterization of Particulates and Suspensions, Ed. S.G. Malghan, NIST, 180-200, (1993)

210. Hunter, J.L. and Dardy, H.D. "Ultrahigh-frequency ultrasonic absorption cell", J. Acoust. Soc. Am., 36, 10, 1914-1917, (1964)

211. Hunter, R.J., "The Electroacoustic Characterization of Colloidal Suspension", in Ultrasonic and Dielectric Characterization Techniques for Suspended Particulates, ed. V. Hackley and J. Texter, Am. Ceramic Soc., Ohio, 25-47, (1998)

212. Hunter, R.J., Liversidge lecture, J. Proc. R. Soc. New South Wales, 121, 165-178 (1988)

213. Hunter, R.J. "Foundations of Colloid Science", Oxford University Press, Oxford, (1989)

214. Hunter, R.J. "Review. Recent developments in the electroacoustic characterization of colloidal suspensions and emulsions", Colloids and Surfaces, 141, 37-65 (1998)

215. Hunter, R.J. "Zeta potential in Colloid Science", Academic Press, NY (1981)

216. Huruguen, J.P., Zemb, T., and Pileni, M.P., Progr. Coll. Polym. Sci. 89, 39 (1992).

217. Hyman, A. and Walsh, J.J. "Philosophy in the Middle Ages", Indianapolis, Hackett Publishing Co, (1973)

218. Inoue, M.; Yoshino, K.; Moritake, H.; Toda, K. "Viscosity measurement of nematic liquid crystal using shear horizontal wave propagation in liquid crystal cell", Japanese Journal of Applied Physics, Part 1, VOL. 40, NO. 5B, 3528-33 (2001)

219. Irani, R.R. and Callis, C.F. "Particle Size: Measurement, Interpretation and Application", John Wiley & Sons, NY-London, (1971)

220. Isakovich, M.A. Zh. Experimental and Theoretical Physics, 18, 907 (1948)

221. Issacs, E.E., Huang, H., Babchin, A.J. and Chow, R.S. "Electroacoustic method for monitoring the coalescence of water-in-oil emulsions", Colloids and Surfaces, 46, 177-192 (1990)

222. Issacs, E.E., Huang, H., Chow, R.S. and Babchin, A.J. "Coalescence behavior of water-in-oil emulsions", Particle technology and surface phenomena in Minerals and Petrolium", Ed. M.K. Sharma and G.D. Sharma, Plenum Press, NY (1991)

223. Jackopin, L. and Yeager, E., "Ultrasonic Relaxation in Manganese Sulfate Solutions", J. of Physical Chemistry, vol.74, 21, 3766-3772, (1970)

224. James, M., Hunter, R.J. and O'Brien, R.W. "Effect of particle size distribution and aggregation on electroacoustic measurements of Zeta potential", Langmuir, 8, 420-423 (1992)

225. James, R.O., Texter, J. and Scales, P.J. "Frequency dependence of Electroacoustic (Electrophoretic) Mobilities", Langmuir, 7, 1993-1997 (1991)

226. James, R.O.. "Calibration of Electroacoustic (ESA) Apparatus using lattices and oxides colloids", in S.B. Malghan (Ed.) Electroacoustics for Characterization of Particulates and Suspensions, NIST, 67-92 (1993)

227. Javanaud, C., Gladwell, N.R., Gouldby, S.J., Hibberd, D.J., Thomas, A. and Robins, M.M. "Experimental and theoretical values of the ultrasonic properties of dispersions: effect of particle state and size distribution", Ultrasonics, 29, 330-337, (1991)

228. Jain, M. K.; Schmidt, S.; Grimes, C. A. "Magneto-acoustic sensors for measurement of liquid temperature, viscosity and density", Applied Acoustics, VOL. 62, NO. 8, 1001-11 (2001)

229. Johnson, D., Koplik, J. and Dashen, R. "Theory of dynamic permeability and tortuosity in fluid saturated porous media", J. Fluid Mech vol. 176, 379-402, (1987)

230. Jong-Rim Bae; Jeong-Koo Kim; Meyung-Ha Yi, "Ultrasonic velocity and absorption measurements for polyethylene glycol and water solutions", Japanese Journal of Applied Physics, Part 1, VOL. 39, NO. 5B, 2946-7 (2000)

231. Jones, G.M., Povey, M.J.W. and Campbell, J. "Measurement and control of asphaltene agglomeration in hydrocarbon liquids", US Patent 5,969,237 (1999)

232. Kabanov, A.V., *Makromol. Chem., Macromol. Symp.* 44, 253 (1991).

233. Kamphuis, H., Jongschaap, R.J. and Mijnlieff, P.F. "A transient-network model describing the rheological behaviour of concentrated dispersions", Rheological Acta, 23, 329-344 (1984)

234. Karamushka, V.I., Ulberg, Z.R., Gruzina, T.G. and Dukhin, A.S. "ATP-Dependent gold accumulation by living Chlorella Cells", Acta Biotechnologica, 11, 3, 197-203 (1991)

235. Katchalsky, A. and Curran, P.F. "Non-equilibrium thermodynamics in biophysics", Harvard University press, Cambridge (1967)

236. Kessler, L.W. and Dunn, F. "Ultrasonic investigation of the conformal changes of bovine serum albumin in aqueous solution", J. Phys. Chem., 73, 12, 4256-4262 (1969)

237. Kijlstra, J., van Leeuwen, H.P. and Lyklema, J. "Effects of Surface Conduction on the Electrokinetic Properties of Colloid", Chem. Soc., Faraday Trans., 88, 23, 3441-3449 (1992)

238. Kijlstra, J., van Leeuwen, H.P. and Lyklema, J. "Low-Frequency Dielectric Relaxation of Hematite and Silica Sols", Langmuir, (1993)

239. Kinsler, L., Frey, A., Coppens, A., and Sanders, J. "Fundamentals of Acoustics", J. Wiley & Sons, NY, 547 p, (2000)

240. Kippax, P. and Higgs, D. "Application of ultrasonic spectroscopy for particle size analysis of concentrated systems without dilution", American Laboratory, 33, 23, 8-10, (2001)

241. Kirchhoff, "Vorlesungen uber Mathematische Physik", (1876)

242. Kirchhoff, Pogg.Ann., vol.CXXXIV, p.177, (1868)

243. Kikuta, M. "Method and apparatus for analyzing the composition of an electro-deposition coating material and method and apparatus for controlling said composition", US Patent 5,368,716 (1994)

244. Knauss, C. J.; Leppo, D.; Myers, R. R. , "Viscoelastic measurement of polybutenes and low viscosity liquids using ultrasonic strip delay lines", Journal of Polymer Science, Polymer Symposia, USA, NO. 43, 179-86 (1973)

245. Kol'tsova, I.S. and Mikhailov, I.G. "Scattering of Ultrasound Waves in Heterogeneous Systems", Soviet Physics-Acoustics, 15, 3, 390-393 (1970)

246. Kondoh, J.; Hayashi, S.; Shiokawa, S., "Simultaneous detection of density and viscosity using surface acoustic wave liquid-phase sensors", Japanese Journal of Applied Physics, Part 1 (Regular Papers, Short Notes & Review Papers) , VOL. 40, NO. 5B, 3713-17 (2001)

247. Kondoh, J.; Saito, K.; Shiokawa, S.; Suzuki, H. "Simultaneous measurement of liquid properties using multichannel shear horizontal surface acoustic wave microsensor", Japanese Journal of Applied Physics, Part 1, VOL. 35, NO. 5B, 3093-6, (1996)

248. Kondoh, J.; Saito, K.; Shiokawa, S.; Suzuki, H. , " Multichannel shear horizontal surface acoustic wave microsensor for liquid characterization", 1995 IEEE Ultrasonics Symposium. Proceedings. An International Symposium (Cat. No.95CH35844) , Ed.: Levy, M.; Schneider, S. C.; McAvoy, B. R. , vol.1, 445-9 (1995)

249. Kong, L., Beattie, J.K. and Hunter, R.J. "Electroacoustic study of concentrated oil-in-water emulsions", JCIS, 238, 70-79 (2001)

250. Kornbrekke, R.E., Morrison, I.D. and Oja, T. "Electrokinetic measurements of concentrated suspensions in apolar liquids", in Electroacoustics for Characterization of Particulates and Suspensions, Ed. S.G. Malghan, NIST, 92-111, (1993)

251. Kosmulski, M. and Rosenholm, J.B. "Electroacoustic study of adsorption of ions on anatase and zirconia from very concentrated electrolytes", J. Phys. Chem., 100, 28, 11681-11687 (1996)

252. Kosmulski, M. "Positive electrokinetic charge on silica in the presence of chlorides", JCIS, 208, 543-545 (1998)

253. Kosmulski, M., Gustafsson, J. and Rosenholm, J.B. "Correlation between the Zeta potential and Rheological properties of Anatase dispersions", JCIS, 209, 200-206 (1999)

254. Kosmulski, M., Durand-Vidal, S., Gustafsson, J. and Rosenholm, J.B. "Charge interactions in semi-concentrated Titania suspensions at very high ionic strength", Colloids and Surfaces A, 157, 245-259 (1999)

255. Kostial, P. "Surface Acoustic Wave Control of the Ion Concentration in Water", Applied Acoustics, 41, 187-193 (1994)

256. Kozak M.W. and Davis J.E. "Electrokinetic phenomena in fibrous porous media", JCIS, 112, 2, 403-411 (1986)

257. Krishnan, R. "Role of particle size and ESA in performance of solvent based gravure inks", in Electroacoustics for Characterization of Particulates and Suspensions, Ed. S.G. Malghan, NIST, 263-274, (1993)

258. Kruyt, H.R. "Colloid Science", Elsevier: Volume 1, Irreversible systems, (1952)

259. Kuwabara, S. "The forces experienced by randomly distributed parallel circular cylinders or spheres in a viscous flow at small reynolds numbers", J. Phys. Soc. Japan, 14, 527-532 (1959)

260. Lamb, H. "Hydrodynamics", Sixth Edition, Dover Publications, NY, 1932

261. Landau, L.D. and Lifshitz, E.M., "Electrodynamics of Continuous Media", London, Pergamon Press, (1960)

262. Larson, J.R. "Advances in liquid toners charging mechanism: Electroacoustic measurement of liquid toner charge", in Electroacoustics for Characterization of Particulates and Suspensions, Ed. S.G. Malghan, NIST, 301-315, (1993)

263. Lec, R., Vetelino, J.F., Clarke, P., Roy, A., and Turner, J. "Prototype microwave acoustic fluid sensors", Ultrasonic Symposium, 543 (1988)

264. Levine, S. and Neale, G.H. "The Prediction of Electrokinetic Phenomena within Multiparticle Systems.1.Electrophoresis and Electroosmosis.", J. of Colloid and Interface Sci., 47, 520-532 (1974)

265. Li, Yongcheng., Park, C.W. "Effective medium approximation and deposition of colloidal particles in fibrous and granular media", Adv. In Colloid and Interface Sci., 87, 1-74 (2000)

266. Liftshitz, E.M., "Theory of molecular attractive forces", Soviet Phys. JETP 2, 73 (1956)

267. Lischonski, K. "Representation and Evaluation of Particle Size Analysis data", Part. Charact., 1, 89-95 (1984)

268. Lin, W.H. and Raptis, A.C. "Thermoviscous effects on acoustic scattering by thermoelastic solid cylinders and spheres", J. Acoust. Soc. Am., 74, 5, 1542-1554 (1983)

269. Litovitz T.A. and Lyon, "Ultrasonic Hysteresis in Viscous Liquids", J. Acoust. Soc. Amer., vol.26, 4, 577- 580, (1954)

270. Ljunggren, S. and Eriksson, J.C. "The lifetime of a colloid sized gas bubble in water and the cause of the hydrophobic attraction", Colloids and Surfaces, 129-130, 151-155 (1997)

271. Lloyd, P. and Berry, M.V. "Wave propagation through an assembly of spheres", Proc. Phys. Soc., London 91, 678-688 (1967)

272. Loeb, A.L., Overbeek, J.Th.G. and Wiersema, P.H. "The Electrical Double Layer around a Spherical Colloid Particle", MIT Press (1961)

273. Loewenberg, M. and O'Brien, R.W. "The dynamic mobility of nonspherical particles", JCIS, 150, 1, 159-168 (1992)

274. Lyklema, J., "Fundamentals of Interface and Colloid Science", vol. 1-3, Academic Press, London-NY, (1995-2000).

275. Lubomska, M. and Chibowski, E. "Effect of Radio Frequency Electric Fields on the Surface Free Energy and Zeta Potential of Al_2O_3 ", Langmuir, 17, 4181-4188 (2001)

276. Ma, Y., Varadan, V.K., and Varadan, V.V. "Comments on ultrasonic propagation in suspensions", J. Acoust. Soc. Amer., 87, 2779-2782 (1990)

277. Mackenzie, K.V. "Reflection of Sound from Coastal Bottoms", J.Acoust.Soc.Amer.,vol.32, 2, 221-231, (1960)

278. Malkin, E. S. ; Dukhin, A. S. "Interaction of Dispersed Particles in an Electric Fields and Linear Concentration Polarization of the Double Layer", Kolloid. Zh., 5, 801-810, (1982)

279. Margulies, T.S. "Multiphase continuum theory for sound wave propagation through dilute suspensions of particles", J.Acoust.Soc.Am., 96, 1, 319-331 (1994)

280. Marlow,B.J., Fairhurst,D. and Pendse,H.P., "Colloid Vibration Potential and the Electrokinetic Characterization of Concentrated Colloids", Langmuir, 4,3, 611-626 (1983)

281. Marlow, B.J, Fairhurst, D. and Schutt, W. "Electrophoretic Fingerpronting an the Biological Activity of Colloidal Indicators", Langmuir, 4, 776 (1988)

282. Marlow, B.J. and Rowell, R.L. "Electrophoretic Fingerprinting of a Single Acid Site Polymer Colloid Latex", Langmuir, 7, 2970-2980 (1991)

283. Marlow, B.J. and Rowell, R.L. "Acoustic and Electrophoreric mobilities of Coal Dispersions", J.of Energy and Fuel, 2, 125-131 (1988)

284. Marlow, B.J., Oja, T. and Goetz, P.J., "Colloid Analyser", US Patent 4,907,453 (1990)

285. Marshall, L., Goodwin, J.W., and Zukoski, C.F. "The effect of Electric Fields on the Rheology of Concentrated Suspensions", J. Chem. Soc., Faraday Trans., (1988)

286. Mason, W.P. "Dispersion and Absorption of Sound in High Polymers", in Handbuch der Physik., vol.2, Acoustica part1, (1961)

287. Mason, W.P., Baker, W.O. McSkimin, H.J. and Heiss, J.H. "Measurement of shear elasticity and viscosity of liquids at ultrasonic frequencies", Physical Review, 75, 6, 936-945, (1949)

288. Maxwell, "On the Viscosity or Internal Friction of Air and other Gases", Phil. Trans. vol 156, p.249, (1866)

289. Maxwell, J.C. "Electricity and Magnetism", Vol.1, Clarendon Press, Oxford (1892)

290. Meister, R., and Laurent, R., "Ultrasonic absorption and velocity in water containing algae in suspension", J. Acous. Soc. Am., 32, 5, 556-559 (1960)

291. McCann, C. "Compressional Wave Attenuation in Concentrated Clay Suspensions", Acustica, 22, 352-356 (1970)

292. McClements, J.D. and Povey, M.J. "Ultrasonic velocity as a probe of emulsions and suspensions", Adv. Colloid Interface Sci., 27, 285-316 (1987)

293. McClements, D.J. "Comparison of Multiple Scattering Theories with Experimental Measurements in Emulsions", The Journal of the Acoustical Society of America, vol.91, 2, pp. 849-854, (1992)

294. McClements, D.J. and Povey, M.J.W. "Scattering of Ultrasound by Emulsions", J. Phys. D: Appl. Phys., 22, 38-47 (1989)

295. McClements, D.J. "Ultrasonic Characterization of Emulsions and Suspensions", Adv. Colloid Int. .Sci., 37, 33-72 (1991)

296. McClements, J.D. "Characterization of emulsions using a frequency scanning ultrasonic pulse echo reflectometer", Proc. Inst. Acoust., 13, 71-78 (1991)

297. McClements, J.D. "Ultrasonic Determination of Depletion Flocculation in Oil-in-Water Emulsions Containing a Non-Ionic Surfactant", Colloids and Surfaces, 90, 25-35 (1994)

298. McClements, J.D. and Powey, M.J.W., "Scattering of ultrasound by emulsions", J. Phys. D: Appl. Phys., 22, 38-47 (1989)

299. McClements, J.D. "Principles of ultrasonic droplet size determination in emulsions:, Langmuir, 12, 3454-3461 (1996)

300. McClements, J.D. and Coupland, J.N. "Theory of droplet size distribution measurement in emulsions using ultrasonic spectroscopy", Colloids and Surfaces A, 117, 161-170 (1996)

301. McClements, J.D., Hermar, Y. and Herrmann, N. "Incorporation of thermal overlap effects into multiple scattering theory", J. Acous. Soc. Am., 105, 2, 915-918 (1999)

302. McClements, D.J. "Ultrasonic characterization of food emulsions", in Ultrasonic and Dielectric Characterization Techniques for Suspended Particulates, Ed. V.A. Hackley and J. Texter, Amer. Ceramic Soc., 305-317, (1998)

303. McLeroy, E.G. and DeLoach, A. "Sound Speed and Attenuation from 15 to 1500 kHz Measured in Natural Sea-Floor Sediments", J. Acoust. Soc. Amer., vol.44, 4, pp. 1148-1150 (1968).

304. Midmore, B.R. and Hunter, R.J., J. Colloid Interface Science, 122, 521 (1988)

305. Mie, G., "Beitrage zur Optik truber Medien speziell kolloidaler Metallosungen", Ann Phys., 25, 377-445 (1908)

306. Mitaku, S.; Ohtsuki, T.; Enari, K.; Kishimoto, A.; Okano, K, "Studies of ordered monodisperse polystyrene latexes. I. Shear ultrasonic measurements", Japanese Journal of Applied Physics, VOL. 17, NO. 2, 305-13 (1978)

307. Mitchell, S.K., and Focke, K.C. "New measurements of compressional wave attenuation in deep ocean sediments", J. Acoust. Soc. Am, 67, 5, 1582-1589 (1980)

308. Moritake, H.; Takahashi, K.; Yoshino, K.; Toda, K., "Viscosity measurement of ferroelectric liquid crystal", Japanese Journal of Applied Physics, Part 1, VOL. 35, NO. 9B, 5220-3 (1996)

309. Morkun, V.S., Khorolsky, V.P., Protsuto, V.S. and Potapov, V.N. "Method and apparatus for measuring parameters of solid phase of slurries", US Patent 5,058,432 (1991)

310. Morse, P.M. "Vibration and Sound", Acoustical Society of Amer. Publications; ISBN 0883 182874, (1991)

311. Morse, P.M. and Uno Ingard, K., "Theoretical Acoustics", 1968 McGraw-Hill, NY, 1968, Princeton University Press, NJ, 925 p., (1986)

312. Morfey, C.L. "Sound attenuation by small particles in a fluid", J.Sound Vib., 8, 1, 156-170 (1968)

313. Moudgil, B.M. and Damodaran, R. "Surface Charge Characterization of minerals by electroacoustic measurements", in Electroacoustics for Characterization of Particulates and Suspensions, Ed.S.G.Malghan, NIST, 219-238, (1993)

314. Murpy, S.R., Garrison, G.R. and Potter, D.S. "Sound Absorption at 50 to 500 kHz from Transmission Measurements in the Sea", J. Acoust. Soc. Amer., vol.30, 9, pp. 871-875, (1958).

315. Murtsovkin, V.A. and Muller, V.M. "Inertial Hydrodynamic Effects in Electrophoresis of Particles in an Alternating Electric Field", J.of Colloid and Interface Sci., 160, 2, 338-346 (1993)

316. Murtsovkin, V.A. and Muller, V.M. "Steady-State Flows Induced by Oscillations of a Drop with an Adsorption Layer", J.of Colloid and Interface Sci., 151, 1 150-156 (1992)

317. Myers, D.F. and Saville, D.A. "Dielectric Spectroscopy of Colloidal Suspensions", J. Colloid and Interface Sci., 131, 2, 448-460 (1988)

318. Nicholson, J.D., Doherty, J.V., and Clarke, J.H.R. *in* "Microemulsions" (I.D. Rob, Ed.), p. 33, Plenum Press, New York, (1982)

319. Nomoto, O. and Kishimoto, T. "Velocity and Dispersion of Ultrasonic Waves in Electrolytic Solutions", Bulletin of Kobayashi Institute, vol.2, 2, 58-62, (1952)

320. Nomura, T.; Saitoh, A.; Horikoshi, Y. "Measurement of acoustic properties of liquid using liquid flow SH-SAW sensor system", Sensors and Actuators B (Chemical), VOL. B76, NO. 1-3, 69-73 (2001)

321. O'Brien, R.W. "Electro-acoustic effects in a dilute suspension of spherical particles", Preprint, School of Mathematics, The University of New South Wales, (1986)

322. O'Brien, R.W., Midmore, B.R., Lamb, A. and Hunter, R.J. "Electroacoustic studies of Moderately concentrated colloidal suspensions", Faraday Discuss. Chem. Soc., 90, 1-11 (1990)

323. O'Brien R.W., Rowlands W.N. and Hunter R.J. "Determining charge and size with the Acoustosizer", in S.B. Malghan (Ed.) Electroacoustics for Characterization of Particulates and Suspensions, NIST, 1-21 (1993)

324. O'Brien, R.W. "Determination of Particle Size and Electric Charge", US Patent 5,059,909, Oct.22, (1991).

325. O'Brien, R.W. Particle size and charge measurement in multi-component colloids", US Patent 5,616,872, (1997).

326. O'Brien, R.W. "Electro-acoustic Effects in a dilute Suspension of Spherical Particles", J. Fluid Mech., 190, 71-86 (1988)

327. O'Brien, R.W. "Electroosmosis in Porous Materials", Journal of Colloid and Interface Science", 110, 2, 477-487 (1986)

328. O'Brien, R.W. and White, L.R. "Electrophoretic mobility of a spherical colloidal particle", J.Chem.Soc.Faraday Trans., II, 74, 1607-1624 (1978)

329. O'Brien, R.W., "The solution of electrokinetic equations for colloidal particles with thin double layers", J.Colloid Interface Sci, 92, 204-216 (1983)

330. O'Brien, R.W., Wade, T.A., Carasso, M.L., Hunter, R.J., Rowlands, W.N. and Beattie, J.K. "Electroacoustic determination of droplet size and zeta potential in concentrated emulsions", JCIS, (1996)

331. O'Brien, R.W., Cannon, D.W. and Rowlands, W.N. "Electroacoustic determination of particle size and zeta potential", JCIS, 173, 406-418 (1995)

332. O'Konski, C.T., "Electric Properties of Macromolecules v. Theory of Ionic Polarization in Polyelectrolytes", J. Phys. Chem, 64, 5, 605-612 (1960)

333. Ohshima, H. "A simple expression for Henry's function for retardation effect in electrophoresis of spherical colloidal particles", JCIS, 168, 269-271 (1994)

334. Ohshima, H., Healy, T.W., White, L.R. and O'Brien, R.W. "Sedimentation velocity and potential in a dilute suspension of charged spherical colloidal particles", J. Chem. Soc. Faraday Trans., 2, 80, 1299-1377 (1984)

335. Ohshima, H. "Dynamic Electrophoretic Mobility of Spherical Colloidal Particles in Concentrated Suspensions", J. of Colloid and Interface Sci., 195, 137-148 (1997)

336. Ohshima, H. "Electrokinetic phenomena in concentrated dispersion of charged mercury drops", J. of Colloid and Interface Sci., 218, 535-544 (1999)

337. Ohshima, H. and Dukhin, A.S. "Colloid Vibration Potential in a Concentrated Suspension of Spherical Colloidal Particles", JCIS, 212, 449-452 (1999)

338. Oja, T., Petersen, G., and Cannon, D. "Measurement of Electric-Kinetic Properties of a Solution", US Patent 4,497,208, (1985)

339. Osseo-Asare, K. and Khan, A. "Chemical-mechanical polishing of tungsten: An electrophoretic mobility investigation of alumina-tungstate interactions", Electrochemical Soc. Proc., 7, 139-144 (1998)

340. Overbeek, J.Th.G., Verhiekx, G.J., de Bruyn, P.L and Lakkerkerker, J. Colloid Interface Sci., 119, 422 (1987)

341. Pai, R., "Techniques for measuring the composition (oil and water content) of emulsions-a state of the art review", Colloids and Surfaces A, 84, 141-193 (1994)

342. Pallas-Areny, R. and Webster, J.G. "Ultrasonic based sensors", Sensors, 16-19, (1992)

343. Pedersen, P.C., Lewin, P.A. and Bjorno, L. "Application of time-delay spectroscopy for calibration of ultrasonic transducers", IEEE Transaction on Ultrasonics, 35, 2, 185-203 (1988)

344. Pellam, J.R. and Galt, J.K. "Ultrasonic propagation in liquids: Application of pulse technique to velocity and absorption measurement at 15 Megacycles", J. of Chemical Physics, 14, 10 , 608-613 (1946)

345. Pendse, H.P., Bliss T.C. and Wei Han "Particle Shape Effects and Active Ultrasound Spectroscopy", Ultrasonic and Dielectric Characterization Techniques for Suspended Particulates, Edited by V.A. Hackley and J. Texter, American Ceramic Society, Westerville, OH, 165-177, (1998)

346. Pendse, H.P, Strout, T.A. and Shanna, A. A. "Theoretical Consideration in Acoustophoretic Analysis of Concentrated Colloids". National Institute of Standards and Technology Special Publication 856, USA Department of Commerce, 23-39 (1993)

347. Pileni, M. P., Zemb, T. and Petit, C., *Chem. Phys. Lett.* 118, 4, 414 (1985).

348. Pinkerton, J.M.M. "A pulse method for measurement of ultrasonic absorption in liquids", Nature, 160, 128-129 (1947)

349. Poisson, "Sur l'integration de quelques equations lineaires aux differnces prtielles, et particulierement de l'equation generalie du mouvement des fluides elastiques", Mem., de l'Institut, t.III, p.121, (1820)

350. Poisson, Journal de l'ecole polytechnique, t.VII, (1808)

351. Polat, H., Polat, M. and Chandler, S. "Characterization of coal slurries by ESA technique", in Electroacoustics for Characterization of Particulates and Suspensions, Ed. S.G. Malghan, NIST, 200-219, (1993)

352. Pollinger, J.P. "Oxide and non-oxide powder processing applications using electroacoustic characterization", in S.B. Malghan (Ed.) Electroacoustics for Characterization of Particulates and Suspensions, NIST, 143-161 (1993)

353. Povey, M., "The Application of Acoustics to the Characterization of Particulate Suspensions", in Ultrasonic and Dielectric Characterization Techniques for Suspended Particulates, ed. V. Hackley and J. Texter, Am. Ceramic Soc., Ohio, (1998)

354. Povey, M.J.W. "Ultrasonic Techniques for Fluid Characterization", Academic Press, San Diego, (1997)

355. Prek, M. , "Experimental determination of the speed of sound in viscoelastic pipes", International Journal of Acoustics and Vibration, VOL. 5, NO. 3 , 146-50 (2000)

356. Price, W.J. "Ultrasonic measurements on Rochelle salt crystals", Physical Review, 75, 6, 946-952 (1949)

357. Probstein, R.F., Sengun, M.R. and Tseng, T.C. "Bimodal model of concentrated suspension viscosity for distributed particle sizes", J. Rheology, 38, 4, 811-828, (1994)

358. Prugne C.; van Est J.; Cros B.; Leveque G.; Attal J. "Measurement of the viscosity of liquids by near-field acoustics", Measurement Science and Technology, 9/11, 1894-1898, (1998)

359. Radiman, S., Fountain, L.E., Toprakcioglu, C., de Vallera, A., and Chieux, P., *Progr. Coll. Polym. Sci.* 81, 54 (1990).

360. Ratinskaya, A. "On sound attenuation in emulsions", Soviet Physics-Acoustics, 8, 2, 160-164 (1962)

361. Rasmusson, M. "Volume fraction effects in Electroacoustic Measurements", JCIS, 240, 432-447 (2001)

362. Rayleigh, J.W "On the Application of the Principle of Reciprocity to Acoustics", Royal Society Proceedings, vol XXV, p. 118, (1876)

363. Rayleigh, J.W "On the scattering of Light by small particles", Phil. Mag., (1871)

364. Rayleigh, J.W. "The Theory of Sound", Volume 1, Macmillan & Co., London, (1926)

365. Rayleigh, J.W. "The Theory of Sound", Vol.2, Macmillan and Co., NY, second edition 1896, first edition (1878).

366. Rayleigh, J.W. "Acoustical Observations", Phil. Mag., vol IX, p. 281, (1880)

367. Rayleigh, J.W. "On the Light from the Sky", Phil. Mag., (1871)

368. Reynolds, O. Proceedings of the Royal Society, vol.XXII, p.531, (1874)

369. Ricco, A.J., and Martin, S.J. "Acoustic wave viscosity sensor", Appl. Phys. Lett., 50, 21 1474-1476 (1987)

370. Richardson, E.G. "Acoustic Experiments Relating to the Coefficients of Viscosity of Various Liquids", Faraday Soc. Discussion, vol.226, pp. 16-24 (1954)

371. Riche, L.; Levesque, D.; Gendron, R.; Tatibouet, J., "On-line ultrasonic characterization of polymer flows."- Nondestructive Characterization of Materials VI, 37-44 (1994)

372. Rider,P.F. and O'Brien,R.W., "The Dynamic Mobility of Particles in a Non-Dilute Suspension", J. Fluid. Mech., 257, 607-636 (1993)

373. Rider, P.F., "Sound Wave Mobilities in a Nondilute Suspension of Spheres", J. Colloid and Interface Sci., 172, 1-13 (1995)

374. Riebel, U. et al. "The Fundamentals of Particle Size Analysis by Means of Ultrasonic Spectrometry" Part. Part. Syst. Charact., vol.6, pp.135-143, (1989)

375. Riebel, U. "Method of and an apparatus for ultrasonic measuring of the solids concentration and particle size distribution in a suspension", US patent 4,706,509 (1987)

376. Riebel, U. and Krauter, U. "Ultrasonic extinction and velocity in dense suspensions", Characterization Techniques for Suspended Particulates, ed. V. Hackley and J. Texter, Am. Ceramic Soc., Ohio, 97-111, (1998)

377. Riebel, U. "Process and apparatus for measuring density and mass flow", US patent 6,202,494 (2001)

378. Rogers, P.H. and Williams, A.O. "Acoustic field of circular plane piston in limits of short wavelength or large radius", J. Acoust. Soc. Am., 52, 3, 865-870 (1972)

379. Rose, L.A., Baygents, J.C. and Saville, D.A. "The Interpretation of Dielectric Response Measurements on Colloidal Dispersions using dynamic Stern Layer Model", J. Chem. Phys., 98, 5, 4183-4187 (1992)

380. Rossi, B. "Optics", Addison-Wesley, Reading, MA, 510 p. (1957)

381. Rowlands, W.N., O'Brien, R.W. "The dynamic mobility and dielectric response of kaolinite particles", JCIS, 175, 190-200 (1995)

382. Rowlands, W.N., O'Brien, R.W., Hunter, R.J., Patrick, V. "Surface properties of aluminum hydroxide at high salt concentration", JCIS, 188, 325-335 (1997)

383. Russel, W.B., Saville, D.A. and Schowalter, W.R. "Colloidal Dispersions", Cambridge University Press, (1989)

384. Rutgers, A.J. and Rigole, W. "Ultrasonic vibration potentials in colloid solutions, in solutions of electrolytes and pure liquids", Trans.Faraday Soc., 54, 139-143 (1958)

385. Sachiko, S.; Toyoe, M; Yoshihiko, T, "Measuring instrument for liquid viscosity by surface acoustic wave", Patent Japan 02006728 JP, (1990)

386. Sastry, G.L.N. "Effect of dielectric constant on EDTA-Metal Chelates. An Ultrasonic Study", Indian J. of Pure & Applied Phys., 21, 320-322 (1983)

387. Sayer, T.S.B. "Electroacoustic measurements in concentrated pigment dispersions", Colloids and Surfaces A, 77, 39-47 (1993)

388. Sawatzky R.P. and Babchin, A.J. "Hydrodynamics of electrophoretic motion in an alternating electric field", J. Fluid. Mech., 246, 321-334 (1993)

389. Scales, P.J. and Jones, E. "The effect of particle size distribution on the accuracy of electroacoustic mobilities", Langmuir, 8, 385-389 (1992)

390. Scott, D.M., Boxman, A. and Jochen, C.E. "Ultrasonic measurement of sub-micron particles", Part. Part. Syst. Character., 12, 269-273 (1995)

391. Scott, D.M. "Industrial applications of in-line ultrasonic spectroscopy", Ultrasonic and Dielectric Characterization Techniques for Suspended Particulates, Edited by V.A. Hackley and J. Texter, American Ceramic Society, Westerville, OH, 155-165, (1998)

392. Sewell, C.T.J., "The extinction of sound in a viscous atmosphere by small obstacles of cylindrical and spherical form", Phil. Trans. Roy. Soc., London, 210, 239-270 (1910)

393. Sette, D. "Ultrasonic studies" in "Physics of simple liquids", North-Holland Publ., Co., Amsterdam, (1968)

394. Sharma, A., and Pendse, H.P., "Ultrasonic spectroscopy measurement system for on-line sensing of concentrated particulate suspensions", Ultrasonic and Dielectric Characterization Techniques for Suspended Particulates, Edited by V.A. Hackley and J. Texter, American Ceramic Society, Westerville, OH, 177-191, (1998)

395. Sheen, S. H.; Reimann, K. J.; Lawrence, W. P.; Raptis, A. C. "Ultrasonic techniques for measurement of coal slurry viscosity", - IEEE 1988 Ultrasonics Symposium. Proceedings (IEEE Cat. No.88CH2578-3) Ed.: McAvoy, B. R., vol.1, 537-41, (1988)

396. Sheen, Shuh-Haw; Lawrence, William P.; Chien, Hual-Te; Raptis, Apostolos C. "Method for measuring liquid viscosity and ultrasonic viscometer", US PATENT NUMBER- 05365778, (1994)

397. Sherman, N.E. "Ultrasonic velocity and attenuation in aqueous kaolin dispersions", J. Colloid Interface Sci., 146, 2, 405-414 (1991)

398. Shilov, V.N., Zharkih, N.I. and Borkovskaya, Yu.B. "Theory of Nonequilibrium Electrosurface Phenomena in Concentrated Disperse System.1.Application of Nonequilibrium Thermodynamics to Cell Model.", Colloid J., 43,3, 434-438 (1981)

399. Shilov, V.N. and Dukhin A.S. "Sound-induced thermophoresis and thermodiffusion in electric double layer of disperse particles and electroacoustics of concentrated colloids." Langmuir, submitted.

400. Shortley, G. and Williams, D. "Elements of Physics", Prentice Hall Inc., NJ, (1963)

401. Shumway, "Sound Speed and Absorption Studies of Marine Sediments by a resonance Method", Geophysics, vol. 25, no. 3, pp. 659-682 (1960)

402. Singh, V. R.; Dwivedi, S. "Ultrasonic detection of adulteration in fluid foods", IEEE Engineering in Medicine and Biology Society and 14th Conference of the Biomedical Engineering Society of India. An International Meet (Cat. No.95TH8089), 1/73-4 , (1995)

403. Sinha, D.N. "Noninvasive identification of fluids by swept-frequency acoustic interferometry" US Patent, 5,767,407 (1998)

404. Smith, K.L. and Fuller, G.G. "Electric Field Induced Structure in Dense Suspensions", J. of Colloid and Interface Sci., 155, 1, 183-190, (1993)

405. Smoluchowski, M., in Handbuch der Electrizitat und des Magnetismus", vol.2, Barth, Leipzig (1921).

406. So, J. H.; Esquivel-Sirvent, R.; Yun, S. S.; Stumpf, F. B., "Ultrasonic velocity and absorption measurements for poly(acrylic acid) and water solutions", Journal of the Acoustical Society of America, VOL. 98, NO. 1, 659-60 (1995)

407. Sokolov, M., "On an acoustic method for complex viscosity measurement", Transactions of the ASME. Series E, Journal of Applied Mechanics, VOL. 41, NO. 3, 823-5 (1974)

408. Solovyev, V.A., Montrose, C.J.,Watkins, M.H. and Litovitz, T.A. "Ultrasonic Relaxation in Etnanol-Ethyl Halide Mixtures", J. of Chemical Physics, vol.48, 5, (1968), pp. 2155-2162

409. Sowerby, B.D. "Method and apparatus for determining the particle size distribution, the solids content and the solute concentration of a suspension of solids in a solution bearing a solute", US Patent 5,569,844 (1996)

410. Soucemarianadin, A.; Gaglione, R; and Attane, P. "High-frequency acoustic rheometer, and device for measuring the viscosity of a fluid using this rheometer" PATENT NUMBER- 00540111/EP B1, DESG. COUNTRIES- DE; ES; GB; IT; NL; SE, (1996)

411. Soucemarianadin, A; Gaglione, R; Attane, P., "High-frequency acoustic rheometer and device to measure the viscosity of a fluid using this rheometer", US PATENT NUMBER- 05302878, (1994)

412. Stakutis, V.J., Morse, R.W., Dill, M. and Beyer, R.T. "Attenuation of Ultrasound in Aqueous Suspensions", J. Acous. Soc. Amer., 27, 3 (1955)

413. Stieler. T., Scholle. F-D., Weiss. A., Ballauff. M., and Kaatze, U. "Ultrasonic spectrometry of polysterene latex suspensions. Scattering and configurational elasticity of polymer chains", Langmuir, 17, 1743-1751 (2001)

414. Stintz, M., Hinze, F. and Ripperger, S. "Particle size characterization of inorganic colloids by ultrasonic attenuation spectroscopy", in Ultrasonic and Dielectric Characterization Techniques for Suspended Particulates, Ed. V.A. Hackley and J. Texter, Amer. Ceramic Soc., 219-229, (1998)

415. Stokes, "Dynamic Theory of Diffraction", Camb. Phil. Trans., IX, (1849)

416. Stokes, "On a difficulty in the Theory of Sound", Phil.Mag., Nov. (1848)

417. Stoll, R. "Theoretical Aspects of Sound Transmission in Sediments", J. Acoust. Soc. Amer., vol.68, 5, 1341-1350, (1980)

418. Strout, T.A., "Attenuation of Sound in High-Concentration Suspensions: Development and Application of an Oscillatory Cell Model", A Thesis, The University of Maine, (1991)

419. Sulun, O., "Determination of elastic constants by measuring ultra-sound velocity in Al/sub 2/(SO/sub 4/)/sub 3/ and Ce/sub 2/(SO/sub 4/)/sub 3/ solutions" Revue de la Faculte des Sciences de l'Universite d'Istanbul, Serie C (Astronomie, Physique, Chimie), VOL. 39-41, NO. 1-3, . 18-31 (1974-1976)

420. Swarup, S.; Chandra, S. "Measurement of Ultrasonic Velocity in Resin Solutions." Acta Polym., 40, 8, 526-529 (1989)

421. Swarup, S.; Chandra, S., "Ultrasonic velocity measurements in acetone solutions of some standard epoxy resins and xylene solutions of fumaric resins", Indian Journal of Physics, Part B, VOL. 61B, NO. 6, 515-21 (1987)

422. Takeda, Shin-ichi and Goetz. P.J. "Dispersion/Flocculated Size Characterization of Alumina Particles in Highly Concentrated Slurries by Ultrasound Attenuation Spectroscopy", Colloids and Surfaces, 143, 35-39 (1998)

423. Takeda, Shin-ichi, "Characterization of Ceramic Slurries by Ultrasonic Attenuation Spectroscopy", Ultrasonic and Dielectric Characterization Techniques for Suspended Particulates, Edited by V.A. Hackley and J. Texter, American Ceramic Society, Westerville, OH, (1998)

424. Takeda, Shin-ichi, Chen, T. and Somasundaran P. "Evaluation of Particle Size Distribution for Nanosized Particles in Highly Concentrated Suspensions by Ultrasound Attenuation Spectroscopy", Colloids and Surfaces, (1999)

425. Temkin S. "Elements of Acoustics", 1st ed., John Wiley & Sons, NY (1981)

426. Temkin, S. "Sound Propagation in Dilute Suspensions of Rigid Particles" The Journal of the Acoustical Society of America, vol. 103, 2, pp.838-849, (1998)

427. Temkin, S. "Sound Speed in Suspensions in Thermodynamic Equilibrium" Phys. Fluids, vol. 4, 11, pp.2399-2409, (1992)

428. Temkin, S. "Viscous attenuation of sound in dilute suspensions of rigid particles", The Journal of the Acoustical Society of America, vol. 100, 2, pp.825-831, (1996)

429. Texter, J., "Electroacoustic characterization of electrokinetics in concentrated pigment dispersions", Langmuir, 8, 291-305 (1992)

430. Thorburn, W.M. "The Myth of Occam's Razor", Mind, 27, 345-353 (1918)

431. Tian Hao, "Electrorheological fluids", Adv. Materials, 13, 24, 1847-1857 (2001)

432. Treffers, R. and Cohen, M., "High resolution spectra of cool stars in the 10- and 20-micron region", Astrophys. J., 188, 545-552 (1974)

433. Trinh, E. H.; Ohsaka, K , "Measurement of density, sound velocity, surface tension, and viscosity of freely suspended supercooled liquids", International Journal of Thermophysics , USA, VOL. 16, NO. 2, 545-55 (1995)

434. Trinh, E., Apfel, R.E. "Method for the measurement of the sound velocity in metastable liquids, with an application to water", J. Acoust. Soc. Amer., 63, 3, 777-780 (1978)

435. Tsang, L., Kong, J.A. and Habashy, T. "Multiple scattering of acoustic waves by random distribution of discrete spherical scatterers with the quasicrystalline and Percus-Yevick approximation", J. Acoust. Soc. Amer., 71, 552-558 (1982)

436. Twersky, V., "Acoustic bulk parameters in distribution of pair-correlated scatterers", J. Acoust. Soc. Amer., 64, 1710-1719 (1978)

437. Tyndall, J., "Light and Electricity", D. Appleton and Com., NY, (1873).

438. Tyndall, J., "Sound", Phil. Trans., 3rd addition, (1874)

439. Ulberg, Z. R. ; Dukhin, A. S. "Electrodiffusiophoresis- Film Formation in AC and DC Electrical Fields and Its Application for Bactericidal Coatings", Progress in Organic Coatings, 1, 1-41 (1990).

440. Urick R.J. "Absorption of Sound in Suspensions of Irregular Particles", J. Acoust. Soc. Amer., 20, 283 (1948)

441. Urick, R.J. "A Sound Velocity Method for determining the Compressibility of Finely Divided Substances", J. Appl. Phys., 18, 983 (1947)

442. Uusitalo, S.J., von Alfthan, G.C., Andersson, T.S., Paukku, V.A., Kahara, L.S. and Kiuru, E.S. "Method and apparatus for determination of the average particle size in slurry", US Patent 4,412,451 (1983)

443. Van Tassel, J. and Randall, C.A. "Surface chemistry and surface charge formation for an alumina powder in ethanol with the addition HCl and KOH", JCIS 241, 302-316 (2001)

444. Valdes, J.L. "Acoustophoretic characterization of colloidal diamond particles", in Electroacoustics for Characterization of Particulates and Suspensions, Ed.S.G.Malghan, NIST, 111-129, (1993)

445. Valdes, J.L., Patel, S.S., Chen Y.L. "Reaction chemistries in dynamic colloidal systems", in Ultrasonic and Dielectric Characterization Techniques for Suspended Particulates, Ed. V.A. Hackley and J. Texter, Amer. Ceramic Soc., 275-290, (1998)

446. Ven, van de T.G.M., "Colloidal Hydrodynamics", Academic Press, (1989)

447. Ven, van de. T.G.M. and Mason, S.G., "The microrheology of colloidal dispersions", J. Colloid Interface Sci., 57, 505 (1976)

448. Verwey, E.J.W. and Overbeek, J.Th.G., "Theory of the Stability of Lyophobic Colloids", Elsevier (1948)

449. Vidmar, P. and Foreman, T."A Plane-Wave Reflection Loss Model Including Sediment Rigidity", J.Acoust.Soc.Amer., vol.66, 6, (1979), pp. 1830-1835.

450. Vorobyeva, T.A., Vlodavets, I.N. and S.S.Dukhin, Kolloid.Zh. USSR, 32, 189, (1970)

451. Wagner, K.W., Arch. Elektrotech., 2, 371 (1914)

452. Walldal, C., Akerman, B. "Effect of ionic strength on the dynamic mobility of polyelectrolytes", Langmuir, 15, 5237-5243 (1999)

453. Waterman, P.S. and Truell, R., "Multiple Scattering of Waves", J.Math.Phys., 2, 512-537 (1961)

454. Watson, J. and Meister, R. "Ultrasonic absorption in water containing plankton in suspension", J. Acoust. Soc. Am., 35, 10, 1584-1589 (1963)

455. Webster's Unabridged Dictionary, Second Edition, Simon & Schuster, (1983)

456. Wedlock, D.J., McConaghy, C.J. and Hawksworth, S. "Automation of ultrasonic velocity scanning for concentrated dispersions", Colloids and Surfaces A, 77, 49-54 (1993)

457. Wilson, R., Leschonski, K., Alex, W., Allen, T., Koglin, B., Scarlett, B. "BCR Information", Commission of the European Communities, EUR 6825, (1980)

458. Wines, T.H., Dukhin A.S. and Somasundaran, P. "Acoustic spectroscopy for characterizing heptane/water/AOT reverse microemulsion", JCIS, 216, 303-308 (1999)

459. Wood, A.B. "A Textbook of Sound", Bell, London, (1940)

460. Wood, A.B. and Weston, D.E. "The Propagation of Sound in Mud", Acustica, vol.14, pp.156-162 (1964)

461. Yeager, E., Bugosh, J., Hovorka, F. and McCarthy, J., "The application of ultrasonics to the study of electrolyte solutions.II.The detection of Debye effect", J. Chem. Phys., 17, 4, 411-415 (1949)

462. Yeager, E., and Hovorka, F., "The application of ultrasonics to the study of electrolyte solutions.III.The effect of acoustical waves on the hydrogen electrode", J. Chem. Phys., 17, 4, 416-417 (1949)

463. Yeager, E., and Hovorka, F., "Ultrasonic waves and electrochemistry.I. A survey of the electrochemical applications of ultrasonic waves", J. Acoust. Soc. Am., 25, 3, 443-455 (1953)

464. Yeager, E., Dietrick, H., and Hovorka, F., J. "Ultrasonic waves and electrochemistry. II. Colloidal and ionic vibration potentials", J. Acoust. Soc. Am., 25, 456 (1953)

465. Zana, R., and Yeager, E., "Determination of ionic partial molal volumes from Ionic Vibration Potentials", J. Phys. Chem, 70, 3, 954-956 (1966)

466. Zana, R., and Yeager, E., "Quantative studies of Ultrasonic vibration potentials in Polyelectrolyte Solutions", J. Phys. Chem, 71, 11, 3502-3520 (1967)

467. Zana, R., and Yeager, E., "Ultrasonic vibration potentials in Tetraalkylammonium Halide Solutions", J. Phys. Chem, 71, 13, 4241-4244 (1967)

468. Zana, R., and Yeager, E., "Ultrasonic vibration potentials and their use in the determination of ionic partial molal volumes", J. Phys. Chem, 71, 13, 521-535 (1967)

469. Zana, R., and Yeager, E., "Ultrasonic vibration potentials", Mod. Aspects of Electrochemistry, 14, 3-60 (1982)

470. Zinin, P.V. "Theoretical Analysis of Sound Attenuation Mechanisms in Blood and Erythrocyte Suspensions", Ultrasonics, 30, 26-32, (1992)

471. Zholkovskij, E., Dukhin, S.S., Mischuk, N.A., Masliyah, J.H., Charnecki, J., "Poisson-Bolzmann Equation for Spherical Cell Model: Approximate Analytical Solution and Applications", Colloids and Surfaces, accepted

472. Zukoski, C.F. and Saville, D.A. "Electrokinetic properties of particles in concentrated suspensions", J. Colloid Interface Sci., 115, 422-436 (1987)

473. Zukoski, C.F. and Saville, D.A. "The interpretation of electrokinetic measurements using a dynamic model of the Stern layer", J. Colloid Interface Sci., 114, 1, 32-44 (1986)

474. Zukoski, C.F. and Saville, D.A. "An experimental test of electrokinetic theory using measurements of electrophoretic mobility and electrical conductivity", J Colloid Interface Sci., 107, 2, 322-333 (1985)

475. Zulauf, M., and Eicke, H.-F., J. Phys. Chem., 83, 4, 480 (1979).

Index

A
Accuracy, 232, 289
Acoustic Impedance, 82, 240
Acoustic Sensor, 213
Acoustic Spectrometer, 181, 193, 210, 211, 288, 289
Acoustic Spectroscopy, 278, 288
Acoustics, 1, 2, 3, 6, 7, 8, 9, 21, 62, 63, 75, 77, 94, 150, 196, 251, 256, 264, 288, 291
AcoustoSizer, 162
air bubbles, 145, 219, 311, 312
Alba, 148, 209, 210
Aldrich, 252
Alekssev, 308
Allegra, 6, 8, 101, 115, 116, 187, 318
alumina, 186, 196, 267, 271, 275, 276
alumina, Sumitiomo, 186, 187, 193, 267, 297
alumina-zirconia, 267, 269, 271, 272, 274, 275
Analysis of Attenuation, 224
anatase, 247, 252, 253, 254, 255, 256, 297, 309, 310, 314, 315
Anderson, 136
Andrea, 208, 209, 210, 221
Andreae, 8
Anson, 9, 22, 23, 109, 129, 146, 187
Anthony, 287
AOT, 188, 189, 258, 259, 260
Apfel, 205
Archimedes, 175
area-based, 182
area-weighted, 187
Array, 220
asphaltines, 256, 320
ATH, 283, 284, 285, 286, 287
Atkinson, 112, 136
Attenuation, 78, 79, 80, 83, 88, 90, 91, 92, 94, 105, 126, 134, 135, 182, 183, 193, 194, 195, 207, 216, 221, 222, 233, 250, 254, 258, 260, 261, 263, 276, 279, 280, 281, 283, 284, 285, 287, 290, 292, 293, 294, 312

B
Babchin, 197, 257, 313
Babick, 148, 321
barodiffusion, 156
BCR silica, 108, 109, 185, 187, 193, 231
Beattie, 257
Bell Labs, 288
Berry, 136
Bessel, 141
Beyer, 115
Bhardwaj, 205
Bimodal, 289
bimodality, 63, 226, 293, 320

Biot, 87, 318
bitumen-in-water, 197, 257
Block, 214
Bohren, 63, 226
Boltzman, 157
Booth, 8, 153
Brinkman, 51, 58
Brownian, 6, 26, 45
Bruggeman, 51, 58
Bugosh, 8, 153, 157, 308
Busby, 112, 193, 316

C
Cabot, 185, 187, 230, 289, 290, 291, 294, 295
calculating the PSD, 228, 269
Cannon, 9, 153, 197
Cant, 238, 239
CAPA-700, 189
Carasso, 227, 314, 315
Carhart, 6, 8, 101, 115
Carnie, 154, 156, 162
Case, 123, 175
Ceramics, 4, 267, 297
Challis, 187, 210, 221, 319
CHDF, 187
Chemical-Mechanical Polishing (CMP), 4
Chivers, 9, 22, 23, 109, 129, 146, 187
Chow, 197, 257
Chromatographic, 187
CMP (Chemical-Mechanical Polishing), 231, 287, 288, 291, 294, 295, 296, 304, 315, 320
coagulation, 43, 45
Colloid Chemistry, 296
Colloid Loss, 216, 217
Colloid Vibration Current, 4, 9, 154, 155, 156, 158, 159, 160, 162, 163, 164, 165, 166, 167, 168, 169, 170, 172, 173, 175, 198, 199, 200, 201, 206, 234, 235, 237, 238, 239, 241, 298, 300, 304, 305, 306, 308, 309, 310, 311, 312, 315, 316
Colloid Vibration Potential, 4, 153, 155, 158, 159, 164, 165, 170, 171, 175, 196, 197, 206, 235, 257, 312, 313
Colloidal Dynamics, 3, 153, 162, 210
compressibility, 21, 75, 76, 87, 93, 94, 130, 137, 143, 146, 205, 318
core-shell model, 46, 51, 114, 115, 118, 129, 130, 131, 189, 319, 320
Coulter, 62, 198, 224
CVI Probe, 234
CVI Probe, 235, 297, 307
CVP for monodisperse, 197

D
Dahlberg, 197
Dann, 205
Darby, 91
Dashen, 88

Debye, 3, 4, 6, 8, 35, 36, 38, 44, 45, 53, 153
Debye-Hückel, 36
Deff, 41
Deinega, 308
demulsifiers, 257
Denizot, 8, 153
Denko, 189
Derouet, 8, 153
Dielectric, 2, 92
Dielectric Spectroscopy, 92
Dielectrophoresis, 2
dipole-dipole interaction, 61, 62
dispergating, 267
Dispersion Science, 17
Dispersion Technology, 3, 7, 129, 146, 153, 181, 193, 198, 201, 210, 211, 223, 234, 235, 247, 288, 297, 299, 307, 309, 310
Dispersion Technology Material Database, 129, 130, 146
DLVO, 43, 44
Dohner, 209
Droplet, 259, 260
DSP, 213
Dukhin, 7, 9, 37, 38, 41, 53, 55, 111, 112, 118, 120, 155, 164, 166, 168, 171, 172, 181, 187, 209, 210, 308, 315, 320
Dukhin-Semenikhin, 55
Dukhin-Shilov, 308
duPont, 182, 248, 289, 290

E
ECAH, 6, 8, 101, 102, 104, 106, 110, 114, 115, 116, 117, 118, 127, 144, 187, 210
ECC, 295, 296
Eigen, 6, 8, 89, 92, 93
El-Aasser, 257, 313
Electric Double Layer, 33, 35, 36, 37, 38, 39, 41, 42, 44, 45, 53, 55, 56, 156, 160, 161, 164, 166, 169, 171, 172, 173, 197, 307, 308, 309, 310, 311
electroacoustics, 2, 4, 5, 6, 7, 8, 9, 12, 13, 20, 24, 25, 26, 27, 28, 42, 50, 86, 89, 103, 121, 153, 154, 155, 156, 157, 158, 160, 162, 163, 170, 174, 175, 181, 196, 197, 198, 200, 205, 206, 234, 235, 237, 238, 257, 288, 297, 305, 306, 307, 308, 309, 310, 311
electrochemical, 25, 28, 33, 43
electrocoagulation, 58
electro-deposition, 205
electro-diffusion, 157
electrodynamic, 24
electro-hydrodynamic, 60
electrokinetics, 3, 8, 9, 12, 38, 41, 46, 52, 53, 56, 57, 58, 101, 102, 103, 168, 239, 252, 288, 295, 308
electromagnetic, 63, 64
electro-mechanical, 1, 2
electro-neutral, 41, 166

electroosmosis, 2, 12, 38, 54, 55, 61, 158, 308
electrophoresis, 2, 6, 52, 53, 54, 55, 58, 154, 156, 157, 159, 160, 161, 162, 163, 197, 198, 308
electro-rheology, 58, 61
electro-rotation, 2
electrostriction, 158
Electro-viscosity, 2
electro-zone, 62, 224
Emulsions, 4, 256
Enderby, 8, 153, 154, 155, 156, 162, 196
Enderby-Booth, 153, 154, 155, 156, 162, 196
Ennis, 154, 156, 162, 197
Epstein, 6, 8, 101, 114, 115, 116, 316
Erwin, 209
ESA (Electric Sonic Amplitude), 4, 9, 24, 153, 154, 155, 156, 157, 160, 162, 163, 173, 176, 197, 198, 201, 206, 234, 257, 297, 309, 312, 313, 314, 315, 316
Euler, 6
Extinction, 63, 64, 104

F
Fairhurst, 8, 153, 196
Faraday, 25
Faran, 114, 136, 193, 316
far-field, 139, 140
Federal Institute for Material Research, 297
Fine, 297
flocculation, 9, 19, 46, 247, 248, 252, 255, 256, 257, 300, 310, 319
Focke, 115, 316, 318
Foldy, 8, 311
Fourier, 210
fractionation, 31, 62, 224, 225, 226
Fresnel, 8, 65, 66, 209

G
Galt, 8, 208
gamma-radiation, 205
George, 92
Gibbs, 33
glycerin-water, 188
Goetz, 7, 9, 111, 112, 118, 120, 155, 181, 187, 209, 210, 257, 313, 315, 320
Gravimetric, 21, 26
Gunnison, 197
Gushman, 209

H
Halse, 205
Hamaker, 45
Hampton, 111, 118, 317
Hamstead, 205
Happel, 8, 48, 49, 50, 57, 123, 168
Harker, 9, 319
Hartmann, 115, 316
Hawley, 6, 8, 101, 115, 116, 187, 317, 318
Helos, 189

Hemar, 112, 118, 129, 189
Henry, 2, 3, 7, 53, 54, 200, 298
Henry-Ohshima, 200
Hermans, 6, 8, 153, 157, 196
Hertz, 83
hexadecane, 187, 189, 256, 319, 320
hexametaphosphate, 45, 46, 267, 300, 301, 302, 303, 304
Holmes, 187, 210, 221, 319
Hookean, 52, 132, 133, 134, 255, 256
Horiba, 189
Hovorka, 8
Huang, 257, 314
Huffman, 63, 226
Hunter, 55, 90, 157, 197, 198, 257, 312
Huygens, 6
hydrodynamically, 38, 125, 187

I

ill-defined problem, 62, 63, 87, 107, 224, 225, 226
incompressibility, 47
Ingard, 27, 75, 87
Interferometry, 206, 207
inter-particle, 103, 109, 125, 256
Ion Vibration Current, 24, 157, 158, 304, 305, 306, 307, 308, 310, 311
Ion Vibration Potential, 153, 157, 158, 308
Irani, 29
Isaacs, 257
Isakovich, 5, 8, 127, 128, 129
Isobutyl, 82
iso-electric point, 229, 230, 248, 249, 251, 252, 253, 254, 255, 256, 297, 298, 300, 305, 306, 309, 310, 316

J

Jacobin, 93
James, 197, 198, 313
Japan Fine Ceramics Center, 297
Johnson, 88
Jones-Dole, 25

K

ka, 11, 101, 102, 108, 109, 115, 136, 141, 142, 187, 188
kaolin-water, 300
KCl, 25, 93, 239, 252, 253, 254, 267, 289, 299, 307
Kikuta, 205, 282
Kirchhoff, 5, 7
Kong, 257, 315
Koplik, 88
Kosmulski, 252, 309, 310, 314, 315
Kostial, 205
Kruyt, 55
Kuvabara, 49, 57
Kuwabara, 8, 48, 49, 50, 168

Kytomaa, 112, 136

L

Lame, 116
Langmuir, 247
Laplace, 5, 7, 166, 172
latex/water, 20
latices, 17, 26, 27, 112, 127, 128, 131, 146, 147, 256, 264
Legendre, 116, 117
Leschonski, 29, 224
Levine, 8, 57, 154, 166, 167, 168, 197, 198
Levine-Neale, 57
Lindner, 193
Lloyd, 136
Loeb-Dukhin-Overbeek, 37
Loewenberg, 162
lognormal PSD, 31, 32, 33, 191, 206, 227, 228, 229, 230, 266, 269, 270, 271, 273, 274, 278, 292, 293, 295, 320
long-wavelength, 139, 160
Lyklema, 17, 19, 33, 34, 35, 41
lyophilic, 43, 46
lyophobic, 43, 46

M

Ma, 136
Macroscopic, 63, 102, 225
macroviscosity, 26
Malvern, 3, 7
Marlow, 6, 8, 45, 111, 153, 174, 196, 197, 198, 312
Martin, 205
Matec, 3, 7, 153, 197, 257, 297, 309, 313, 314, 315, 316
Maxwell, 2, 7, 42, 56, 57, 58, 156, 157, 158, 161, 164, 171, 172
Maxwell-Wagner, 42, 56, 57, 58, 156, 157, 158, 161, 164, 171, 172
McCann, 132
McClements, 7, 9, 112, 118, 128, 136, 187, 189, 206, 207, 256, 318, 319, 320
McLeroy, 111, 118, 119, 317
Mercer, 136
microelectrophoresis, 12, 160, 197, 198, 257, 297, 298, 306, 308, 312
microemulsions, 63, 146, 188, 256, 257, 258, 259, 260, 264
microviscosity, 26, 87, 193, 256
Midmore, 55
Mie, 66
Mikhailov, 114, 318
monodisperse PSD, 19, 32, 47, 50, 105, 124, 127, 168, 170, 193, 197, 228
Morse, 8, 27, 75, 87, 106, 108, 112, 115, 136, 137, 141, 143, 144, 194, 196

multi-phase systems, 266, 267, 269, 270, 271, 272, 273, 275, 276, 277, 278
multiple-scattering, 102
Murtsovkin, 59, 61

N
Navier-Stokes, 47
Neale, 8, 57
nepers, 79, 83, 240
Neumann, 141
Newton, 5, 6, 7, 119
Newtonian, 21, 22, 52, 251
Next Mask, 219, 220
NIST, 297
Nobel Prize, 6, 89, 92
Non-aqueous, 4
non-associated non-polar, 81
non-conducting, 39, 40, 42, 61, 62, 169
non-destructive characterization, 205
non-equilibrium, 1, 33, 43
non-linearity, 184, 197, 198
non-spherical, 112
non-stationary, 102, 118
non-symmetrical PSD, 229
Norrish, 92

O
Occam, 228, 230
Ohshima, 6, 53, 54, 154, 155, 200
oil-in-water, 33, 189, 257, 260, 261, 262, 263, 278, 280
Oja, 9
on-line, 10
Onsager, 50, 57, 158, 159
Osseo-Asare, 288
Overbeek, 37, 55

P
particle-free supernate, 310
particle-particle interactions, 44, 48, 52, 58, 61, 101, 102, 104, 106, 110, 112, 113, 114, 115, 117, 118, 124, 127, 128, 129, 153, 154, 160, 162, 181, 184, 187, 198, 199, 278
Pascal, 21
PCC-alumina, 270, 272, 274, 275
PCC-silica, 271, 272, 276, 277
Pellam, 8, 208
Pen Kem, 153
Pendse, 7, 8, 153, 196
Percus-Yevick, 162
permittivity, 24, 38, 56, 88, 158, 168, 207
pH, 45, 206, 220, 229, 230, 239, 248, 249, 250, 251, 252, 253, 254, 255, 256, 267, 275, 289, 295, 296, 297, 298, 299, 300, 302, 303, 304, 306, 309, 310, 311
photo-centrifugation, 189
photomicrography, 257
pH-titrations, 310

Pinkerton, 208
Pogorzelski, 68
Polarization, 41, 42, 165
polydispersity, 10, 26, 28, 29, 50, 62, 118, 120, 121, 124, 134, 144, 163, 164, 168, 169, 170, 172, 196, 197, 198, 225, 290, 313
Porter, 92
Portland cement, 212
Povey, 7, 9, 187, 256, 318
precipitated calcium carbonate (PCC), 267, 268, 269, 270, 271, 272, 273, 274, 275, 276, 277, 278, 300, 301, 314, 320
precision (reproducibility), 67, 221, 268, 284, 291
Prediction, 207, 225, 277
pressure, 21, 23, 24, 47, 57, 60, 76, 77, 80, 81, 82, 83, 84, 85, 87, 102, 103, 119, 120, 122, 123, 133, 137, 138, 139, 143, 144, 145, 146, 154, 158, 159, 173, 174, 175, 205, 209, 238, 240
PSD, 29, 31, 32, 33, 62, 63, 107, 125, 224, 225, 226, 227, 228, 229, 230, 231, 232, 239, 253, 256, 258, 264, 265, 266, 269, 270, 271, 272, 273, 274, 275, 276, 277, 278, 288, 289, 291, 292, 293, 294, 316, 320
pulse-echo, 208
pyncnometer, 94, 148

Q
quasi-stationary, 158, 159

R
Randall, 198, 298, 321
Rasmussen, 197, 201
Rayleigh, 2, 3, 6, 7, 8, 10, 66, 87, 104, 105, 106, 108, 109, 118, 136, 137, 138, 140, 142, 188, 209
Rayleigh-Gans, 66
Rexolite, 236
Reynolds, 3, 7
RF, 212, 213, 218, 235, 237, 238
Rheology, 3, 9, 21, 22, 46, 77
Ricco, 205
Richardson, 112, 188, 316
Richrdson, 193
Rider, 154, 162
Riebel, 7, 9, 209, 210
Rigole, 196, 312
Rosenholm, 252, 309, 310, 314
Rouse-Bueche-Zimm, 131
Rowell, 45, 197, 312
Rowlands, 310, 314
Rutgers, 6, 153, 196, 312
rutile, 110, 112, 182, 183, 184, 185, 196, 198, 200, 201, 223, 229, 247, 248, 249, 250, 251, 252, 253, 278, 283, 284, 285, 286, 287, 297, 312, 313, 315
rutile-ATH, 285, 286, 287
S

Samatzky, 197, 257
Saville, 54
Scales, 198, 313
Scattering, 7, 8, 11, 63, 67, 102, 104, 106, 112, 139, 140, 144, 145, 193
SDEL, 156, 158
Sedimentation Current, 2, 164, 165
semi-conducting, 28
Sensitivity, 289
Sensor Loss, 215, 216, 217
Sette, 206, 208
Shilov, 5, 6, 8, 9, 38, 57, 58, 155, 156, 163, 166, 167, 308
Shilov-Zharkich, 167
Showa, 189
Shugai, 154, 156, 162
Sigmund, 193
Sigmund Lindner GmbH, 193
Signal Level, 216
Signal Processor, 215, 216, 218
signal-to-noise, 221, 307, 308
Sihna, 207
Silica, 187, 231, 240
silica, Cabot, 185, 187, 230, 289, 290, 291, 294, 295
silica, Geltech, 185, 187, 233, 267, 271, 275, 276, 277, 289, 290, 291, 292, 293, 294, 295
silica, Ludox, 149, 182, 183, 184, 185, 187, 196, 198, 199, 221, 222, 223, 233, 239, 240, 289, 290, 291, 292, 293, 294, 296, 306, 307, 308, 312
silica, Minusil, 216, 221, 223, 231, 232
Silicon, 186, 187
Smoluchowski, 53, 54, 55, 56, 58, 155, 156, 158, 159, 160, 163, 164, 200
sols, 28, 196
sonication, 258, 288, 289, 300, 301
Sound, 1, 3, 7, 8, 9, 77, 93, 104, 146, 148, 149, 150, 184, 224, 240
sound speed, 5, 7, 22, 26, 28, 75, 76, 77, 80, 81, 82, 83, 84, 88, 89, 90, 91, 92, 93, 94, 101, 114, 117, 118, 119, 120, 121, 123, 133, 134, 146, 148, 149, 150, 157, 167, 168, 170, 175, 184, 185, 205, 206, 207, 208, 216, 218, 219, 220, 221, 222, 224, 239, 240, 241, 257, 266, 300, 311, 318, 319
Sowerby, 205
space-time, 115
Stakutis, 115, 316
Standards, 17
Stern, 34, 35, 36, 45, 55, 197, 315
Stokes, 2, 5, 7, 12, 47, 78, 79, 80, 131, 251, 252, 280
Streaming current/potential, 2
stress-strain, 131
structure, 6, 18, 21, 34, 36, 37, 38, 39, 81, 87, 91, 94, 101, 114, 131, 134, 163, 164, 189, 191, 207, 256, 307

sub-population, 231, 288, 289
sunflower-in-water, 257
sunscreen, 264, 278, 279, 280, 281
supernate, 310
surface-bulk, 297, 298, 299, 300
surfactants, 303
Sympatec, 3, 7, 189

T
Takeda, 288, 320
Taylor, 167
Temkin, 75, 102, 117, 118
Temperature, 82
Texter, 198, 313
thermodiffusion, 156
Thermodynamic, 22, 118
toluene-in-water, 257
tone-burst, 210, 212, 218
Total Vibration Current, 305, 310
Transducer Loss, 216, 217
transmembrane potential, 33
Truel, 136
Tsai, 68
Tsang, 136
Tsukuba, 189
TVI, 305, 306, 310
Twersky, 136
Tyndall, 2, 3, 5, 7

U
ultra-centrifuge, 63, 225
Urick, 118, 188, 316
Uusitalo, 7, 8, 209

V
Van der Waals, 43, 44, 45
Van Tassel, 198, 298, 321
viscoelastic, 22, 52

W
Wagner, 42, 56, 57, 58, 156, 157, 158, 161, 164, 171, 172
water-ethanol, 88, 91
water-in-car oil, 258, 259
water-in-crude oil, 197, 257
water-in-heptane microemulsion, 188, 259, 260
water-in-hexadecane, 256
water-in-oil, 33, 135, 136
Waterman, 136
wavenumber, 101, 120, 121, 133, 134, 185
Wedlock, 187
weight-basis, 182
wettability, 312
Wines, 188, 320

X
X-ray, 2, 188, 226

Y
Yeager-Zana, 6, 8, 93, 153, 158, 196, 308, 312

Z

Zharkikh, 8, 57, 58, 166
zirconia, 186, 267, 268, 269, 270, 271, 272, 274, 275, 276, 277, 298, 299, 315, 320
Zirconia-alumina, 270, 276, 277
Zukoski, 54